Solubility of Gases in Liquids

Solubility of Gases in Liquids

A Critical Evaluation of Gas/Liquid Systems in Theory and Practice

P. G. T. Fogg
and
W. Gerrard

JOHN WILEY & SONS
Chichester · New York · Brisbane · Toronto · Singapore

Copyright © 1991 by John Wiley & Sons Ltd.
Baffins Lane, Chichester
West Sussex PO19 1UD, England

All rights reserved.

No part of this book may be reproduced by any means,
or transmitted, or translated into a machine language
without the written permission of the publisher.

Other Wiley Editorial Offices

John Wiley & Sons, Inc., 605 Third Avenue,
New York, NY 10158-0012, USA

Jacaranda Wiley Ltd, G.P.O. Box 859, Brisbane,
Queensland 4001, Australia

John Wiley & Sons (Canada) Ltd, 22 Worcester Road,
Rexdale, Ontario M9W 1L1, Canada

John Wiley & Sons (SEA) Pte Ltd, 37 Jalan Pemimpin 05-04,
Block B, Union Industrial Building, Singapore 2057

Library of Congress Cataloging-in-Publication Data:
Fogg, Peter G. T.
  Solubility of gases in liquids : a critical evaluation of
gas/liquid systems in theory and practice / P. G. T. Fogg and
W. Gerrard.
     p.   cm.
  Includes bibliographical references and index.
  ISBN 0 471 92925 5
  1. Gases—Solubility. I. Gerrard, William. II. Title.
QD543.F633   1990
541.3′422—dc20   90-45670
                                                      CIP

British Library Cataloguing in Publication Data:
Fogg, P. G. T.
 Solubility of gases in liquids.
 1. Gases. Solubility
 I. Title II. Gerrard, William *1900–*
 541.342
 ISBN 0 471 92925 5

Typeset in Great Britain by
Latimer Trend & Co. Ltd, Plymouth, Devon

Printed and bound in Great Britain by
Biddles Ltd, Guildford, Surrey

To our grandchildren

Dr William Gerrard died two weeks before his ninetieth birthday. He was able to discuss the completed manuscript but he did not live to see its publication. William Gerrard had an infectious delight in aspects of chemistry and many former colleagues and students have been inspired by his unforgettable enthusiasms, scholarship and devotion to hard work. May this book serve as a memorial to his interest in the solubility of gases.

September 1990   Peter Fogg

CONTENTS

Preface		xi
Chapter 1	Introduction	1
	1.1 Aim of the work	1
	1.2 Methods of presenting data	2
Chapter 2	Methods of Measuring the Solubility of Gases in Liquids	7
	2.1 General principles	7
	2.2 Gas-bubbler methods	10
	2.3 Volumetric methods	10
	2.4 Thin film methods	14
	2.5 Chromatographic methods	15
	2.6 Vapour pressure methods	17
	2.7 Measurements at high pressures	17
Chapter 3	Raoult's Law, Henry's Law and the Reference Line	20
	3.1 The relationship between Raoult's law and the reference line	20
	3.2 Difficulties associated with Henry's law	24
	3.3 Duhem–Margules equation and Henry's law	26
Chapter 4	Making Use of Available Data	30
	4.1 Sources of data	30
	4.2 Empirical equations and the thermodynamic quantities	31
	4.3 Criteria for evaluating published data	32
	4.4 Predicting gas solubilities	35
Chapter 5	Solubility of Sulfur Dioxide	39
	5.1 General behaviour	39
	5.2 Solubility of sulfur dioxide in water and in aqueous solutions	41

	5.3 Solubility of sulfur dioxide in non-aqueous solvents	44
	5.4 Tables summarising the solubility of sulfur dioxide	75
Chapter 6	Solubility of Ammonia and the Amines	86
	6.1 Physical properties of ammonia and amines gaseous at 298 K and 1.013 bar	86
	6.2 Solubility of ammonia and amines in water	88
	6.3 Solubility of ammonia in organic solvents	91
	6.4 Solubility of methanamine in organic solvents	98
	6.5 Solubility of N-methylmethanamine in organic solvents	102
	6.6 Solubility of N,N-dimethylmethanamine in organic solvents	105
	6.7 Solubility of ethanamine and higher amines	110
Chapter 7	Solubilities of Methane and Other Gaseous Hydrocarbons	113
	7.1 Physical properties of hydrocarbons with boiling points below 298 K at 1.013 bar	113
	7.2 Solubility of methane	115
	7.3 Solubility of ethene	133
	7.4 Solubility of ethane	138
	7.5 Solubility of ethyne (acetylene)	146
	7.6 Solubility of propane, butane and 2-methylpropane	155
	7.7 Solubilities of other gaseous hydrocarbons	165
Chapter 8	Solubility of Chlorine	173
	8.1 General behaviour	173
	8.2 Solubility of chlorine in water and aqueous solutions	174
	8.3 Solubility of chlorine in non-aqueous solvents	177
	8.4 Tables summarizing the solubility of chlorine	182
Chapter 9	Solubilities of the Hydrogen Halides	187
	9.1 General behaviour	187
	9.2 Solubility of hydrogen halides in water	188
	9.3 Solubility of hydrogen halides in non-aqueous solvents	191
Chapter 10	Solubility of Hydrogen Sulfide	218
	10.1 General behaviour	218
	10.2 Solubility of hydrogen sulfide in aqueous solvents	219
	10.3 Solubility of hydrogen sulfide in non-aqueous solvents	223

	10.4 Tables summarising the solubility of hydrogen sulfide	233
Chapter 11	Solubility of Carbon Dioxide	241
	11.1 General behaviour	241
	11.2 Solubility of carbon dioxide in aqueous solvents	242
	11.3 Solubility of carbon dioxide in organic solvents	246
Chapter 12	Solubility of the Oxides of Nitrogen	265
	12.1 General properties of oxides of nitrogen, stable and gaseous at 298 K	265
	12.2 Solubility of nitrous oxide	265
	12.3 Solubility of nitric oxide	270
	12.4 Solubility of nitrogen dioxide/dinitrogen tetroxide	273
Chapter 13	Solubilities of Carbon Monoxide, Nitrogen, Oxygen and Hydrogen	274
	13.1 General behaviour	274
	13.2 Solubility of carbon monoxide	274
	13.3 Solubility of nitrogen	281
	13.4 Solubility of oxygen	292
	13.5 Solubility of hydrogen	300
Chapter 14	Solubilities of Other Gases	315
Index of Solvents		321
Subject Index		325

PREFACE

Many processes concerned with the environment, medicine, industry and the chemical laboratory involve the dissolution of gases in liquids. Quantitative information on such phenomena as the dissolution of hydrogen sulfide in crude oil during oil drilling operations or the dissolution of greenhouse gases in sea water are of great importance to modern society. There are many facets to the study of the solubility of gases and it is not easy to gain a general overview of gas–liquid equilibria from individual research papers giving new data, from review articles or from general textbooks of physical chemistry.

This book gives a critical and detailed survey of the solubility in a wide range of liquids of all gases in common use. Additional information on other substances usually considered to be gases is also included. The reader will readily obtain an overall picture of the differing gas solubilities under conditions often encountered in chemical plant and in the laboratory.

The earlier chapters of the book cover the basic theoretical and practical aspects of the measurement of solubilities of gases. The limitations in the reliability of the available data are discussed. Ways of predicting approximate solubilities of gases are indicated. Those requiring more specialised details of recent developments in the theoretical treatment of the subject are directed to the appropriate comprehensive works.

The later chapters of the book contain many tables of solubility data for dissolution in aqueous and non-aqueous solvents. Detailed literature references to the sources of the data are given.

It is difficult to understand the significance of extensive numerical data without a visual aid. The authors believe that the behaviour of gas–liquid systems can be comprehended most readily by means of simple diagrams linked together by reference lines based upon the vapour pressure of the gas in question. Graphs showing the variation of solubility with pressure or with temperature form an essential part of the book. Not only do these graphs indicate general patterns of gas solubility but they also highlight instances in which there are discrepancies between solubilities measured in different laboratories. In some cases the graphs assist in identifying bad data.

PREFACE

This volume was partly prompted by two earlier books published by Gerrard, *Solubility of Gases and Liquids*, published in 1976, and *Gas Solubilities, Widespread Applications*, published in 1980. Gerrard developed the concept of a reference line in these earlier volumes and included many diagrams showing the behaviour of gas–liquid systems. Much work on the solubility of gases has been published during the last ten years and the authors have been able to make use of more extensive material than was available for the earlier volumes.

The present volume contains smoothing equations for gas solubilities where these are appropriate. Smoothing equations are invaluable for the drawing of graphs by computer techniques and for the estimation of solubilities at particular temperatures and pressures. It is important to bear in mind the possibility of errors in the original data on which such smoothing equations are based, especially if this data is from one source only. Smoothing equations are useful for interpolation of data but can often be unreliable when they are extrapolated beyond the range of the original data.

Some of the smoothing equations are taken from the published literature. Minor changes have been made where necessary to give uniform presentation. Equations which carry no literature reference have been derived by Fogg for inclusion in this book.

The book is intended for use by chemists and chemical engineers in industrial and research laboratories requiring a ready access to numerical data and references to specialised papers in the literature. It will also be of interest to final year students requiring a background knowledge of phenomena associated with gas–liquid equilibria. It is hoped that the highlighting of inconsistencies and gaps in the corpus of published data will encourage further experimental and theoretical work.

Both authors are members of the IUPAC Solubility Data Commission and have obtained much stimulation from contact with their colleagues at meetings of the Commission. They have relied heavily on volumes in the *Solubility Data Series* devoted to the solubility of individual gases which have been published for the Commission.

The authors are particularly appreciative of the help given by Dr Ronald Brown, a former colleague of the authors at the Polytechnic of North London. Dr Brown has carefully checked the text and advised us on the presentation of material to avoid inconsistencies.

They are very grateful also for the help given to them by Heyden & Son. This book could never have been produced without their patience, friendly co-operation and advice.

Peter Fogg
William Gerrard

Chapter 1
INTRODUCTION

1.1 AIM OF THE WORK

Data on the solubilities of gases in liquids are measured in many ways. The aim of this book is to explain the use and limitations of such data. The basis upon which data may be evaluated for reliability is explained.

Sufficient data relating to gases of especial industrial importance are included to indicate general solubility trends. Graphs are often the most convenient and easily comprehended way of conveying information. The graphs which are presented form an essential part of the book and enable the general behaviour of gas-liquid systems to be readily understood.

Equilibria involving two components distributed between a gas phase and a liquid phase have been studied for many years and data have been expressed in many forms. There is no sharp distinction between the study of liquid-liquid equilibria and that of gas-liquid equilibria so an arbitrary division has to be made. The systems to be treated in this book will consist of those in which one component is either above its critical temperature or has a vapour pressure above 1.013 bar (1 atm) at the temperature of study. The other component will exist as a liquid at the temperature of study. For convenience the one will be referred to as the *gas* and the other as the *solvent*. However, this does not necessarily imply that the former is the minor component and the latter the major component of the liquid phase. There have been many measurements in which the mole fraction of the gas in the liquid phase was greater than 0.5.

According to the phase rule there are two degrees of freedom associated with a two phase, two component system. However, there are many variables associated with such a system: for example, composition of the liquid phase, composition of the gas phase, densities of the phases, partial pressures of the components in the gas phase. Sometimes approximations have to be introduced when a value of one of these variables is calculated from an experimental measurement of another variable.

There have been few studies of the rate at which equilibrium is established between a liquid phase and a gas phase. The data under discussion are intended to relate to states of thermodynamic equilibria, but one must always bear in mind that equilibria may not have been attained in some of the older experimental measurements of gas solubilities. Even when one can be certain that literature data are reliable, applying such data to industrial and other situations may lead to errors if equilibrium is not reached. Despite these caveats there is little doubt of the importance of having access to accurate gas solubility data. The significance of such data must be properly understood before they can be correctly used in an industrial context.

1.2 METHODS OF PRESENTING DATA

The solubility of a gas in a solvent can be expressed in many different ways. Accurate interconversion from one measure of solubility to another is not always possible.

1.2.1 Mole fraction solubilities.

The mole fraction solubility, x_g, of a gas dissolved in a solvent is given by:

$$x_g = \frac{n_g}{n_g + n_s}$$

where n_g and n_s are the number of moles of gas and of solvent respectively. If the system contains more than two components then:

$$x_g = \frac{n_g}{\Sigma n}$$

The scale of mole fractions extends from 0 to 1, and this enables graphical plotting over the whole of the possible concentration range.

1.2.2 Mole ratio solubilities

The mole ratio solubility, N_g, is the ratio of the number of moles of gas to that of solvent

$$N_g = \frac{n_g}{n_s}$$

The lower the concentration the closer the value of the mole fraction to the mole ratio. It is sometimes convenient to assume that mole fraction and mole ratio are approximately equal at low concentrations when n_g is small in comparison with n_s. The difference increases with concentration, and the assumption that the two are equal can lead to errors.
For a two component system it follows that:

$$N_g = \frac{x_g}{1 - x_g}$$

and hence:

$$\frac{N_g - x_g}{N_g} = x_g$$

$$\frac{\text{mole ratio} - \text{mole fraction}}{\text{mole ratio}} = \text{mole fraction}$$

Hence for a mole fraction of gas of 0.1 there is a 10% difference between the mole ratio and the mole fraction (expressed as a percentage of the mole ratio). For a mole fraction of 0.01 the difference is only 1% which may be comparable with experimental errors in the measurement of a solubility. In practice, mole fraction can usually be equated with the mole ratio if either is less than 0.01.

Mole ratio concentrations extend from zero to infinity. This scale of concentration therefore cannot be used for graphical plots over the whole of the concentration range.

1.2.3 Ostwald coefficient

The Ostwald coefficient, L, is usually defined as the ratio of the volume of gas absorbed to the volume of the absorbing liquid, both volumes being measured at the same temperature. It is often assumed that this volume is independent of pressure when the pressure is of the order of 1 bar and the temperature is greater than the boiling point of the gas under 1.013 bar (1 atm). This is equivalent to assuming the validity of Henry's law (mole ratio form).

This assumption may lead to significant errors. The Ostwald coefficient becomes a more reliable measure of the solubility when pressure and temperature are specified.

Sometimes the Ostwald coefficient is taken to be equal to the ratio of the concentration (weight or moles per unit volume) of gas in the liquid phase to that in the gas phase. This definition could be exactly equivalent to that given above only if the gas obeyed the ideal gas laws and there were no change in volume of the liquid phase when the gas dissolved.

This can be shown as follows: Let

d_g = density of the gas under the conditions of measurement of the solubility.
d_s = density of the pure solvent under the same conditions.
d_l = density of the liquid solution under the conditions of measurement.

L = ratio of the volume of gas absorbed to the volume of the absorbing liquid, both volumes being measured at the same temperature.
M_g = relative molecular mass of the dissolved gas.

It follows that:
Moles of dissolved gas/volume of solvent

$$= \frac{Ld_g}{M_g}$$

Moles of dissolved gas/weight of solvent

$$= \frac{Ld_g/M_g}{d_s}$$

Moles of dissolved gas/weight of solution

$$= \frac{Ld_g/M_g}{d_s + Ld_g}$$

Moles of dissolved gas/volume of solution

$$= \frac{Ld_g/M_g}{(d_s + Ld_g)/d_l}$$

Moles of gas per unit volume in the gas phase

$$= \frac{d_g}{M_g}$$

Ratio, L_c, of concentrations of gas in liquid and gas phases,

= mole per unit volume of gas in solution/mole per unit volume in the gas phase

$$= \frac{L}{(d_s + Ld_g)/d_l}$$

L_c can be equal to L only if $(d_s + Ld_g)/d_l$ is unity. This can be true only if dissolution of gas causes no change in the volume of the liquid phase.

There may be appreciable differences between the apparent values of the

Ostwald coefficient defined in these two ways. Dissolution of gas may cause appreciable changes of volume of the liquid phase if the gas is very soluble under particular conditions of temperature and pressure.

It follows from the equations given above that the difference between L and L_c expressed as a percentage of L, i.e. $[(L - L_c)/L] \times 100$, can be found from the relationships:

$$\frac{L - L_c}{L} = \frac{L - L/[(d_s + Ld_g)/d_l]}{L}$$

$$= 1 - \frac{1}{(d_s + Ld_g)/d_l}$$

The densities of various aqueous solutions of hydrochloric acid at 293.15 K are given in the literature.[3] These data may be used to calculate the percentage difference between the value of L and that of L_c for the particular partial pressures of HCl with which these solutions are in equilibrium.

% by wt of HCl	d_l	$d_s + Ld_g$	$d_s + Ld_g$	$\frac{L - L_c}{L} \times 100$
	g cm^{-3}	g cm^{-3}	d_l	
0.5	1.0007	1.0032	1.0025	0.25
1.0	1.0031	1.0083	1.0052	0.52
5.0	1.0228	1.0507	1.0274	2.66
10.0	1.0476	1.1091	1.0587	5.54
20.0	1.0980	1.2478	1.1364	12.00
30.0	1.1492	1.4260	1.2408	19.41
40.0	1.1977	1.6637	1.3890	28.01

The possibility of significant discrepancies between Ostwald coefficients defined in different ways must always be borne in mind. More rigorous treatments of the definitions of the Ostwald coefficient have been given by Battino[1] and Gerrard.[2]

1.2.4 Henry's law constant

This law implies that the concentration of gas in the liquid phase is proportional to the partial pressure of the dissolved gas. The proportionality constant is the Henry's law constant. Problems associated with Henry's law are discussed in Section 3.2.

1.2.5 Bunsen coefficient

The Bunsen coefficient, α, is the volume of gas, reduced to 273.15 K and 1.013 bar, which is absorbed by unit volume of solvent at a stated temperature under a partial pressure of gas of 1.013 bar. In the estimation of mole fraction solubilities from Bunsen coefficients allowance should be made if possible for any approximations, such as assumption of ideal behaviour of the gas, which have been made in relating the original experimental measurements to the reported coefficient.

1.2.6 Absorption coefficient

The absorption coefficient, β, is usually defined as the volume of gas, reduced to 273.15 K and 1.013 bar, which is absorbed by unit volume of solvent at a stated temperature under a total pressure (partial pressure of gas + partial pressure of solvent) of 1.013 bar.

1.2.7 Kuenen coefficient

The Kuenen coefficient, S, is the volume of gas, reduced to 273.15 K and 1.013 bar, which is absorbed by 1 gram of solvent at a stated temperature when the partial pressure of gas is 1.013 bar.

1.2.8 Other methods of expressing gas solubilities

A gas solubility may be expressed as the weight percentage or as the number of moles per unit volume of solution. It may also be expressed as the number of moles of gas per unit weight of solvent.

REFERENCES

1. Battino R. *Fluid Phase Equilib.* 1984, 15, 231.
2. Gerrard, W. *Gas Solubilities—Widespread Applications*, Pergamon Press, Oxford, 1980, Chapter 1.
3. *Handbook of Chemistry and Physics*, 70th ed. C.R.C. Press Inc., Boca Raton, Florida, U.S.A., 1989.

Chapter 2
METHODS OF MEASURING THE SOLUBILITY OF GASES IN LIQUIDS

2.1 GENERAL PRINCIPLES

The solubility of a gas in a liquid depends on temperature and on the partial pressure of the gas. Solubility measurements require the determination of a temperature, a pressure and the corresponding composition of the liquid phase when thermodynamic equilibrium has been established. Detailed reviews of experimental techniques and methods of treatment of data have been published.[1-8]

The temperature is often quoted to $\pm 0.1°C$. This is usually a measure of the accuracy to which the measuring instrument can be read. Errors in the calibration of the instrument may be a source of error in solubility values, especially at elevated temperatures as in the case of solubilities of gases in molten salts. Errors will also arise if the liquid phase is not of uniform temperature.

The vapour phase in a gas absorption experiment contains at least two components—the gas under test and solvent vapour. Sometimes a third component is present—a diluent, often dry air or nitrogen. The total pressure of the vapour phase can usually be measured accurately, but estimation of the partial pressure of the gas under test can cause great problems and be a significant cause of error. The proportion in the vapour phase of the gas under test is sometimes estimated by chemical analysis, and the partial pressure of this gas calculated from this proportion and the total pressure. However, very accurate estimation of the partial pressure by this method requires a knowledge of deviations of the mixture from Dalton's law of partial pressures.

In practice, the total pressure of the system during the absorption process is usually measured and reported, although it is sometimes stated that measurements were made at an undefined 'barometric' pressure. This leads to an uncertainty of about $\pm 3\%$ in the total pressure.

Often the gas phase in contact with the liquid is assumed to consist solely of the gas under test, neglecting the presence of solvent vapour. This may be justified if the vapour pressure of the pure solvent at the temperature of measurement is less than about 1% of the total pressure. The lower the partial pressure of the gas under test, the higher the proportion of the total pressure contributed by the partial pressure of the solvent.

Some workers assume that the partial pressure of the solvent is unaffected by the presence of dissolved gas and therefore equal to the vapour pressure of pure solvent at a specified temperature. The partial pressure of the gas under test at a measured total pressure can then be found by difference. For measurements carried out at a total pressure equal to barometric pressure, a measured solubility is often corrected to correspond to a partial pressure of gas of 1.013 bar, on the assumption that mole fraction solubility is approximately proportional to partial pressure of gas over a small pressure range.

A better approximation allows for the decrease in partial pressure of the solvent due to dissolved gas. Hildebrand[9] carried out such an approximation in the following manner.

Suppose the mole fraction solubility x_g of the gas at a total pressure P_t is made up of partial pressures of gas and solvent vapour, and the corresponding vapour pressure of pure solvent is P_s. The mole fraction of solvent in the liquid phase is then $(1-x_g)$. If the vapour pressure of the solvent follows Raoult's law then:

partial pressure of solvent

$$= (1-x_g)P_s$$

Hence partial pressure of gas in the gas phase

$$= P_t - (1-x_g)P_s$$

The mole fraction solubility x'_g of the gas, under a partial pressure of P_t is then given by:

$$x'_g = x_g P_t / [P_t - (1-x_g)P_s]$$

assuming that the mole fraction solubility can be taken to be approximately proportional to the partial pressure, over a small pressure range.

Under some circumstances there may be little change in the mole fraction solubility of a gas with change in its partial pressure. This may occur at higher partial pressures if the gas is very soluble at lower pressures. Hydrogen chloride behaves in this way when it is dissolved in many solvents containing oxygen. The variation with total pressure of the mole fraction solubility of hydrogen chloride in 1-ethoxybutane[9] at 270.25 K is shown in Fig. 2.1. The vapour pressure of pure solvent at this temperature is about 0.023 bar so this total pressure is close to the partial pressure of the hydrogen chloride. At 1.013 bar there is only an 0.46% change in mole fraction solubility for a change in partial pressure of gas of 0.023 bar. This change is

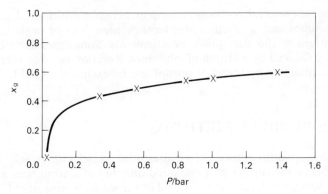

Figure 2.1. Mole fraction solubility of hydrogen chloride in 1-ethoxybutane at 270.25 K.

likely to be less than the percentage errors in the original measurements, and any attempt to correct for the partial pressure of the solvent is likely to increase the error.

The difference between the mole fraction solubility at a total pressure of 1.013 bar and that at a partial pressure of 1.013 bar may be insignificant for other systems which have similar solubility patterns. This may be the case even if the pure solvent has appreciable vapour pressure at the temperature of measurement. Attempts to correct for the vapour pressure of the solvent by one or other of the simple methods mentioned above can, in this situation, increase the error in an estimation of the solubility at a partial pressure of 1.013 bar.

There is little quantitative knowledge of the rates at which absorption of different gases occur. The establishment of absorption equilibrium between liquid phase and gas phase may be slow, especially if the liquid phase has a high viscosity. The decision as to when such an equilibrium is established is usually an empirical matter—the experiment is continued until maximum absorption seems to be achieved. Solvents are usually stirred during absorption. Failure to achieve absorption equilibrium is likely to be the chief cause of discrepancies between solubilities reported by different workers for the same system under apparently identical conditions. Unwanted gases can affect absorption. Preliminary degassing of solvent and use of a very pure sample of gas under test usually speeds up the establishment of absorption equilibrium. There is the possibility of unwanted gases accumulating above the surface of the solvent and reducing the proportion of the gas under test in the immediate vicinity of the liquid.

Under some circumstances liquids may become supersaturated with gas. Even when a gas is bubbled through a liquid with the surface of the liquid subject to barometric pressure there may be some degree of supersaturation, because the pressure at the bottom of an inlet tube is greater than the pressure at the surface of the liquid.

Irreversible chemical reaction may occur between solvent and gas under test. Sometimes such a reaction may be slow enough to be neglected under the conditions of the absorption measurements. Sometimes chemical reaction can be reduced by addition of inhibitors. Reaction between alkanes and chlorine is inhibited by traces of phenol and by exclusion of light.

2.2 GAS-BUBBLER METHODS

The simple apparatus consists of an absorption vessel with an inlet tube, to pass gas below the surface of a measured quantity of solvent, and an outlet tube. The vessel and its contents are held in a constant temperature bath. Gas under test, either in a pure state or mixed with a diluent gas of low solubility, is bubbled through the liquid until no further gas is absorbed. The total pressure above the liquid is equal to barometric pressure and is the sum of the partial pressures of gas under test, solvent and diluent gas (if this is present). The quantity of gas absorbed is found by chemical analysis or from the increase in weight of the absorption vessel and its contents.

Mixing the gas under test with a diluent gas of low solubility enables the measurement of solubilities over a range of partial pressures. Another method of obtaining solubilities over a range of partial pressures with this type of apparatus is to saturate the solution at a total pressure (sum of partial pressures of solute gas and of solvent) equal to barometric pressure, connect the vessel to a vacuum line fitted with a manometer, and pump away some of the absorbed gas from the solution.[11] The new total pressure above the solution is then measured, and the quantity of gas remaining in solution is measured. The process of partial evacuation and measurement of pressure can than be repeated (Fig. 2.2).

2.3 VOLUMETRIC METHODS

In this type of apparatus a fixed quantity of gas at a measured pressure comes into contact with a measured quantity of solvent with appropriate agitation to ensure that equilibrium is established between gas and liquid phases. Temperatures are controlled by an air or liquid thermostat. The volumes of gas phase before and after absorption are measured at the same pressure, usually barometric pressure. The amount of gas absorbed is found from the difference between the measurement of initial and final volumes of gas, with appropriate allowance for dead volume in capillary tubes, taps and absorption vessel.

There have been many designs of volumetric apparatus since this technique was used by Ostwald. A simple apparatus of this type is shown in Fig. 2.3. This consists of a calibrated gas burette, B, and magnetically stirred absorption vessel, C, connected by capillary tubing to a vacuum line, A, and source of gas. The temperature of the gas burette and the absorption vessel must be controlled. Gas under test is introduced over mercury into the gas

METHODS OF MEASURING THE SOLUBILITY OF GASES IN LIQUIDS

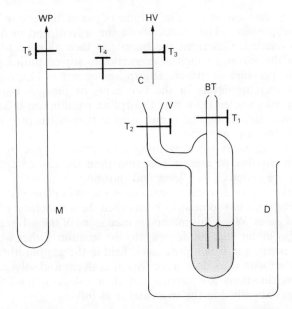

Figure 2.2. Bubbler tube apparatus as used by Gerrard: M, manometer; WP, water pump; HV, high vacuum pump; BT, bubbler tube; D, Dewar flask; T_1–T_5, taps; C, cone to fit socket V.

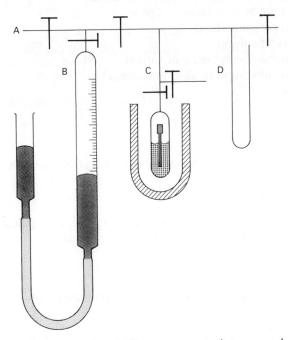

Figure 2.3. Gas burette apparatus for solubility measurements: A, to gas supply; B, gas burette; C, bulb holding solvent and magnetic stirrer and surrounded by constant temperature bath; D, manometer.

burette from a reservoir of gas. The volume of gas in the burette is measured at barometric pressure. The spaces above the solvent and in the capillary tubes are evacuated. Gas from the burette is then allowed to come into contact with the solvent, which is magnetically stirred until equilibrium is reached. The pressure at which absorption occurs is controlled by the difference in mercury levels in the two arms of the gas burette and is measured by manometer D. When absorption equilibrium is attained, the volume of gas in the gas burette is measured at barometric pressure with the two levels of mercury the same.

Designs of apparatus for measuring absorption of gas by volumetric methods, but capable of greater precision than the simple apparatus described here, are reviewed by Clever and Battino.[2]

Many workers have advocated saturating the gas in the gas burette with solvent vapour. This has sometimes been called the *wet* method of measuring solubilities of gases. When this method is used some of the solvent vapour, as well as the gas under test, condenses into the solution. Other workers have used the *dry* method in which pure gas is held in the gas burette but the gas phase in contact with the solution consists of both gas and solvent vapour. If various approximations are carried out then calculating solubility from volumes of gas absorbed in the two cases is as follows:

Volume of gas absorbed	$= V_g$
Total pressure of gas plus solvent vapour	$= P_t$
Partial pressure of solvent vapour	$= P_s$
Partial pressure of gas	$= P_g$
Density of solvent	$= d_s$
Relative molecular mass of solvent	$= M_s$
Volume of solvent in absorption vessel	$= V_s$

Wet method

Moles of gas dissolved at a partial pressure of P_g (ideal gas approximation)

$$= P_g V_g / RT$$

Moles of gas dissolved at a partial pressure of P_t (assuming that absorption is proportional to pressure over a small pressure range)

$$= P_t V_g / RT$$

Moles of solvent condensed (ideal gas approximation)

$$= P_s V_g / RT$$

Volume of solvent condensed

$$= \frac{P_s V_g / RT}{d_s / M_s}$$

Moles of gas absorbed per unit volume of solvent

$$= \frac{P_t V_g / RT}{V_s + (P_s V_g / RT)/(d_s / M_s)}$$

The volume of solvent which condenses from the gas phase depends upon the magnitude of V_g. The more soluble the gas the greater the volume of solvent which condenses and the greater the errors introduced by incorrect estimation of P_s. If the volume of solvent which condenses from the gas phase is small compared with the volume of solvent originally in the absorption vessel then any errors in estimating the partial pressure of the solvent will cause very small errors in calculating absorption per unit volume of solvent.

Dry method

Moles of gas absorbed at a partial pressure of P_g over the solution (ideal gas approximation)

$$= P_t V_g / RT$$

Moles of gas absorbed at a partial pressure of P_t over the solution (assuming that absorption is proportional to pressure over a small pressure range)

$$= \frac{P_t (P_t V_g / RT)}{P_g}$$

The partial pressure, P_g, of the gas over the solution

$$= P_t - P_s$$

Moles of gas absorbed per unit volume of solvent

$$= \frac{P_t (P_t V_g / RT)}{V_s P_g}$$

$$= \frac{P_t (P_t V_g / RT)}{V_s (P_t - P_s)}$$

The effect of approximations in the estimation of P_s upon the error in the solubility depends, in this case, on the relative values of P_t and P_s.

2.4 THIN FILM METHODS

Recently the importance of making special provision to ensure equilibrium between liquid and gas phases has been recognized. Apparatus in which a thin film of liquid comes into contact with the gas has been developed by various groups.[2,12-16]

The Benson and Krause apparatus[16] consists of an equilibrator composed of two concentric glass spheres with inner and outer volumes of about 1 dm^3 and 2 dm^3, respectively. Solvent is pumped from the liquid sample bulb into the space between the inner and outer spheres. It then falls as a thin film and circulates back to the sample bulb. The gas fills the inner sphere and the part of the space between the two spheres not occupied by liquid. Equilibrium between gas and liquid phases is established at the surface of the thin film of liquid (Fig. 2.4).

After an appropriate time a volume of the liquid phase is sealed off in a liquid phase sample bulb and analysed. A sample of the gas phase can also be sealed in a gas phase sample bulb and analysed. Equilibrium is reached when there is no further change in composition of the phases with circulation of liquid.

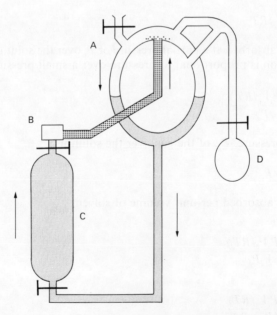

Figure 2.4. The Benson & Krause falling film apparatus: A, spherical equilibrator; B, centrifugal pump; C, liquid phase sample bulb; D, vapour phase sample bulb.

2.5 CHROMATOGRAPHIC METHODS

Gas-liquid chromatography may be used to analyse liquid or gas phases which have come into equilibrium during a determination of gas solubility. It is then being used simply as an analytical tool. Alternatively, the technique gives information about solution processes taking place within the chromatography column itself.[17-21]

The process of gas-liquid chromatography involves dissolving a gas or vapour in a thin liquid film inside the chromatographic column. The greater the solubility the greater the tendency of the gas or liquid to be retained by the column and, consequently, the longer the retention time and greater the retention volume under a particular set of conditions. If a small sample of the gas or vapour under test is injected into the carrier gas stream then the partial pressure of the sample is low and consequently the solution in the stationary phase is very dilute. The nature of the solution may be close to that of an infinitely dilute solution.

If the Henry's law constant, H, is defined as:

$$H = \lim_{x_g \to 0} \frac{f_g}{x_g}$$

where f_g is the fugacity of the solute gas and x_g is the mole fraction of the solute gas in the solvent (i.e. the stationary phase), then it can be shown that:

$$H = \frac{n_1 R T y_2 z}{V_N}$$

where n_1 is the number of moles of solvent, y_2 the fugacity coefficient of the solute, z the compressibility factor of the gas, and V_N the net retention volume.

When the quantity of solute is about one microgram and the total gas pressure is close to barometric pressure then the above equation may be simplified to:

$$H = \frac{n_1 R T}{V_N}$$

The net retention volume, V_N, is related to the retention time t_s by the equation:

$$V_N = (t_s - t_m) U j$$

where t_m is the column dead time or the retention time for a gas which is not absorbed by the stationary phase (e.g., argon); U is the flow rate of the gas measured at the outlet of the column; j is a correction factor related to the inlet pressure P_i and the outlet pressure P_o and given by the relationship:

$$j = 1.5 \frac{(P_i/P_o)^2 - 1}{(P_i/P_o)^3 - 1}$$

The retention time should correspond to an infinitely small sample size. If retention times show any detectable dependence upon sample size then the value to be used in the above equation should be found by extrapolating to zero sample size.

In principle the solubility of a gas or vapour in a solvent may be found by preparing a chromatography column having a known weight of the solvent under test as stationary phase. The retention time of the gas under test, inlet and outlet pressure, and flow rate of carrier gas are then measured, and the limiting value of the Henry's law constant for the particular temperature calculated.

The solubilities of several gases in one solvent may be measured rapidly by this method, enabling comprehensive surveys of solubilities to be carried out. Lenoir et al.,[19] for example, have published the solubilities of 12 gases in 18 solvents with some systems investigated at three temperatures. Unfortunately, solubilities measured by this method sometimes differ from measurements by other methods which have been published in the literature and it is generally agreed that the method is unreliable for some systems.

It is recognized that adsorption of solute gas at the gas-solvent interface may be a cause of error. The extent to which this occurs in the case of a particular gas will vary from solvent to solvent. In addition, adsorption at solid-gas interfaces may also occur. The extent of this will depend upon the nature of the support phase and upon the material from which the column tubing is constructed.

Sweeney[20] has reported a limiting value of 29.8 bar for the Henry's law constant for dissolution of carbon dioxide in tributyl phosphate at 298.15 K. He compares this value with values of 29.5, 28.5 and 29.8 bar reported by different authors from measurements by other methods. In this case the agreement with traditional methods of measuring solubility is excellent, with a difference of about 0.5% between the average of the three values from other authors and the value from chromatographic measurements. There is poor agreement between values for hydrogen sulfide in 1-methyl-2-pyrrolidinone at 298.15 K. Literature values obtained by traditional methods of measurement are 6.91 and 7.75 bar. Sweeney obtained a value of 5.81 which is apparently too low by at least 16%.

Lenoir et al.[19] have shown that chromatographic measurements of Henry's law constant for systems of alkanes deviate by an average of 3% from values from other sources and that the maximum deviation does not exceed 6.5% They admit, however, that they could not find enough reliable data for comparison with the Henry's law constants for polar systems.

2.6 VAPOUR PRESSURE METHODS

Gas solubilities at various pressures have been found from the total vapour pressure of solutions of known composition. The partial pressure of the solute gas is then equal to the total vapour pressure minus the partial vapour pressure of the solvent. This method can be used at pressures above barometric pressure with a suitably designed apparatus. Wright and Maass[22] prepared solutions of hydrogen sulfide at various concentrations and measured the vapour pressures of these solutions using a sensitive glass diaphragm manometer with the internal pressure very nearly balanced against the external pressure.

2.7 MEASUREMENTS AT HIGH PRESSURES

Many mixtures of gases and liquids have been studied at high pressures. Apparatus for such studies is usually constructed largely of steel. There is often provision to withdraw samples of the liquid phase and the gas phase while the system is under pressure.

A description and schematic diagram of apparatus used in this case for measurement of the solubility of hydrogen sulfide in water have been given by Lee and Mather[23] (see Fig. 2.5). The cell consisted of a high pressure liquid

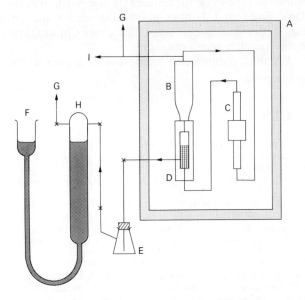

Figure 2.5. Apparatus for high pressure measurements of gas solubilities: A, air bath; B, vapour reservoir; C, magnetic pump; D, high pressure liquid level gauge; E, liquid sampler; F, mercury leveller; G, vent line; H, gas collecting burette; I, to gas chromatograph.

level gauge of capacity 200 cm^3 mounted below a vapour reservoir of capacity 250 cm^3. The cell was charged with about 100 cm^3 of water. Hydrogen sulfide at the required pressure was circulated through the cell by a magnetic pump. Pressure was measured by a Heise bourdon tube gauge and temperature by thermopiles. The cell and pump were housed in an air-bath.

Small samples of the liquid phase could be withdrawn through a valve into a sampling vessel under barometric pressure. Gas evolved from the liquid in this vessel was collected in a gas burette. The residual liquid in the vessel after evolution of this gas was analysed by gas chromatography. There was also provision for withdrawing samples of the gas phase from the cell directly into a chromatograph.

Gas-liquid equilibria at high pressures can also be studied by the bubble and dew point method. A measured quantity of gas and liquid solvent are pumped under pressure into a small glass cell. The temperature is increased until the mixture in the cell is seen to be completely in the gas phase. The mixture is then slowly cooled and the temperature noted when the first droplets of liquid appear. This is the dew point for the particular mixture and pressure. After further cooling the phase boundary disappears and the system becomes one liquid phase. This is the bubble point. The dew point corresponds to a point on the phase boundary between the one phase gas region and the two phase gas + liquid region. The bubble point corresponds to a point on the boundary between the one phase liquid region and the two phase gas + liquid region. At the temperature and pressure of the bubble point the solubility of the gas is the same as the composition of the mixture as a whole. Dew point and bubble point coincide at the critical point. Measurements of dew and bubble points for a series of mixtures of gas and solvent give the complete phase diagram.

Kurata and Kohn[24] published a description of apparatus for measuring dew and bubble points.

REFERENCES

1. Battino, R.; Clever, H.L. *Chem. Rev.* 1966, 66, 395.
2. Clever, H.L.; Battino, R. in *Solutions and Solubilities*, Ed. M.R.J. Dack, J.Wiley and Sons, New York, 1975, Chapter 7.
3. Hildebrand, J.H.; Prausnitz, J.M.; Scott, R.L. *Regular and Related Solutions*, Van Nostrand Reinhold, New York, 1970, Chapter 8.
4. Markham, A.E.; Kobe, K.A. *Chem. Rev.* 1941, 63, 449.
5. Wilhelm, E.; Battino, R. *Chem. Rev.* 1973, 73, 1.
6. Wilhelm, E.; Battino, R.; Wilcock, R.J. *Chem. Rev.* 1977, 77, 219.
7. Kertes, A.S.; Levy, O.; Markovits, G.Y. in *Experimental Thermochemistry, Vol II*, Ed. B. Vodar and B. LeNaindre, Butterworth, London, 1974, Chapter 15.
8. Wilhelm, E.; *CRC Crit. Rev. Anal. Chem.* 1985/86, 16, 129.
9. Taylor, N.W.; Hildebrand, J.H. *J. Amer. Chem. Soc.* 1923, 45, 682.
10. Kapoor, K.P.; Luckcock, R.G.; Sandbach, J.A. *J. Appl. Chem. Biotech.* 1971, 21, 97.
11. Gerrard, W. *Solubility of Gases and Liquids*, Plenum Press, New York, 1976.
12. Morrison, T.J.; Billet, F. *J. Chem. Soc.* 1948, 2033.
13. Battino, R.; Evans, F.D.; Danforth, W.F. *J. Am. Oil Chem. Soc.* 1968, 45, 830.
14. Hayduk, W.; Walter, E.B.; Simpson, P. *J. Chem. Eng. Data* 1972, 17, 59.

15. Dymond, J.; Hildebrand, J.H. *Ind. Eng. Chem. Fundam.* 1967, 6, 130.
16. Benson, B.B.; Krause, D., Jr.; Peterson, M.A. *J. Solution Chem.* 1979, 8, 655.
17. Conder, J.R.; Young, C.L. *Physicochemical Measurements by Gas Chromatography*, Wiley, Chichester, U.K. 1979.
18. Ng, S.; Harris, H.G.; Prausnitz, J.N. *J. Chem. Eng. Data* 1969, 14, 482.
19. Lenoir, J.Y.; Renault, P.; Renon, H. *J. Chem. Eng. Data* 1971, 16, 340.
20. Sweeney, C.W. *Chromatographia* 1984, 18, 663.
21. Maffiolo, G.; Lenoir, J.Y.; Renon, H. *Chem. Eng. Sci.* 1970, 25, 1847.
22. Wright, R.H.; Maass, O. *Can. J. Research* 1932, 6, 94.
23. Lee, J.I.; Mather, A.E. *Ber. der Bunsen-Gesellschaft* 1977, 81, 1020.
24. Kurata, F.; Kohn, J.P. *Petroleum Processing* 1956, 11 (12), 57.

Chapter 3
RAOULT'S LAW, HENRY'S LAW AND THE REFERENCE LINE

3.1 THE RELATIONSHIP BETWEEN RAOULT'S LAW AND THE REFERENCE LINE

Early work on the vapour pressure of solutions was carried out by von Babo and by Wüllner. The latter concluded from his work on solutions of salts (1856–60) that the vapour pressure lowering was proportional to the concentration of the solution. Such work was extended by Raoult (1887–88) who measured the vapour pressure over dilute solutions of relatively non-volatile solutes in volatile solvents, e.g. diethyl ether. He proposed the relationship:

$$\frac{P^\circ - P}{P^\circ} = \frac{n_2}{n_1 - n_2} = x_2$$

which may be written:

$$P = P^\circ x_1$$

where P° is the vapour pressure of the pure solvent, P the vapour pressure of the solution, x_1 the mole fraction of the solvent in the solution and x_2 the mole fraction of the solute in the solvent.

Raoult demonstrated that the relationship was obeyed approximately by a number of systems. It has subsequently been shown to give the vapour pressure of the solvent over many solutions of non-volatile solutes having concentrations up to about 5 mole% with an accuracy of about 1%. It has also been found that the equations usually give the partial vapour pressure of the major component in a completely miscible mixture of two volatile components when either of these components forms at least 95 mole% of the mixture. The relationship is known as 'Raoult's law'.

Mixtures of two volatile components do not usually follow the relationship to a very high degree of accuracy over the whole of the concentration range. Sometimes the deviations are very marked. Nevertheless, the term 'Raoult's

Law' has persisted even though the equations associated with the law are of restricted validity.

An ideal solution is often defined as one in which the solvent obeys the relationship:

$$P_1 = P_1^\circ x_1$$

over the whole of the concentration range from $x_1 = 0$ to 1. The solute, if it is volatile, is assumed to obey a similar relationship:

$$P_2 = P_2^\circ x_2$$

over the whole of the concentration range from $x_2 = 0$ to 1. Usually, the major component is termed 'solvent' and the minor component 'solute', but the terms may be interchanged as convenient.

The conceptual model of an 'ideal solution' has provided a basis for the development of theories of solutions just as the conceptual model of an 'ideal gas' has provided a basis for developments of theories of gases. Like all conceptual models, it has also been abused when the limitations of the model have been neglected. Again, like other conceptual models in all fields of human enquiry, it has provided a standard against which reality may be measured. The equations mentioned above are also called 'Raoult's law' when they form the basis of the conceptual model of an 'ideal solution'.

It follows from the above that the phrase 'Raoult's law' is used in two distinctly different senses. In one sense, the phrase refers to an experimentally discovered approximate relationship of limited application between the vapour pressure of the solvent and the mole fraction concentration of either solvent or solute in a dilute binary solution. In another sense it means a mathematical relationship between partial vapour pressures and mole fractions of components which is exactly true, by definition, for a certain conceptual model, providing a reference standard for reality. There is some justification for dropping the term 'Raoult's law' when the experimentally found variation of partial pressure of a component with variation of mole fraction of that component is being compared with the variation expressed by the mathematical relationship:

$$P = P^\circ x$$

When a plot of this equation is superimposed on a graph of experimental values of mole fractions plotted against experimental values of partial pressures for constant temperature, and hence constant P°, the line is better termed a 'reference line' than a 'Raoult's law' line. The same applies when the line for a fixed value of P, but varying temperature and hence varying value of P°, is superimposed on experimentally found variation of mole fraction with variation of temperature at constant partial pressure. Use of the term *reference line* gets away from any suspicion that experimental results *ought* to fit on the line and *obey the law*.

Reference lines for a gas depend only on the nature of the gas and not on the solvent and are a measure of the tendency of a gas to pass into a liquid phase. Reference lines are especially helpful for pictorially representing variation of solubility with pressure or temperature and readily enable the comparison of behaviour of one system with another.

At constant temperature a reference line on a plot of mole fraction solubility, x_g, against partial pressure of gas, P_g, is a straight line connecting the origin with the point corresponding to $x_g = 1$ and $P_g = $ vapour pressure of liquefied gas. The reference line for constant pressure follows a curve (see Figs 3.1 and 3.2). If both pressure and temperature are considered to be variables then a reference surface can be constructed (see Fig 3.3).

Above the critical temperature a liquid and a gas become identical fluid phases and the value of P° is an imaginary quantity. Solubilities of gases, measured below and above critical temperatures, show no discontinuities at or near critical temperatures. Values of P° below the critical temperature can be extrapolated to temperatures above the critical to provide hypothetical values for constructing reference lines.

A simple method of extrapolating the vapour pressure of a liquid is to assume that vapour pressure varies with temperature according to the Clausius-Clapeyron equation which is of the type

$$\ln(P/\text{bar}) = A + B/(T/\text{K})$$

where A and B are constants.

The Clausius-Clapeyron equation may be used in the form:

$$\ln(P/\text{bar}) = \frac{-\Delta H}{R}\left[\frac{1}{T/\text{K}} - \frac{1}{T_b/\text{K}}\right]$$

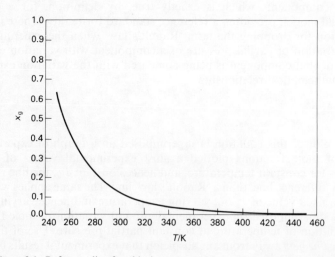

Figure 3.1. Reference line for chlorine at a partial pressure of 1.013 bar (1 atm).

RAOULT'S LAW, HENRY'S LAW AND THE REFERENCE LINE

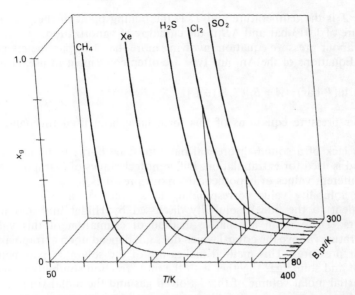

Figure 3.2. Reference lines for a partial pressure of 1.013 bar for gases of differing boiling points (boiling points measured at 1.013 bar).

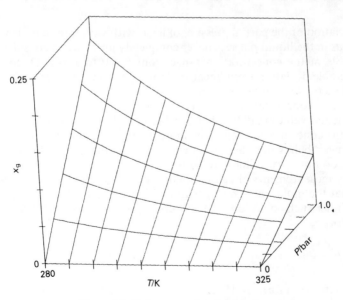

Figure 3.3. Reference surface for chlorine.

where T is the temperature and T_b is the boiling point of the gas under a pressure of 1.013 bar and ΔH is the enthalpy of vaporization.

A vapour pressure equation involving more than two parameters may be used. Equations of the Antoine type are often convenient to use, i.e.

$$\ln(P/\text{bar}) = A + B/[(T/\text{K}) + C]$$

Vapour pressure equations of this form have been used throughout this book.

The lack of a sound theoretical basis must be borne in mind whatever method is used for extrapolation. Different techniques of extrapolation will give different values of hypothetical vapour pressures.

The solubility value corresponding to a point on a reference line is equivalent to the 'ideal' solubility discussed by Hildebrand and his co-workers. There is an alternative method of estimation of this value at temperatures above the critical which does not depend upon extrapolation of vapour pressures. This method makes use of the solubility parameters of solvent and solute gas together with the mole fraction solubility of the gas, the partial molal volume of the dissolved gas and the molal volume of the solvent.[1]

The use of reference lines in the interpretation of the solubility of gases has been extensively discussed by Gerrard.[2,3]

3.2 DIFFICULTIES ASSOCIATED WITH HENRY'S LAW

The variation of the partial pressure of a gas with variation of mole fraction, x_g, of gas in the liquid phase is never completely linear when considered over the whole of the concentration range from $x_g = 0$ to $x_g = 1$. Often there is considerable deviation from linearity. However, it may be expedient to treat a small sector of a curve as a straight line.

For low concentrations of gas in the liquid phase the variation of experimental values may fit a linear relation within experimental error. The bigger the concentration range the more likely is the detection of deviations from linear behaviour. Some systems, such as the sulfur dioxide + water system, do not show a linear relationship, even at low pressures.

Many measurements of gas solubilities have been made under conditions such that the mole fraction of gas in the liquid phase did not exceed about 0.01. Under such conditions it is usually expedient to assume a relationship of the form:

$$P_g = x_g H$$

This equation is a form of 'Henry's law' where H is the Henry's law constant.

If this relationship is used either to predict, for a particular gas, a value of P_g from x_g, or of x_g from P_g, the error becomes greater the greater the value of x_g and of P_g. For a particular value of P_g the error is greater the higher the boiling point of the gas. Whereas it may be a satisfactory equation for predicting the solubility of oxygen at 298 K and a partial pressure of 2 bar, it will be completely unsatisfactory for predicting the solubility of sulfur dioxide at this temperature and pressure (see Fig 5.2).

The mole ratio solubility, N_g, is related to the mole fraction solubility, x_g, by the relationship:

$$N_g = x_g/(1 - x_g)$$

At low concentrations it is sometimes convenient to assume that:

$$N_g = x_g$$

Henry's law is sometimes written in the form:

$$P_g = N_g H$$

The higher the mole ratio solubility, N_g, the less accurate the equation becomes.

Henry's law constants are often treated as limiting values of the ratio of partial pressure to mole fraction solubility, i.e.

$$\lim_{x_g \to 0} \frac{P_g}{x_g} = H$$

In much recent work[4] they are taken to be limiting values of the ratio of fugacity to mole fraction solubility, i.e.

$$\lim_{x_g \to 0} \frac{f_g}{x_g} = H$$

Henry's law constants, H, given in the literature are evaluated in a variety of ways. In some cases H is taken to be the slope of the regression line of a plot of experimental values of P_2 against x_g or N_g measured up to a certain partial pressure which is often about 1.013 bar. In other cases it is calculated from the ratio of P_g to x_g or N_g for a single value of P_g. It is also found by extrapolating P_g/x_g or f_g/x_g to zero pressure.

Unfortunately, authors may not indicate exactly how values of the constant were evaluated. Often there is no precise indication of the individual pressures or range of pressures at which measurements were made. There is then no way of finding the pressure range over which Henry's law is a useful approximation, and valuable information is lost.

3.3 DUHEM-MARGULES EQUATION AND HENRY'S LAW

The fugacities of two volatile components, 1 and 2, in a two component mixture are related to the mole fractions of these components, x_1 and x_2, by the Duhem-Margules[5,6] equation:

$$\frac{d\ln f_1}{d\ln x_1} = \frac{d\ln f_2}{d\ln x_2}$$

If the partial pressures are less than about 1.013 bar then it is usually satisfactory to equate fugacities with partial pressures and to write the equation in the form:

$$\frac{d\ln P_1}{d\ln x_1} = \frac{d\ln P_2}{d\ln x_2} \tag{1}$$

Solutions to this equation put forward by Margules[6] are of the form:

$$P_1 = P_1^\circ x_1 \exp(A x_2^2) \tag{2}$$
$$P_2 = P_2^\circ x_2 \exp(A x_1^2) \tag{3}$$

These equations often give a good approximation to the variation of partial vapour pressure with mole fraction of a component in a two component mixture.

In the case of a solution of a gas in a solvent the equations can be written as:

$$P_g = P_g^\circ x_g \exp(A x_s^2) \tag{4}$$
$$P_s = P_s^\circ x_s \exp(A x_g^2) \tag{5}$$
$$P_{total} = P_g^\circ x_g \exp(A x_s^2) + P_s^\circ x_s \exp(A x_g^2) \tag{6}$$

where 'g' refers to the gas and 's' to the solvent. P_g° and P_s° are the vapour pressures of pure liquefied gas and pure solvent, respectively.

If units of pressure (bar, kPa, mmHg or atm, etc.) are introduced then Equation (4) may be written:

$$\ln \frac{P_g/\text{bar}}{x_g} = \ln(P_g^\circ/\text{bar}) + A x_s^2 \tag{7}$$

As $x_g \to 0$ then $x_s \to 1$
In the limit:

$$\ln \left[\frac{P_g/\text{bar}}{x_g}\right]_{x_g \to 0} = \ln(P_g^\circ/\text{bar}) + A \tag{8}$$

If the limiting value, $H°$, of the Henry's law constant is defined as the limiting value of P_g/x_g as the partial pressure of the gas tends to zero then Equation (7) may be written:

$$\ln(H°/\text{bar}) = \ln(P_g°/\text{bar}) + A \qquad (9)$$

Subtracting Equation (9) from (7) and rearranging gives:

$$\ln \frac{P_g/\text{bar}}{x_g} = \ln(H°/\text{bar}) - A(1 - x_s^2) \qquad (10)$$

Equations (7) to (9) have wider validity if they are written so that fugacities replace partial pressures, but these simple forms are often accurate enough for use at pressures below 1.013 bar.

Equation (10) is similar in form to the Krichevsky-Il'inskaya equation[7,8] which may be written as:

$$\ln \frac{f_g/\text{bar}}{x_g} = \ln(H°/\text{bar}) - A(1 - x_s^2) + \frac{P_g V^*}{RT} \qquad (11)$$

where V^* is the partial molar volume of the solute gas in the liquid phase at infinite dilution.

If the partial vapour pressure of the solute gas is less than about 1.013 bar then the term $(P_g V^*/RT)$ may be sufficiently close to zero to be neglected. In addition, it may be possible to equate the fugacity of the gas with its partial pressure. If these approximations are carried out then Equation (11) becomes identical with Equation (10).

If a system approximately follows Equation (10) then a plot of $(1 - x_s^2)$ against $\ln((P_g/\text{bar})/x_g)$ is close to a straight line of slope $(-A)$ and intercept ln $H°$. This often provides a convenient method of calculating limiting values of Henry's law constant from mole fraction solubilities measured at various pressures.

Equations (10) and (4) sometimes fit experimental points fairly closely over the whole of the mole fraction solubility range if measurements have been confined to pressures below about 1.013 bar. This is less likely to be the case for systems studied at higher pressures.

Lorimer[9] measured the solubility of sulfur dioxide in acetone at low temperatures over almost the whole of the solubility range. Experimental values of mole fraction solubility for various partial pressures are shown in Fig 3.4 together with a point corresponding to pure sulfur dioxide ($x_g = 1$). The limiting value of Henry's law constant, $H°$, from a plot of $(1 - x_s^2)$ against $\ln(P_g/x_g)$ is 0.03577 bar and the value of A is -2.7111. The variation of x_g with variation of P_g from Equation (10) is also shown in the diagram. In this case there is very good correlation between the experimental relationship of pressure with solubility and that given by the equation.

The solubility of ammonia in hexanedinitrile over ranges of temperature

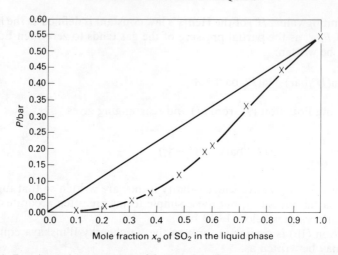

Figure 3.4. Experimental measurements of the variation of the mole fraction solubility of sulfur dioxide in 2-propanone at 250 K, the reference line and the solubility curve from the Duhem-Margules equation (crosses show experimental points).

and pressures was published by Freidson et al.[10] The limiting value of Henry's law constant, estimated as described above, is 5.032 bar. Measurements at 283.15 K in the pressure range 0.089 to 0.959 bar are fitted by the equation:

$$\ln \frac{P_g/\text{bar}}{x_g} = \ln 5.032 + 0.6087(1 - x_s^2) \tag{12}$$

A plot of mole fraction solubility of ammonia against partial pressure is shown in Fig 3.5 together with variation of x_g with P_g corresponding to this equation.

The solubility was also measured in the pressure range 2.60 to 5.69 bar. These measurements do not fit the above equation.

Experimental measurements on this system may also be compared with Equation (4). If the experimental measurement of mole fraction solubility at one partial pressure and the value of the vapour pressure of liquid ammonia are substituted into equation (4), the corresponding value of A may be calculated. This value of A may then be used to estimate variation of mole fraction solubility with partial pressure over a pressure range. The mole fraction solubility of ammonia at 283.2 K and 0.959 bar and a value of the vapour pressure of liquid ammonia of 8.082 bar lead to a value for A of -0.4243. The equation for variation of partial pressure with variation of mole fraction of ammonia is then:

$$P_g/\text{bar} = 8.082 x_g \exp(-0.4243 \, x_s^2) \tag{13}$$

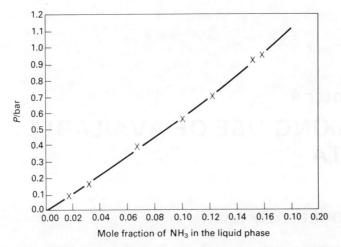

Figure 3.5. Experimental measurements of the variation of the mole fraction solubility of ammonia in hexanedinitrile at 283 K and the solubility curve from the Duhem-Margules equation.

The curve corresponding to this equation is indistinguishable from the curve given by Equation (10) at pressures to 1 atm. The two curves do not coincide at high pressures because of the different methods of estimating the parameters.

In general, Equations (4) and (10) are useful for some, but not all, systems to indicate the general shape of the curve relating partial pressure with mole fraction solubility. They can be extrapolated over limited pressure or mole fraction ranges but are likely to lead to errors at pressures above barometric. Relationships involving Henry's law constant are discussed more rigorously by Wilhelm.[11]

REFERENCES

1. Gjaldbaek, J. Chr.; Andersen, E.K. *Acta Chem. Scand.* 1954, 8, 1398.
2. Gerrard, W. *Solubility of Gases and Liquids*, Plenum Press, New York, 1976.
3. Gerrard, W. *Gas Solubilities—Widespread Applications*, Pergamon Press, Oxford, 1980.
4. Wilhelm, E.; Battino, R.; Wilcock, R.J. *Chem. Rev.* 1977, 77, 219
5. Duhem, P. *Compt. rend.*, 1886, 102, 1449.
6. Margules, M. *Sitzungsber. Wien. Akad.*, 1895, 104, 1243.
7. Krichevsky, I.R.; Il'inskaya, A.A. *Acta Physicochim. URSS* 1945, 20, 327.
8. Bender, E.; Klein, U.; Schmitt, W.Ph.; Prausnitz, J.M. *Fluid Phase Equilib.* 1984, 15, 241.
9. Lorimer, J.W.; Smith, B.C.; Smith, G.H. *J.C.S. Faraday I*, 1975, 71, 2232.
10. Freidson, G.S.; Furmer, I.E.; Amelin, A.G. *VINITI deposited document* 1543-74, 1974.
11. Wilhelm, E. *CRC Crit. Rev. Anal. Chem.* 1985/86, 16, 129.

Chapter 4
MAKING USE OF AVAILABLE DATA

4.1 SOURCES OF DATA

Gas solubility data are published in a variety of forms and types of publications. They may appear in advertisements to promote a particular solvent produced by a chemical company. They may be included in a paper mainly devoted to a theoretical study of the liquid state or in one chiefly concerned with the kinetics of reactions of dissolved gases. Recent measurements are likely to be more reliable than older ones, but even measurements of solubility made towards the end of the nineteenth century may be useful, especially for comparison with more recent work.

There are often problems in interpreting data given in the literature. It is not always clear whether or not allowance has been made for the partial pressure of the solvent, and hence whether pressures which are quoted correspond to partial pressures of gas or to total pressures. The common practice of reporting Henry's law constants without indicating the pressure at which measurements were made can lead to severe limitations on the usefulness of such constants because of invalidity of approximations implied in this law.

Several reviews and books have been devoted to the solubility of gases in liquids. Reviews devoted to a wide range of gases and solvents were published by Markham and Kobe[1] in 1941, by Battino and Clever[2] in 1966 and by Wilhelm and Battino[3] in 1973. The solubilities of gases in water were reviewed by Wilhelm et al.[4] in 1977.

In 1972 Gerrard pointed out general patterns of behaviour of all gas + solvent systems.[5] These were elaborated in two books published in 1976 and 1980.[6,7] Each of these books contains extensive lists of experimental data, many of which were obtained by Gerrard and his co-workers.

Solubility data for gases are also tabulated in books compiled by Seidell and Linke[8] and by Stephen and Stephen,[9] and are included in the Landolt-Bornstein Tables.[10] Older data are included in the International Critical Tables.[11]

The most recent compilations of data for the solubilities of gases in liquids are those which are being produced by The Commission on Solubility Data in the Analytical Division of the International Union of Pure and Applied Chemistry.[12-24] These are published as volumes in the Solubility Data Series. This series is devoted to solubilities of solids and liquids as well as gases. As far as gases are concerned, single volumes are devoted to individual gases or small groups of gases. Almost all the available data for a particular system are compiled and, whenever possible, the reliability of data is evaluated. There are plans to publish update volumes as more data become available. Volumes dealing with many of the important gases have already been published.

4.2 EMPIRICAL EQUATIONS AND THERMODYNAMIC QUANTITIES

It is often convenient to fit an equation to experimental variation of the solubility of a gas with variation of temperature at constant partial pressure of gas. Often the mole fraction solubility, x_g, is expressed as a two constant equation of the form:

$$\ln x_g = A + B/T \tag{1}$$

where A and B are constants for a particular system and partial pressure. Sometimes, data of sufficient accuracy are available to justify fitting an equation containing more than two constants. Equations of the following form are sometimes used:

$$\ln x_g = A + B/T + C \ln T + DT \tag{2}$$

where A, B, C and D are constants for the particular system and partial pressure of gas. These equations are useful for obtaining interpolated values of solubilities within the range of temperature over which experimental measurements have been made. They are less reliable for obtaining extrapolated values of solubilities outside this temperature range.

Quantities called the 'heat or enthalpy of solution', ΔH, 'entropy of solution', ΔS, and 'Gibbs free energy of solution', ΔG, are often calculated from the constants in such equations as these. In the case of Equation (1) these are found as follows:

$\Delta H = -RB$

$\Delta S = RA$

$\Delta G = -RT \ln x_g = -RAT - RB$

Equation (2) leads to the relationships:

$\Delta H = -RB + RCT + RDT^2$

$$\Delta S = RA + RC\ln(T/K) + RC + 2RDT$$
$$\Delta G = -RAT - RB - RCT\ln(T/K) - RDT^2$$

These quantities are sometimes taken to correspond to the transfer of a mole of gas at a partial pressure of 1.01325 bar in the vapour phase to a hypothetical solution of unit mole fraction.[25] A discussion of the precise relationship between such quantities from solubility measurements and the thermodynamic quantities associated with the transfer of gas from the vapour phase to the liquid phase is outside the scope of this book.

Partial molar enthalpies of solution for transfer of gas from the vapour phase to infinitely dilute solution may be found from the variation of solubility with temperature. This has been discussed by Wilhelm.[26] Recently, it has been possible to make direct calorimetric measurement of the partial molar enthalpies of solution of gases at low concentrations of gas.[27–30] There is good agreement with values from solubility measurements. At 293.15 K the partial molar enthalpy change for dissolution of oxygen in water at infinite dilution from solubility measurements by Benson et al.[31] is -12020 J mol^{-1} and from direct calorimetric measurements by Dec and Gill[28] and by Olofsson et al.[30] is -12030 and -12000 J mol^{-1}, respectively.

4.3 CRITERIA FOR EVALUATING PUBLISHED DATA

Gas solubility data given in the literature are of variable reliability. Often, measurements quoted by different authors are incompatible. There are two kinds of information on gas solubilities, primary measurements and secondary data. The primary measurements are made directly during the actual experiment—the total pressures, the weights or volumes of solvent etc. The secondary data are the parameters calculated from the primary measurements, such as Ostwald coefficients, Henry's law constants, and the mole fraction solubility for a particular partial pressure of gas. It is common practice to present incomplete details of primary measurements. Often, for example, there is no indication of the pressures at which measurements were made when Henry's law constants are reported.

If sufficient information is given there are various criteria by which data may be judged in an attempt to assess the reliability of published work.

Although relevant physical properties, such as boiling point and refractive index, are often quoted as an indication of the purity of the solvent, it is not always stated that the solvent was de-gassed immediately prior to use. Liquids exposed to the atmosphere absorb appreciable quantities of carbon dioxide and, to a lesser degree, other gases in the air. The presence of such impurities can seriously interfere with some methods of measurement of gas solubility which involve a measured quantity of gas in contact with a measured quantity of solvent. This is the case when gas solubilities are measured with a gas burette. There is a danger that in such procedures the impurity gases may come out of solution and reduce the partial pressure of the gas under test in the sample bulb holding the solvent under test.

Such a problem is not likely to arise when gas solubilities are measured by bubbling the gas under test through the solvent. In such situations, any impurity gas usually will be swept out of the system. When solubilities are measured by the bubbling method, a gas under test is sometimes deliberately mixed with air so as to reduce its partial pressure below barometric pressure. One is then dealing with a multicomponent system, and the concentration of the gas under test in the liquid phase may be different from the corresponding concentration for the two component system with the same partial pressure of this gas.

A gas under test should have a high purity, although often authors give no indication of the purity of the gas. Traces of air in this gas are just as likely to interfere with an absorption experiment as are traces of air dissolved in the solvent.

Reaching equilibria between gas and liquid phases may be a slow process, but there are few experimental data for accurate prediction of the time needed for equilibria to be established. In the evaluation of published data one may have to make a subjective judgement as to whether equilibrium is likely to have been established under the conditions of the experiment. When solubilities are measured with a gas burette, gaseous impurities tending to be concentrated in the bulb holding the sample may prevent equilibrium with the bulk of the gas phase. When solubilities are measured by bubbling gas through a solvent the liquid phase may become saturated with the gas under test. At the same time, however, the bubbles of gas passing through the solvent may not become saturated with vapour of this solvent. The actual partial pressure of gas in contact with solvent is then uncertain.

Solubilities of gases in solvents with which irreversible chemical reaction may occur are reported in the literature. In these cases it is important that allowance has been made for this irreversible reaction. Sometimes inhibitors are added. It is necessary to assess whether or not the inhibitor is likely to be effective in stopping the chemical reaction under the conditions of the experiment and also to assess whether the addition of this third component is likely to affect the validity of the measurements as representing the behaviour of a two component system.

The computation of the solubility value or values which are finally presented often depends upon a number of approximations. It is the total pressure of the gas phase which is measured in a gas solubility experiment. This pressure is made up of the partial pressure of solvent and of gas under test. If the published solubility relates to a stated partial pressure of gas then allowance usually has to be made for the partial pressure of the solvent. This is not necessary in cases where neglect of this partial pressure would introduce an error which was significantly less than the other experimental errors in the experiment. The method used by Hildebrand[32] for correcting mole fraction solubilities, measured at a particular total pressure to give mole fraction solubilities at a stated partial pressure, is often followed by other authors. In this method, it is assumed that the mole fraction solubility of the gas is proportional to partial pressure of gas and that the vapour pressure of the solvent is proportional to the mole fraction of the solvent. These approximations can lead to errors in the corrected solubility, but in many cases the errors are small.

Authors may assume that the gas under test obeys the ideal gas laws under the conditions of the experiment and when secondary data such as Ostwald coefficients are calculated. This is justified in some situations but in others it is not. Recalculation of such secondary data from the primary data using the properties of the real gas may be necessary.

Another approximation which is often carried out is to equate the volume of a final solution with the volume of the original solvent, on the assumption that dissolution of gas causes no change in volume. This can lead to errors in values of Ostwald coefficients which outweigh experimental errors. Such errors increase with the increase in mole fraction of gas in the liquid phase.

Whether or not unjustifiable approximations have been carried out, it is a sound policy to check the arithmetic involved in conversion of primary to secondary data, if this is possible. There are some surprising mistakes in the literature of gas solubility.

Mole fraction solubilities for various partial pressures of gas which are either given as such in the literature or can be calculated from other secondary data may be tested by comparing a plot of mole fraction solubility against partial pressure of the gas with the corresponding reference line. One can then see whether the experimental data bear the relation to the reference line which would be expected for the particular gas + solvent system on the basis of analogy with comparable systems.

In the same way, mole fraction solubilities for the same partial pressure of gas but different temperatures may be plotted against temperature and tested to see whether they bear the relationship with a reference line for constant pressure which would be expected by analogy with other comparable systems.

Some systems have been investigated by more than one worker or group of workers. The greater the number of independent measurements in different laboratories which agree with each other, the greater the chance that reliable measurements have been made, especially if different techniques have been used. Sometimes different workers have made measurements at different temperatures and/or pressures. The pattern of each set of values of mole fraction solubilities relative to appropriate reference lines may then be compared. The greater the similarity in behaviour, the more likely that all measurements are reliable.

Unfortunately, a more common occurrence is for measurements by different workers to be incompatible. One is faced with the much more difficult problem of deciding which measurements, if any, are reliable. Differences of 10% between values presented by two groups are not uncommon, even though each group appears to have worked to a high standard of precision. Under these circumstances one can only go by analogies with similar systems using appropriate reference lines as a semi-quantitative guide.

4.4 PREDICTING GAS SOLUBILITIES

The solubility of a gas at a particular temperature depends on:

(a) the partial pressure of the gas;
(b) the vapour pressure of pure liquefied gas at that temperature;
(c) interactions between molecules in the liquid phase.

In the conceptual model associated with Raoult's law the mole fraction gas solubility, x_g, is taken to be given by:

$$x_g = P_g/P°$$

where P_g is the partial pressure of the gas and $P°$ is the vapour pressure of the liquefied gas at the same temperature. This simple model is inadequate for reliable prediction of gas solubilities. It is useful, however, to base reference lines on the equation above.

The value of $P°$ increases with increase in temperature. At a constant partial pressure, the variation of $P_g/P°$ with variation of temperature corresponds to a constant pressure reference line. At a constant temperature, the variation of $P_g/P°$ with variation of P_g corresponds to a constant temperature reference line. Alternatively, one can define a reference surface corresponding to the values of $P_g/P°$ when both temperature and partial pressure are variables. For a particular gas, reference lines and the reference surface give a measure of the tendency of the gas to pass into the liquid phase, irrespective of the nature of the solvent.

When the partial pressure of the gas is equal to the vapour pressure of liquefied gas, condensation to pure liquefied gas will occur ($x_g = 1$) as long as this partial pressure is maintained. When the partial pressure is zero the mole fraction solubility must also be zero. Mole fraction gas solubilities, x_g, for different partial pressures at a constant temperature which is below the critical temperature, must approach the constant temperature reference line as the partial pressure, P_g, of the gas approaches the vapour pressure, $P°$, of pure liquefied gas. Values of x_g must also approach this reference line as the partial pressure approaches zero.

The behaviour between these extremes depends upon the nature of both the gas and the solvent. In some cases experimental values of x_g fall below the reference line values over the whole of the composition range between $x_g = 0$ and $x_g = 1$. Sometimes all experimental values fall above the reference line. In some cases experimental values may fall below the reference line over part of the range and above the line over the rest of the range.

Many pure solvents have some degree of short range ordering of molecules, especially if hydrogen bonding between these molecules can occur. Dissolution of solute molecules will tend to disrupt such short range microstructure. This structure breaking effect with its associated energy change will tend to reduce solubility.

Increase in temperature in the absence of a solute also causes a reduction in the short range ordering in a liquid. The structure breaking effect of a solute therefore becomes less significant the higher the temperature, because the pure solvent becomes more disordered. The tendency of the short range ordering in a solvent to reduce the solubility of a solute is therefore less marked the higher the temperature.

In practice it is often found that, for a particular partial pressure of solute gas, experimental values of mole fraction solubilities, x_g are less than corresponding values of $P_g/P°$ at lower temperatures but greater than $P_g/P°$ at higher temperatures. The experimental variation of x_g with temperature crosses the constant pressure reference line from the lower solubility to the higher solubility side as the temperature is increased. In other systems, the experimental curve appears to approach the reference line from the lower solubility side, but not to cross it, as the temperature is increased. This type of behaviour may be related to the break up of short range order in the solvent and the lessening of the structure of the solvent to inhibit dissolution of gas.

With some systems, experimental values of mole fraction solubilities, in part of the temperature range, increase with rise in temperature. The thermal break up of the short range ordering of the solvent may be important in determining the behaviour of such systems.

Gas molecules may be solvated to a greater or lesser extent when dissolved in a solvent. This structure making effect causes energy changes favourable to dissolution, in contrast to the structure breaking effect mentioned above. Whether the experimentally observed mole fraction solubility is greater or less than the reference line value will be related to the relative importance of structure making and structure breaking effects. The extent of this solvation leading to structure making will decrease with increase in temperature. In some systems mole fraction solubility is higher than the reference line at lower temperatures but lower than the reference line at higher temperatures. With such systems it appears that the solvation plays a bigger part in determining the pattern of solubility than does the breaking of the structure of the solvent.

The following generalizations about the solubility of a gas may be made.

Increase in partial pressure of a gas always increases its solubility.

At temperatures below the critical temperature of the gas, increase in temperature, at constant partial pressure, usually decreases the solubility. At higher temperatures the solubility may pass through a minimum. At a partial pressure of 1.013 bar the solubility of helium in water[33] passes through a minimum at about 304 K. The minimum value of the solubility of hydrogen sulfide in water[34] is at about 450 K.

It usually follows that the higher the boiling point of a gas under a pressure of 1.013 bar the lower the vapour pressure, $P°$, of the liquefied gas at a particular temperature. It is therefore to be expected that the higher the boiling point of the gas the higher the mole fraction solubility in a given solvent and under a given partial pressure and temperature. Constant pressure reference lines for different gases give an indication of the way in which the variation of solubility with temperature will depend upon the boiling point of the gas (see Fig. 3.2). Specific interactions between gas and solvent molecules may, however, reverse this general trend.

Solvents having similar chemical structure tend to behave in a similar manner in the dissolution of a particular gas. The variations with temperature of x_g for a fixed partial pressure of 1.013 bar (1 atm) are likely to show similar relationships with the reference line for that pressure. If values of x_g for one solvent tend, for instance, to approach the reference line from the lower solubility side, as the temperature is raised, then it is likely that a solvent of similar chemical structure will show similar behaviour.

Strongly hydrogen bonded solvents may be poor solvents for a gas unless there is also a strong tendency for the solvent molecules to be hydrogen bonded to the gas molecules.

Many theoretical papers have been devoted to the development of reliable methods of accurately predicting gas solubilities from bulk or molecular properties of the components. Some of these methods for predicting solubilities in non-polar solvents are derived from Hildebrand's regular solution theory.[35] Sebastian et al.[36] have, for example, developed a method of predicting the solubility of hydrogen in hydrocarbon solvents in the temperature range of 310 to 700 K and pressures to 300 bar. The predicted solubilities are close to the experimental values.

Group contribution methods have also been developed to estimate solubilities. Such methods have been used to predict solubilities of non-polar gases in normal alcohols to an accuracy of about 6% and in alcohol + water mixtures to about 10%.[37]

There is also considerable interest in the use of the scaled particle theory in which, for the purpose of analysis of the energy changes involved, dissolution is taken to consist of two hypothetical stages. The first stage consists of the creation of a cavity in the solvent and the second stage the filling of the cavity with the gas molecule. On the basis of this model, it is possible to predict the temperature at which the solubility of a gas in water reaches a minimum.[38]

Statistical mechanical methods have been used to estimate the solubility of a number of gases with various degrees of agreement with experimental measurements. Developments in these and other methods of estimating gas solubilities have been reviewed by Wilhelm,[26] who has also provided a comprehensive bibliography of recent primary publications. A detailed discussion of modern work in these fields is outside the scope of this book.

REFERENCES

1. Markham, A.E.; Kobe, K.A. *Chem. Rev.* 1941, 28, 519.
2. Battino, R.; Clever, H.L. *Chem. Rev.* 1966, 66, 395.
3. Wilhelm, E.; Battino, R. *Chem. Rev.* 1973, 73, 1.
4. Wilhelm, E.; Battino, R.; Wilcock, R.J. *Chem. Rev.* 1977, 77, 219.
5. Gerrard, W. *J. Appl. Tech. Biotechnol.* 1972, 22, 623.

6. Gerrard, W. *Solubility of Gases and Liquids*, Plenum Press, New York, 1976.
7. Gerrard, W. *Gas Solubilities—Widespread Applications*, Pergamon Press, Oxford, 1980.
8. Seidell, A.; Linke, W.F. *Solubilities of Inorganic and Organic Compounds*, American Chemical Society, Washington, D.C. 4th ed. 1965.
9. Stephen, H.; Stephen, T. *Solubilities of Inorganic and Organic Compounds*, Pergamon Press, Oxford, 1963.
10. Landolt-Bornstein, *Physikalisch-Chemische Tabellen* (a series), Springer, Berlin.
11. *International Critical Tables Vol. III*, McGraw Hill, New York, 1928.
12. Clever, H.L. *Solubility Data Series, Vol. 1, Helium and Neon*, Pergamon Press, Oxford, 1979.
13. Clever, H.L. *Solubility Data Series, Vol. 2, Krypton, Xenon and Radon*, Pergamon Press, Oxford, 1979.
14. Clever, H.L. *Solubility Data Series, Vol. 4, Argon*, Pergamon Press, Oxford, 1980.
15. Young, C.L. *Solubility Data Series, Vol. 5/6, Hydrogen and Deuterium*, Pergamon Press, Oxford, 1981.
16. Battino, R. *Solubility Data Series, Vol. 7, Oxygen and Ozone*, Pergamon Press, Oxford, 1981.
17. Young, C.L. *Solubility Data Series, Vol. 8, Oxides of Nitrogen*, Pergamon Press, Oxford, 1981.
18. Hayduk, W. *Solubility Data Series, Vol. 9, Ethane*, Pergamon Press, Oxford, 1982.
19. Battino, R. *Solubility Data Series, Vol. 10, Nitrogen and Air*, Pergamon Press, Oxford, 1982.
20. Young, C.L. *Solubility Data Series, Vol. 12, Sulfur Dioxide, Chlorine, Fluorine and Chlorine Oxides*, Pergamon Press, Oxford, 1983.
21. Young, C.L.; Fogg, P.G.T. *Solubility Data Series, Vol. 21, Ammonia, Amines, Phosphine, Arsine, Stibine, Silane, Germane and Stannane in Organic Solvents*, Pergamon Press, Oxford, 1985.
22. Hayduk, W. *Solubility Data Series, Vol. 24, Propane, Butane and 2-Methylpropane*, Pergamon Press, Oxford, 1986.
23. Clever, H.L.; Young, C.L. *Solubility Data Series, Vol. 27/28, Methane*, Pergamon Press, Oxford, 1987.
24. Fogg, P.G.T.; Young, C.L. *Solubility Data Series, Vol. 32, Hydrogen Sulfide, Deuterium Sulfide and Hydrogen Selenide*, Pergamon Press, Oxford, 1988.
25. Battino, R.; Clever, H.L. Ref. 20, page xvii.
26. Wilhelm, E. *CRC Crit. Rev. Anal. Chem.* 1985/86, 16, 129.
27. Gill, S.J.; Wadsö, I. *J. Chem. Thermodyn.* 1982, 14, 905.
28. Dec, S.F.; Gill, S.J. *J. Solution Chem.* 1984, 13, 27.
29. Dec. S.F.; Gill, S.J. *Rev. Sci. Instrum.* 1984, 55, 765.
30. Olofsson, G.; Oshodi, A.A.; Qvarnström, E.; Wadsö, I. *J. Chem. Thermodyn.* 1984, 16, 1041.
31. Benson, B.B.; Krause, D.,Jr.; Peterson, M.A. *J. Solution Chem.* 1979, 8, 655.
32. Taylor, N.W.; Hildebrand, J.H. *J. Amer. Chem. Soc.* 1923, 45, 682.
33. Battino, R. Ref. 6, page 1.
34. Fogg, P.G.T. Ref. 24, page 1.
35. Hildebrand, J.H.; Prausnitz, J.M.; Scott, R.L. *Regular and Related Solutions*, Van Nostrand Reinhold, New York, 1970.
36. Sebastian, H.M.; Lin, H.M.; Chao, K.C. *Am. Inst. Chem. Eng. J.* 1981, 27, 138.
37. Tochigi, K.; Kojima, K. *Fluid Phase Equil.* 1982, 8, 221.
38. Schulze, G.; Prausnitz, J.M. *Ind. Eng. Chem. Fundam.* 1981, 20, 177.

Chapter 5

SOLUBILITY OF SULFUR DIOXIDE

5.1 GENERAL BEHAVIOUR

Sulfur dioxide has the following physical properties:

Melting point	= 197.7 K
Boiling point (1.013 bar)	= 263.13 K
Critical temperature	= 430.7 K
Critical pressure	= 78.80 bar
Critical volume	= 0.122 dm^3 mol^{-1}
Density of gas at 273.15 K; 1.013 bar	= 2.927 g dm^{-3}
Relative mol. mass	= 64.06

The vapour pressures, P, of the pure liquid are given in International Critical Tables for 203 K to 430 K and fit Antoine equations of the type

$$\ln(P/\text{bar}) = A - B/[(T/\text{K}) + C].$$

Values of A, B and C are as follows:

Temp. range	A	B	C
203 to 273 K	9.9934	2251.33	−37.74
273 to 373 K	10.5120	2515.25	−23.68
373 to 430 K	13.6447	5230.11	133.15

The reference line for a partial pressure of sulfur dioxide of 1.013 bar is shown in Fig. 5.1.

Sulfur dioxide forms addition compounds with benzene and some other aromatic hydrocarbons, with many organic nitrogen bases and with some compounds containing oxygen. Some of these compounds have been isolated,[1-5] but many melt and may decompose at temperatures below about 273 K. The 1:1 compound with methanol,[6] for example, melts at 202 K and that with 2-propanone[4] at 191 K. The 1:1 compound with 1,4-dioxane[4] melts at 275.7 K but is highly dissociated at its melting point.

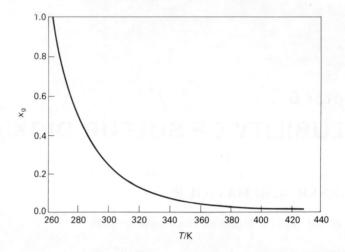

Figure 5.1. Reference line for sulfur dioxide at a partial pressure of 1.013 bar.

Compounds with amines are more stable, for example the 1:1 compound with N-methylbenzenamine,[1] melts at 304 K.

Sulfur dioxide can act as an electron acceptor in the presence of an electron donor. The solubility is high, relative to the reference line, in electron donor solvents such as dimethylformamide, pyridine and benzonitrile. Solubilities may be low, relative to the reference line, in strongly hydrogen bonded solvents such as 1,2-ethanediol and water, although this may not be the case at low partial pressures of sulfur dioxide. Estimations of the variation of solubility of sulfur dioxide with variation of pressure by use of one of the forms of Henry's law can lead to gross inaccuracy, especially at low temperature.

Solubilities in some solvents have been studied by more than one group of workers and consistent and reliable solubility values have emerged. In other cases quoted values are subject to a degree of uncertainty because of inherent errors in the method of measurement or disagreement between groups of workers. Sufficient data are available, however, for the general behaviour of sulfur dioxide in a wide range of solvents to be presented. This enables approximate predictions of the behaviour in solvents for which data are not available.

Although the solubilities of sulfur dioxide have been studied by very many workers, a large proportion of these measurements have been made by a relatively small number of people. The accuracy of the overall picture of the solubilities is very dependent upon the reliability of these measurements.

Gerrard[8,9] measured solubilities in a wide variety of organic solvents at 273 to 298 K and total pressures never greater than 1.067 bar. The absorption vessel of capacity 20–25 cm^3 was fitted with a side arm carrying a tap and ground glass socket and inlet tube carrying a tap and ground glass cone. This was evacuated and weighed. An appropriate quantity of solvent under test was added, air pumped off, and the vessel reweighed to give the weight of solvent. The vessel was then held in a constant temperature bath, and sulfur dioxide gas bubbled through the liquid with both taps open and the total pressure equal to barometric. Taps were then closed, the supply of gas disconnected and the vessel reweighed. The process of passing gas through the liquid and reweighing was repeated until maximum weight was attained. Allowance was made for the weight of gas in the space above the liquid when the weight of sulfur dioxide absorbed was calculated.

Weights of gas absorbed for total pressures less than barometric were measured by connecting the absorption vessel containing solvent saturated with sulfur dioxide at barometric pressure to a manometer and high vacuum system. Portions of the dissolved sulfur dioxide could be pumped away, the total pressure over the solution measured, and the absorption vessel and contents reweighed to enable the weight of dissolved gas to be calculated. The weight of the liquid when the total pressure was close to the vapour pressure of the pure liquid was usually found to be close to the original weight of the solvent, indicating that negligible quantities of solvent had been lost during the measurements.

It was estimated that solubilities were accurate to about 4% with solvents of low volatility.

Sano[10,11] measured the solubility of sulfur dioxide in 105 liquids at 303.2 K. Two methods were used. In one, a measurement was made of the volume of gas absorbed by solvent in a thermostatically regulated and magnetically stirred 50 cm^3 flask. In the other, sulfur dioxide was bubbled through the solvent and absorption measured by iodimetry. Bunsen coefficients were reported after allowing for the partial pressure of the solvent. There was usually close agreement between the two methods.

Makranczy and Patyi, in two papers[12,13] written in association with various other workers, stated that they used a static method described by Bodor *et al.*[14] However, Bodor *et al.* described apparatus for use below 273 K but referred to another paper[15] in which there is a description of apparatus for use above 273 K. In each case the volume of gas absorbed by a given quantity of liquid was measured by a gas burette. Allowance was made for the vapour pressure of the solvent.

5.2 SOLUBILITY OF SULFUR DIOXIDE IN WATER AND IN AQUEOUS SOLUTIONS

Sulfur dioxide, dissolved in water, is in equilibrium with other species in solution. Literature values of solubilities usually refer to bulk solubilities which include the sulfur dioxide equivalent to the sulfur in all species derived from the dissolved sulfur dioxide.

Battino[17] evaluated the available measurements of the solubility of sulfur dioxide in water. He derived a smoothing equation for the mole fraction solubility at a partial pressure of 1.013 bar which was based upon measurements by Hudson,[18] by Beuschlein and Simenson,[19] by Rabe and Harris[20] and by Lavrova and Tudorovskaya.[21] This equation may be written :

$$\ln x_g = -51.445 + 4575.5/(T/K) + 5.6854\ln(T/K)$$

standard deviation in $x_g = 0.00017$
temperature range 283 to 380 K.

The variation of solubility of sulfur dioxide with pressure does not approximate to Henry's law, even at very low pressures (see Fig. 5.2). The lower the concentration of the sulfur dioxide the larger the proportion that is ionized. As a consequence, the lower the partial pressure of sulfur dioxide the greater the relative contribution of the ionization process to the overall dissolution of sulfur dioxide. At 298.2 K under a partial pressure of 1.013 bar, about 10% of the dissolved sulfur dioxide is ionized. The proportion is about 75% under a partial pressure of 0.1013 bar.

Various workers have studied the dissolution of sulfur dioxide in solutions of electrolytes. For comprehensive information, the reader should refer to the compilation and evaluation of solubilities of sulfur dioxide edited by Young.[16]

Fox[22] measured solubilities in solutions of 19 salts at various concentrations. Solubilities were reported as Ostwald coefficients taken to be the volume of gas divided by the volume of solution. It is uncertain, however, whether the 'volume of solution' refers to the solution of the salt before or after addition of sulfur dioxide. This ambiguity limits the utility of these measurements. Solutions of all salts were studied at 298 K, and some were also studied at 308 K. The solubility of sulfur dioxide as measured by the

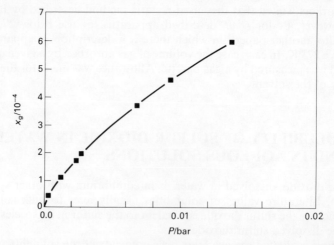

Figure 5.2. Variation with partial pressure of the solubility of sulfur dioxide in water at 298.2 K (Ref. 48).

Ostwald coefficient increased with increase in salt concentration to 3 mol dm^{-3} for solutions of ammonium or potassium chloride, bromide, nitrate and thiocyanate and for solutions of sodium bromide and thiocyanate. Potassium iodide, cadmium iodide and ammonium sulfate solutions, studied to 1.5 mol dm^{-3}, also showed an increase in solubility of gas with increasing salt concentration. Solubilities decreased as salt concentration was increased to 1.5 mol dm^{-3} in the case of cadmium chloride, bromide, and sulfate and also sodium sulfate. Solubilities also decreased as the concentration of sodium chloride was increased to 3 mol dm^{-3}.

Hudson[18] studied the solubility of sulfur dioxide in solutions of potassium chloride and of sodium sulfate at a total pressure equal to barometric pressure. The temperatures were from 293 K to 363 K, in the case of potassium chloride, and to 323 K in the case of sodium sulfate. The weight of sulfur dioxide absorbed by 100 g of water increased at all temperatures as the potassium chloride concentration was increased to 20 g in 100 g of water. Increase in the concentration of sodium sulfate caused the solubility of sulfur dioxide to pass through a maximum. Increase in temperature caused this maximum to occur at higher concentration of salt (see Fig. 5.3).

Jager[23] studied the effect of zinc sulfate on the solubility of sulfur dioxide, as measured by the weight of gas absorbed per dm^3 of solution of salt, from 293 to 373 K. Solubility decreased with increase of zinc sulfate concentration except at 373 K at concentrations of zinc sulfate above 124 g dm^{-3}.

The solubility, under barometric pressure, of sulfur dioxide in aqueous solutions of sulfuric acid of various concentrations was measured by Miles and Fenton[24] at 293 K and by Cupr[25] at 314.2 and 335.2 K. The mole fraction of sulfur dioxide in solution may be calculated without taking ionization into account and defined as:

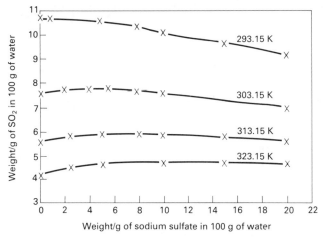

Figure 5.3. Dependence on concentration of salt and temperature of the solubility of sulfur dioxide in aqueous solutions of sodium sulfate at a partial pressure of gas of 1.013 bar.

moles of SO_2/(total moles of $H_2SO_4 + SO_2 + H_2O$)

Cupr's measurements at 314.2 K and 335.2 K show that the mole fraction solubility passes through a minimum with increase in acid concentration (Fig. 5.4). Measurements by Miles and Fenton confirm that, at high acid concentrations, the mole fraction solubility of gas increases with concentration of acid.

Measurements by Lavrova and Tudorovskaya[21] show that mole fraction solubilities in hydrochloric acid pass through a maximum as the acid concentration is increased.

5.3 SOLUBILITY OF SULFUR DIOXIDE IN NON-AQUEOUS SOLVENTS

5.3.1 Solubilities in non-aromatic hydrocarbons

Solubilities in these solvents, in the temperature and pressure ranges in which most measurements have been made, fall well below the reference lines. Mole fraction solubilities for individual compounds when the partial pressure of sulfur dioxide is 1.013 bar are given in Table 5.1. The mole fraction solubilities of all alkanes at this partial pressure are likely to be close to the general pattern indicated in Fig. 5.5.

Solubility in alkanes has been investigated by a number of workers. Makranczy et al.[12] measured solubilities at 298 K and 313 K in alkanes with

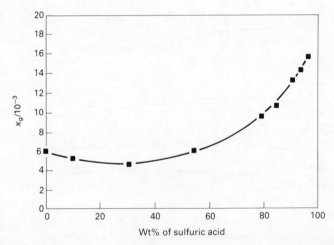

Figure 5.4. Variation in the solubility of sulfur dioxide, at a partial pressure of 1.013 bar and 335.15 K, with concentration of aqueous sulfuric acid.

Table 5.1 Mole fraction solubility of sulfur dioxide in alkanes at a partial pressure of 1.013 bar.

Solvent	T/K	x_g	Ref.
Pentane	298.2	0.0543	12
	313.2	0.0402	12
Hexane	298.2	0.0562	12
	298.2	0.0534	13
	303.2	0.0449	10
	313.2	0.0408	12
Heptane	298.2	0.0567	12
	313.2	0.0417	12
Octane	298.2	0.0583	12
	313.2	0.0429	12
2,2,4-Trimethylpentane	293.2	0.0710	27
Nonane	298.2	0.0602	12
	313.2	0.0441	12
Decane	273.2	0.151	9
	293.2	0.095	9
	298.2	0.0620	12
	313.2	0.0456	12
Undecane	298.2	0.0647	12
	313.2	0.0471	12
Dodecane	298.2	0.0662	12
	313.2	0.0484	12
Tridecane	298.2	0.0682	12
	313.2	0.0500	12
Tetradecane	298.2	0.0702	12
	313.2	0.0514	12
Pentadecane	298.2	0.0722	12
	313.2	0.0531	12
Hexadecane	298.2	0.0745	12
	313.2	0.0545	12
	298.2	0.077	29
	300	0.066	28
	325	0.045	28
	350	0.032	28
	375	0.025	28
	400	0.020	28
	425	0.017	28
	450	0.014	28
	475	0.013	28
Heptadecane	323.2	0.061	29
Cyclohexane	298.2	0.0625	13
	298.2	0.035	30
	303	0.039	10
Decahydronaphthalene	293.2	0.060	9
	293.2	0.048	21
	298.2	0.053	29
	323.2	0.039	29
	303	0.043	10
1,2,3,4-Tetrahydronaphthalene	293.2	0.228	9
	303	0.202	10

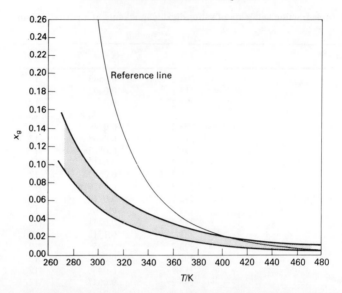

Figure 5.5. General pattern of solubilities of SO$_2$ in alkanes. Mole fraction solubilities, x_g, of SO$_2$ in alkanes for a partial pressure of SO$_2$ of 1.013 bar, estimated from data in the literature, fall in the shaded area of the graph. Mole fraction solubilities at this partial pressure for other temperatures and other alkanes may lie between extrapolations of the upper and lower boundaries of the shaded area. In all cases the mole fraction solubility for a partial pressure of 1.013 bar will approach 1.0 as the temperature is reduced to 263.15 K, the boiling point of SO$_2$ at a pressure of 1.013 bar, provided the system does not split into two liquid layers.

five to sixteen carbon atoms. Values were corrected to correspond to a partial pressure of sulfur dioxide of 1.013 bar. Mole fraction solubilities at 298 K for this partial pressure range from 0.0543 in pentane to 0.0745 in hexadecane.

The solubility in hexane has also been measured by Sano[11] and by Patyi et al.[13] Values of mole fraction solubility for a partial pressure of 1.013 bar are plotted in Fig. 5.6 and fitted to a curve of the form

$$\ln x_g = -9.081 + 1835/(T/\text{K})$$

where x_g is the mole fraction solubility of sulfur dioxide at a partial pressure of sulfur dioxide of 1.013 bar.

Solubilities reported by Sobolov et al.[26] for octane, 2,2,4-trimethylpentane and dodecane at 1.013 bar and five temperatures are an internally consistent set of values but are about half the values suggested by the general pattern of measurements by other workers. There is no obvious reason for the discrepancy.

Nitta et al.[27] measured the mole fraction solubility in 2,2,4-trimethylpentane at various partial pressures and 293 K. The interpolated value for 1.013 bar is 0.069. This may be compared with an extrapolated value of 0.065 for octane from measurements by Macranczy.[12]

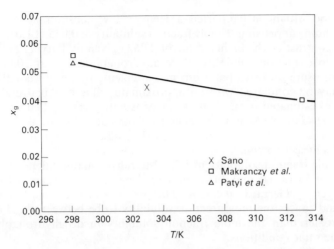

Figure 5.6. Mole fraction solubility in hexane of sulfur dioxide at a partial pressure of 1.013 bar.

Gerrard[9] gave a value of 0.095 for the mole fraction solubility in decane at a partial pressure of 1.013 bar and 293 K. This is appreciably higher than the value of 0.069 by extrapolation of values given by Makranczy (see Fig. 5.7).

The mole fraction solubility in hexadecane at 293 K is 0.068 when estimated from the Henry's law constant given by Tremper and Prausnitz[28] and is 0.077 from that given by Lenoir et al.[29]

Solubilities at partial pressures of sulfur dioxide of 1.013 bar have been measured in cyclohexane by Patyi et al.[13] at 298 K and by Sano[10] at 303 K.

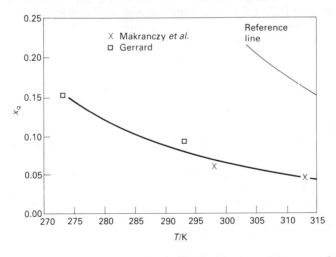

Figure 5.7. Mole fraction solubility in decane of sulfur dioxide at a partial pressure of 1.013 bar.

Benoit and Milanova[30] reported a Henry's law constant at 298 K. The corresponding approximate mole fraction solubility of 0.035 at 1.013 bar is in marked contrast to the higher value of 0.0625 given by Patyi et al.[13]

The mole fraction solubility in decahydronaphthalene at 303 K and a partial pressure of 1.013 bar from measurements by Sano[10] is 0.043. This value may be compared with an approximate value of 0.050 under these conditions estimated from Henry's law constants given by Lenoir et al.[29] A value of 0.048 at 293 K may be calculated from data given by Weissenberger and Hadwiger.[31] This is in contrast to the value of 0.060 at 293 K from Gerrard's measurements.[9]

The mole fraction solubility in 1,2,3,4-tetrahydronaphthalene at a partial pressure of 1.013 bar and 303 K measured by Sano[10] is in accord with the value given by Gerrard[9] at 293 K. The presence of a double bond in this compound causes the mole fraction solubility to be between three and five times greater than the mole fraction solubility in decahydronaphthalene under the same conditions.

5.3.2 Solubilities in aromatic hydrocarbons

Mole fraction solubilities in aromatic hydrocarbons for a partial pressure of sulfur dioxide of 1.013 bar are higher than those in alkanes at the same temperature and pressure. The general region in which values fall in relation to the reference line, is shown in Fig. 5.8.

Measurements of the solubility in benzene at various pressures and temperatures by different groups of workers are consistent to within about 10% and are given in Table 5.2. Measurements by Horiuti[32] ranged from 283 to 333 K and total pressures from 0.38 to 1.35 bar. Ipatieff[33] made measurements to 2.69 bar. Mole fraction solubilities for a partial pressure of 1.013 bar are plotted in Fig. 5.9. The mole fraction solubility in benzene is about three times that in cyclohexane at the same temperature at this partial pressure. As noted above there is a similar difference between the mole fraction solubility in the unsaturated 1,2,3,4-tetrahydronaphthalene and in the saturated decahydronaphthalene.

The variation of mole fraction solubility with partial pressure at 298 K is shown in Fig. 5.10.

The mole fraction solubility in methylbenzene at a partial pressure of 1.013 bar and temperatures from 273 to 333 K may be calculated from solubilities measured by Sano,[10] Lloyd[34] and Gerrard.[8] These form a consistent pattern. (See Table 5.3 and Fig. 5.11.) Lorimer et al.[35] measured the partial pressures of sulfur dioxide over mixtures of methylbenzene and sulfur dioxide at 228, 237 and 250 K, for the whole of the composition range of the liquid phase. It is interesting that a plot of the mole fraction of sulfur dioxide against partial pressure of sulfur dioxide has a reverse sigmoid shape with points above the reference line for low concentrations of gas and below for high concentrations (Fig. 5.12).

The behaviour of sulfur dioxide in 1,3-dimethylbenzene and in 1,3,5-

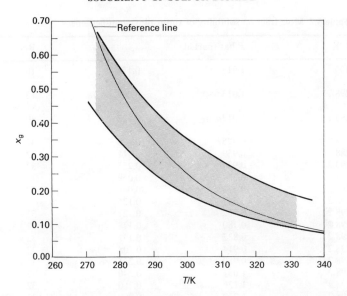

Figure 5.8. General pattern of solubilities of SO$_2$ in aromatic hydrocarbons. Mole fraction solubilities, x_g, of SO$_2$ in aromatic hydrocarbons for a partial pressure of SO$_2$ of 1.013 bar, estimated from data in the literature, fall in the shaded area of the graph. Mole fraction solubilities at this partial pressure for other temperatures and other aromatic hydrocarbons may lie between extrapolations of the upper and lower boundaries of the shaded area. In all cases the mole fraction solubility for a partial pressure of 1.013 bar will approach 1.0 as the temperature is reduced to 263.15 K, the boiling point of SO$_2$ at a pressure of 1.013 bar, provided the system does not split into two liquid layers.

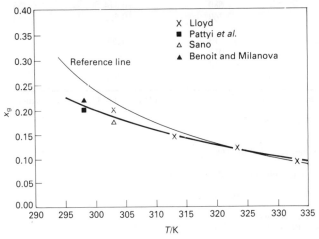

Figure 5.9. Mole fraction solubility in benzene of sulfur dioxide at a partial pressure of 1.013 bar.

Table 5.2 Mole fraction solubility of sulfur dioxide in benzene

T/K	P/bar (partial)	x_g	Ref.
303	1.013	0.176	10
298.2	1.013	0.202	13
298.2	1.013	0.222	30
283.2	0.975	0.376	32
288.2	0.959	0.311	32
293.2	0.939	0.255	32
298.2	0.264	0.059	32
298.2	0.273	0.062	32
298.2	0.547	0.125	32
298.2	0.562	0.127	32
298.2	0.780	0.178	32
298.2	0.832	0.191	32
298.2	0.914	0.208	32
298.2	0.980	0.225	32
298.2	1.051	0.240	32
298.2	1.136	0.260	32
298.2	1.210	0.280	32
298.2	1.258	0.291	32
303.2	0.883	0.172	32
308.2	0.845	0.140	32
313.2	0.800	0.114	32
318.2	0.746	0.092	32
323.2	0.681	0.074	32
328.2	0.606	0.057	32
333.2	0.517	0.044	32
298	2.640	0.605	33
299	1.120	0.231	33
299	1.680	0.314	33
300	2.360	0.508	33
303.2	1.013	0.200	34
313.2	1.013	0.146	34
323.2	1.013	0.126	34
333.2	1.013	0.096	34

Mole fraction solubility for a partial pressure of SO_2 of 1.013 bar fits the equation

$$\ln x_g = -8.844 + 2174/(T/K)$$

standard deviation of $x_g = 0.010$
temperature range 298 to 333 K.

Mole fraction solubility for a total pressure of 1.013 bar fits the equation

$$\ln x_g = -13.891 + 3658/(T/K)$$

standard deviation of $x_g = 0.036$
temperature range 298 to 333 K.

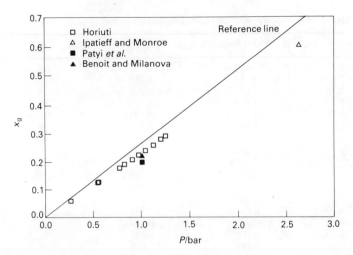

Figure 5.10. Mole fraction solubility in benzene of sulfur dioxide at 298 K.

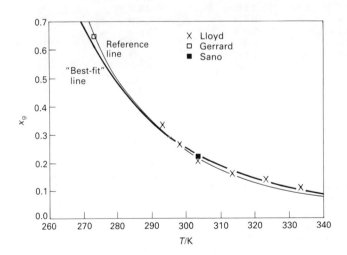

Figure 5.11. Mole fraction solubility in methylbenzene of sulfur dioxide at a partial pressure of 1.013 bar.

Table 5.3 Mole fraction solubility of sulfur dioxide in methylbenzene

T/K	P/bar (partial)	x_g	Ref.
293.2	1.013	0.331	18
298.2	1.013	0.268	18
303.2	1.013	0.204	18
313.2	1.013	0.162	18
323.2	1.013	0.142	18
333.2	1.013	0.110	18
273.2	1.013	0.648	8
227.6	0.010	0.086	35
227.6	0.019	0.195	35
227.6	0.040	0.288	35
227.6	0.058	0.383	35
227.6	0.082	0.513	35
227.6	0.100	0.603	35
227.6	0.109	0.656	35
227.6	0.124	0.762	35
227.6	0.139	0.881	35
227.6	0.156	1.000	35
237.4	0.018	0.083	35
237.4	0.040	0.185	35
237.4	0.074	0.279	35
237.4	0.104	0.363	35
237.4	0.147	0.490	35
237.4	0.175	0.590	35
237.4	0.191	0.637	35
237.4	0.221	0.752	35
237.4	0.248	0.878	35
237.4	0.279	1.000	35
249.8	0.036	0.076	35
249.8	0.085	0.175	35
249.8	0.141	0.263	35
249.8	0.185	0.325	35
249.8	0.269	0.452	35
249.8	0.326	0.558	35
249.8	0.353	0.595	35
249.8	0.408	0.729	35
249.8	0.484	0.869	35
249.8	0.541	1.000	35
303.2	1.013	0.216	10

Mole fraction solubility for a partial pressure of SO_2 of 1.013 bar fits the equation

$$\ln x_g = -10.392 + 2706/(T/K)$$

standard deviation of $x_g = 0.020$
temperature range 273 to 333 K.

Mole fraction solubility for a total pressure of 1.013 bar fits the equation

$$\ln x_g = -11.306 + 2963/(T/K)$$

standard deviation of $x_g = 0.018$
temperature range 273 to 333 K.

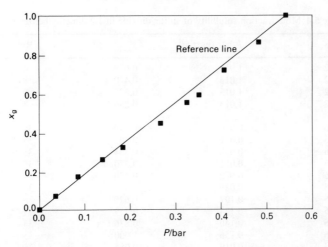

Figure 5.12. Variation with partial pressure of the mole fraction solubility of sulfur dioxide in methylbenzene at 249.8 K.

trimethylbenzene is similar to that in methylbenzene (Tables 5.3 and 5.4 and Figs. 5.13 to 5.16).

5.3.3 Solubilities in compounds containing oxygen

The presence of oxygen in a compound usually favours the dissolution of sulfur dioxide, because oxygen atoms tend to be electron-donors in organic

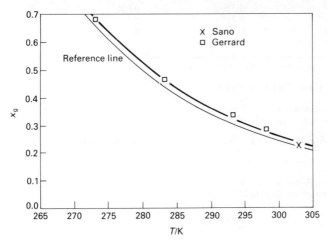

Figure 5.13. Mole fraction solubility in 1,3-dimethylbenzene of sulfur dioxide at a partial pressure of 1.013 bar.

Table 5.4 Mole fraction solubility of sulfur dioxide in 1,3-dimethylbenzene

T/K	P/bar (partial)	x_g	Ref.
273.2	1.013	0.679	8
283.2	1.013	0.470	8
293.2	1.013	0.338	8
298.2	1.013	0.286	8
227.6	0.010	0.089	35
227.6	0.021	0.202	35
227.6	0.032	0.304	35
227.6	0.051	0.370	35
227.6	0.070	0.470	35
227.6	0.089	0.572	35
227.6	0.107	0.625	35
227.6	0.124	0.759	35
227.6	0.140	0.876	35
227.6	0.156	1.000	35
237.4	0.019	0.085	35
237.4	0.040	0.195	35
237.4	0.062	0.296	35
237.4	0.090	0.352	35
237.4	0.124	0.452	35
237.4	0.161	0.557	35
237.4	0.179	0.605	35
237.4	0.222	0.748	35
237.4	0.251	0.873	35
237.4	0.279	1.000	35
249.9	0.037	0.078	35
249.9	0.080	0.183	35
249.9	0.122	0.280	35
249.9	0.159	0.319	35
249.9	0.224	0.416	35
249.9	0.296	0.523	35
249.9	0.320	0.556	35
249.9	0.417	0.719	35
249.9	0.484	0.864	35
249.9	0.544	1.000	35

Mole fraction solubility for a partial pressure of SO_2 of 1.013 bar fits the equation

$$\ln x_g = -10.656 + 2805/(T/K)$$

standard deviation of $x_g = 0.001$
temperature range 273 to 298 K.

Mole fraction solubility for a total pressure of 1.013 bar fits the equation

$$\ln x_g = -10.744 + 2829/(T/K)$$

standard deviation of $x_g = 0.001$
temperature range 273 to 298 K.

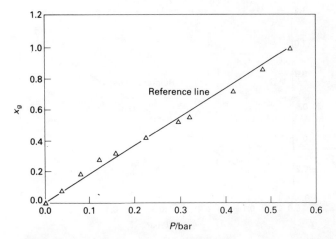

Figure 5.14. Variation with partial pressure of the mole fraction solubility of sulfur dioxide in 1,3-dimethylbenzene at 249.9 K (Ref. 35).

compounds. The solubility of a gas may be reduced if the oxygen atoms take part in hydrogen bonding between solvent molecules. The breaking of such hydrogen bonds to accommodate a gas molecule may be energetically unfavourable.

Mole fraction solubilities in alkanols are higher than those in alkanes from which they are derived. However, in all experimental measurements, the mole fraction solubilities for a partial pressure of 1.013 bar fall below the

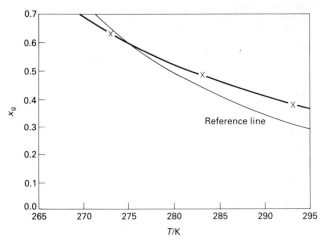

Figure 5.15. Mole fraction solubility in 1,3,5-trimethylbenzene of sulfur dioxide at a partial pressure of 1.013 bar (Ref. 9).

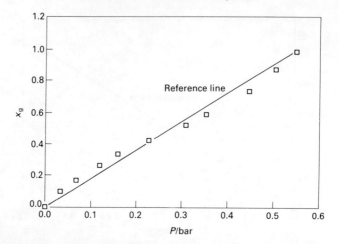

Figure 5.16. Variation with partial pressure of the mole fraction solubility of sulfur dioxide in 1,3,5-trimethylbenzene at 250.0 K (Ref. 35).

corresponding reference line. Experimental values are given in Table 5.5. In the case of straight chain primary alkanols these solubilities show an almost consistent decrease with increase in chain length. The single measurement on 1-hexanol is slightly out of line, suggesting that the reported value is a little too high (Fig. 5.17). The available evidence indicates that solubilities in secondary alkanols and branched chain primary alkanols are similar to those in straight chain alkanols of the same carbon number. Mole fraction

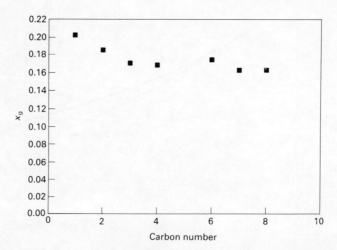

Figure 5.17. Mole fraction solubilities of sulfur dioxide at a partial pressure of 1.013 bar and a temperature of 303 K in straight chain primary alkanols.

Table 5.5 Mole fraction solubility of sulfur dioxide at a partial pressure of 1.013 bar in solvents containing carbon, hydrogen and oxygen

Solvent	T/K	x_g	Ref.
Methanol	298.2	0.241	8
	299.2	0.219	49
	303.2	0.194	11
Ethanol	291.4	0.271	49
	292.7	0.226	38
	293.2	0.272	8
	293.2	0.251	37
	299.2	0.202	49
	303.2	0.185	11
1-Butanol	303.2	0.169	11
1-Hexanol	303.2	0.175	11
1-Octanol	298.2	0.200	8
2-Propanol	303.2	0.155	11
2-Methyl-2-propanol	303.2	0.160	11
1,2-Propanediol	303.2	0.169	11
2,2'-Oxybisethanol	303.2	0.333	11
2-Methoxyethanol	303.2	0.395	11
2-Butoxyethanol	303.2	0.377	11
2-Phenoxyethanol	303.2	0.329	11
Benzenemethanol	303.2	0.219	11
2-Furanmethanol	303.2	0.251	11
Cyclohexanol	303.2	0.139	11
	293.2	0.216	9
Tetrahydrofuran	303.2	0.355	11
1,4-Dioxane	288.2	0.636	8
	303.2	0.487	11
1,1'-Oxybisethane	293.2	0.370[a]	37
1,1'-Oxybisoctane	293.2	0.315	1[c]
2,2'-Oxybispropane (diisopropyl ether)	303.2	0.190	11
Dimethoxymethane	303.2	0.550	11
1,2-Dimethoxyethane	303.2	0.507	11
Cyclohexanone	293.2	0.521	9
	303.2	0.362	11
2-Propanone	298.2	0.483[b]	32
1-Phenylethanone	303.2	0.315	11

[a] At total pressure of 1.183 bar.
[b] At total pressure of 1.141 bar.
[c] Extrapolated.

solubilities, at a partial pressure of 1.013 bar, in monohydric alkanols containing no other functional groups lie in the region indicated in Fig. 5.18.

Measurements of the solubility in methanol and those in ethanol at a partial pressure of 1.013 bar (Figs 5.19 and 5.20) are fairly consistent. Mole fraction solubility in methanol is greater than that in ethanol.

Solubilities have been measured in 1-heptanol to high pressures and various temperatures by Albright et al.[36] Mole fraction solubilities greater than 0.8 have been reached. These show non-linear variation with partial

58 SOLUBILITY OF GASES IN LIQUIDS

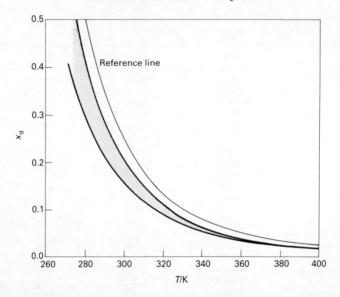

Figure 5.18. General pattern of solubilities of SO_2 in alkanols. Mole fraction solubilities, x_g, of SO_2 in alkanols for a partial pressure of SO_2 of 1.013 bar, estimated from data in the literature, fall in the shaded area of the graph. Mole fraction solubilities at this partial pressure for other temperatures and other alkanols may lie between extrapolations of the upper and lower boundaries of the shaded area. In all cases the mole fraction solubility for a partial pressure of 1.013 bar will approach 1.0 as the temperature is reduced to 263.15 K, the boiling point of SO_2 at a pressure of 1.013 bar, provided the system does not split into two liquid layers.

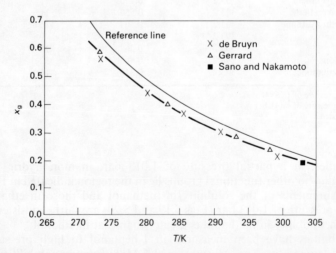

Figure 5.19. Mole fraction solubility in methanol of sulfur dioxide at a partial pressure of 1.013 bar (Refs 8,11,49).

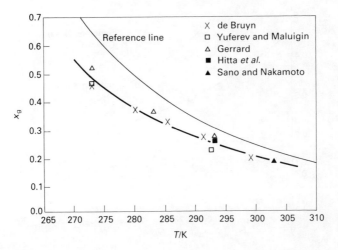

Figure 5.20. Mole fraction solubility in ethanol of sulfur dioxide at a partial pressure of 1.013 bar (Refs 8,11,37,38,49).

pressure (Fig. 5.21). Solubilities in 1-octanol, measured by Gerrard[8] for a partial pressure of 1.013 bar, are consistent with measurements by Sano and Nakamoto[11] (Fig. 5.22).

Mole fraction solubilities in 1,2-ethanediol,[11,29,39] 1,2-propanediol[11] and 1,2,3-propanetriol[11] are similar to those in ethanol but that in 2,2'-oxybisethanol[11] is very high (0.33 at 303.2 K), because of the presence of the ether linkage. The presence of an ether linkage also results in a high mole fraction

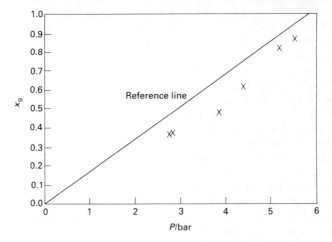

Figure 5.21. Variation with partial pressure of the mole fraction solubility of sulfur dioxide in 1-heptanol at 310.9 K (Ref. 35).

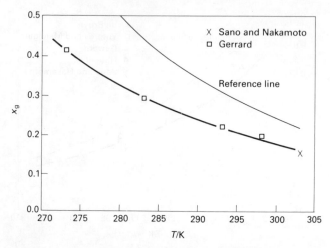

Figure 5.22. Mole fraction solubility in 1-octanol of sulfur dioxide at a partial pressure of 1.013 bar.

solubility in 2-methoxyethanol, 2-butoxyethanol and 2-phenoxyethanol (methyl, butyl and phenyl Cellosolves).[11]

Measurements at 303 K and 1.013 bar indicate that solubilities in the two aromatic alcohols, benzenemethanol[11,29] and 2-furanmethanol,[11] are higher than in aliphatic alcohols. Values are close to that of methylbenzene under the same conditions.

Most liquids containing the ether linkage are good solvents for sulfur dioxide because of their tendency to be electron donors. Solubility in 1,1'-oxybisethane has been measured in the range 273 to 293 K by Nitta et al.[37] and by Yufarev and Maluigin.[38] Total pressures were reported. Sano[10] reported a Bunsen coefficient at 303 K. Solubilities lie above the reference line but values corrected to a partial pressure of 1.013 bar are subject to uncertainty because of the high vapour pressure of the solvent (Fig. 5.23).

The mole fraction solubilities in dibutyl ether have been measured[1] at 293.2 K over a partial pressure range of 0.103 to 0.633 bar (Fig. 5.24). Values are lower than for diethyl ether but still lie above the reference line. The single value for diisopropyl ether, reported by Sano[10], is unusually low. No other values for this system are available for comparison. This low value could be due to steric effects preventing close approach of the sulfur dioxide molecules to the ether linkage.

Dimethoxymethane and 1,2-dimethoxyethane contain two ether linkages and the corresponding mole fraction solubilities are high.[10]

Mole fraction solubilities in the aromatic ethers, methoxybenzene,[10] ethoxybenzene[8] and 1,1'-(methoxymethylene)bisbenzene[10] (i.e. benzyl ether), are similar to those in aliphatic ethers. The introduction of a second methoxy group in a benzene ring increases the mole fraction solubility, but the

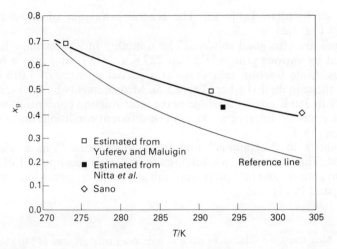

Figure 5.23. Approximate values of the mole fraction solubility in 1,1'-oxybisethane of sulfur dioxide at a partial pressure of 1.013 bar.

reported solubility in 1,3-dimethoxybenzene is markedly greater than that in 1,4-dimethoxybenzene.[10]

The cyclic ethers, tetrahydrofuran[10,30] and tetrahydro-2-furanmethanol,[11] behave in a similar manner to aliphatic ethers but the presence of two ether links in 1,4-dioxane greatly increases solubility.[8,10,27]

Solubilities in various cyclic and linear compounds containing an ether

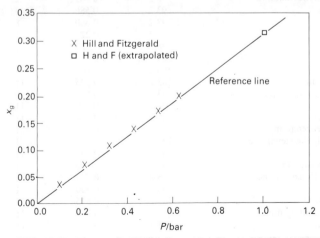

Figure 5.24. Variation with partial pressure of the mole fraction solubility of sulfur dioxide in 1,1'-oxybisbutane at 293.15 K.

linkage are listed in Table 5.6. The general behaviour of mono-ethers is shown in Fig. 5.25.

Ketones are also good solvents. The solubility in 2-propanone has been measured by various groups[10,32,35,40] at 227 K to 313 K and over a range of pressures. Mole fraction solubilities at a partial pressure of 1.013 bar are close to those in diethyl ether (Table 5.5). Measurements by Lorimer et al.[35] from 227 to 250 K cover the whole of the mole fraction concentration range. Measurements by different workers under different conditions are consistent with each other.

Solubilities in 2-octanone[36] and in cyclohexanone[9,10] have also been measured. The mole fraction solubilities for a partial pressure of 1.013 bar in these two ketones and in 2-propanone fall above the reference line within the shaded band in Fig. 5.26.

Table 5.6 Mole fraction solubilities of sulfur dioxide at a partial pressure of 1.013 bar in solvents containing nitrogen

Solvent	T/K	x_g	Ref.
Nitrobenzene	298.2	0.311	34
	298.2	0.293	36[a]
	298.2	0.305	8
	298.2	0.314	30
	303.2	0.333	10
1-Methyl-2-nitrobenzene	298.2	0.299	34
Nitromethane	303.2	0.347	10
Benzonitrile	298.2	0.333	8
	303.2	0.337	10
Acetonitrile	303.2	0.479	10
Benzeneacetonitrile	303.2	0.439	10
N,N-Dimethylformamide	298.2	0.589	42
	293.2	0.586	43
	298.2	0.600	8
Formamide	303.2	0.222	11
Pyridine	298.2	0.609	8
	303.2	0.581	8[a]
	303.2	0.606	11
Methylpyridine	303.2	0.595	11
1-Methyl-2-pyrrolidinone	303.2	0.603	10
Quinoline	303.2	0.549	11
Benzenamine	303.2	0.534	11
N-Methylbenzenamine	298.2	0.560	1[a]
N,N-Dimethylbenzenamine	298.2	0.610	1[a]
	303.2	0.591	11
N-Ethylbenzenamine	298.2	0.560	1[a]
N,N-Diethylbenzenamine	273.2	0.784	2[b]
	298.2	0.530	1[a]
1-Butanamine	303.2	0.615	11
N,N-Dibutylbutanamine	303.2	0.669	11
2-Aminoethanol	303.2	0.52	11

[a] Extrapolated.
[b] Interpolated.

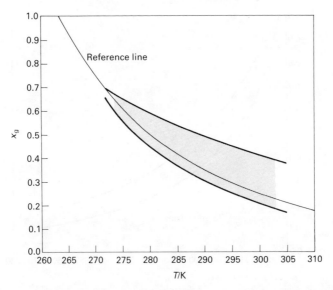

Figure 5.25. General pattern of solubilities of SO_2 in mono-ethers. Mole fraction solubilities, x_g, of SO_2 in mono-ethers for a partial pressure of SO_2 of 1.013 bar, estimated from data in the literature, fall in the shaded area of the graph. Mole fraction solubilities at this partial pressure for other temperatures and other mono-ethers may lie between extrapolations of the upper and lower boundaries of the shaded area. In all cases the mole fraction solubility for a partial pressure of 1.013 bar will approach 1.0 as the temperature is reduced to 263.15 K, the boiling point of SO_2 at a pressure of 1.013 bar, provided the system does not split into two liquid layers.

4-Methyl-1,3-dioxolan-2-one (propylene carbonate) contains a carbonyl group attached to two oxygen atoms. The few data that are available[29,30] indicate a high mole fraction solubility relative to the reference line.

The available mole fraction solubilities in carboxylic acids for a partial pressure of 1.013 bar are, in contrast, below the reference line. Values for acetic acid,[11,27] propanoic acid[11] and hexanoic acid[8,11] lie within a fairly narrow band with mole fraction solubility of gas decreasing with increasing chain length of acid (Fig. 5.27). The single measurement for formic acid[11] is out of line and lies below this band. This low value may be due to extensive hydrogen bonding between formic acid molecules, inhibiting the entry of sulfur dioxide molecules into the liquid.

Mole fraction solubilities in esters,[8–11] at a partial pressure of 1.013 bar, lie in a wide band above the reference line (Fig. 5.28). In the case of esters derived from aliphatic alcohols and aliphatic carboxylic acids, values decrease with increase in chain length of alcohol or acid. However, the mole fraction solubility in benzyl acetate is the same as in methyl acetate and that in benzyl benzoate close to that in ethyl acetate. At this pressure, the mole fraction solubility in tributyl phosphate is very high.[10,41]

The mole fraction solubility in phenol,[11,29] at a partial pressure of gas of 1.013 bar, lies close to the reference line. The single values for 2-chloro-

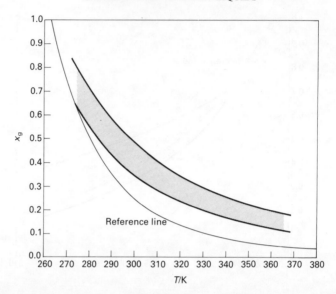

Figure 5.26. General pattern of solubilities of SO_2 in ketones. Mole fraction solubilities, x_g, of SO_2 in ketones for a partial pressure of SO_2 of 1.013 bar, estimated from data in the literature, fall in the shaded area of the graph. Mole fraction solubilities at this partial pressure for other temperatures and other ketones may lie between extrapolations of the upper and lower boundaries of the shaded area. In all cases the mole fraction solubility for a partial pressure of 1.013 bar will approach 1.0 as the temperature is reduced to 263.15 K, the boiling point of SO_2 at a pressure of 1.013 bar, provided the system does not split into two liquid layers.

phenol[11] and 1,2-benzenediol[11] at 303 K lie close to the value for phenol at this temperature. The ether linkage in 2-methoxyphenol enhances the mole fraction solubility relative to that in phenol[11] (Fig. 5.29).

5.3.4 Solubilities in compounds containing nitrogen

Mole fraction solubilities at a partial pressure of 1.013 bar, in compounds containing nitrogen, lie above the reference line. Values for a partial pressure of 1.013 bar are given in Table 5.6. In some cases the solubility is very high. This behaviour is associated with the tendency of compounds containing nitrogen to act as electron donor solvents.

Solubilities in nitrobenzene have been investigated by several workers. Measurements by Gerrard[8] and by Lloyd,[34] corrected to a partial pressure of sulfur dioxide of 1.013 bar, are consistent with each other. They agree with the solubilities extrapolated from measurements at low pressures by Benoit and Milanova,[30] and with those extrapolated from measurements at high pressures by Albright et al.[36] The mole fraction solubility, from measure-

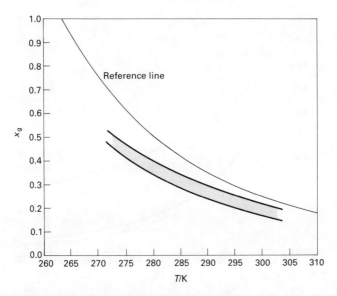

Figure 5.27. General pattern of solubilities of SO_2 in aliphatic monocarboxylic acids. Mole fraction solubilities, x_g, of SO_2 in aliphatic monocarboxylic acids for a partial pressure of SO_2 of 1.013 bar, estimated from data in the literature, fall in the shaded area of the graph. Mole fraction solubilities at this partial pressure for other temperatures and other aliphatic monocarboxylic acids may lie between extrapolations of the upper and lower boundaries of the shaded area. In all cases the mole fraction solubility for a partial pressure of 1.013 bar will approach 1.0 as the temperature is reduced to 263.15 K. the boiling point of SO_2 at a pressure of 1.013 bar, provided the system does not split into two liquid layers.

ments by Sano,[10] is greater than that indicated by other sets of measurements (Fig. 5.30).

Mole fraction solubilities to high pressures, measured by Albright et al.,[36] indicate that the solubility curves for constant temperature cross the corresponding reference lines (see Fig. 5.31).

Lloyd[34] also measured solubility in 1-methyl-2-nitrobenzene at a partial pressure of 1.013 bar and over a temperature range which overlaps the range for measurements of the solubility in nitrobenzene. As would be expected, mole fraction solubilities at this pressure in the substituted nitrobenzene are almost identical with values for nitrobenzene itself at the same temperature.

Measurements by Sano[10] indicate that the mole fraction solubility in nitromethane is 0.347 at 303 K and 1.013 bar, about 25% greater than the solubility in nitrobenzene under these conditions. Under these conditions the mole fraction solubility in benzonitrile is between 0.33 and 0.34, about the same as that in nitromethane. Solubilities in acetonitrile and benzyl cyanide are appreciably greater, 0.479 and 0.439, respectively.[10]

There is good agreement on the solubility in N,N-dimethylformamide (DMF) between various workers.[8,10,30,36,42,43] The mole fraction solubility in

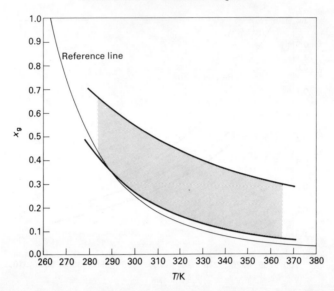

Figure 5.28. General pattern of solubilities of SO_2 in esters of monocarboxylic acids. Mole fraction solubilities, x_g, of SO_2 in esters of aliphatic monocarboxylic acids for a partial pressure of SO_2 of 1.013 bar, reported in the literature, fall in the shaded area of the graph. Mole fraction solubilities at this partial pressure for other temperatures and other esters may lie between extrapolations of the upper and lower boundaries of the shaded area. In all cases the mole fraction solubility for a partial pressure of 1.013 bar will become closer to 1.0 as the temperature is reduced to 263.15 K, the boiling point of SO_2 at a pressure of 1.013 bar, provided the system does not split into two liquid layers.

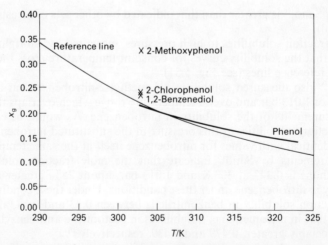

Figure 5.29. Mole fraction solubilities in phenol, 2–chlorophenol, 2–methoxyphenol and 1,2 benzenediol of sulfur dioxide at a partial pressure of 1.013 bar.

Table 5.7 Mole fraction solubilities of sulfur dioxide at a partial pressure of 1.013 bar in solvents containing halogens or sulfur

Solvent	T/K	x_g	Ref.
Trichloromethane	298.2	0.168	45
	303.2	0.118	10
Tetrachloromethane	298.2	0.071	32
	303.2	0.058	10
1,2-Dichloroethane	303.2	0.228	10
1,2-Dichloropropane	303.2	0.167	10
1-Chlorooctane	298.2	0.174	8
1-Bromooctane	298.2	0.163	8
2,2,2-Trichloroethanol	298.2	0.167	8
Chlorobenzene	293.2	0.220	32
	293.2	0.219	8
	298.2	0.187	8
	303.2	0.156	32
	303.2	0.145	10
1,2-Dichlorobenzene	303.2	0.105	32
2-Chlorophenol	303.2	0.244	10
Bromobenzene	298.2	0.174	8
Iodobenzene	298.2	0.153	8
Sulfinylbismethane	293.2	0.638	46
	303.2	0.587	10
Dimethyl sulfate	303.2	0.325	10
Tetrahydrothiophene 1,1-dioxide	303.2	0.482	10

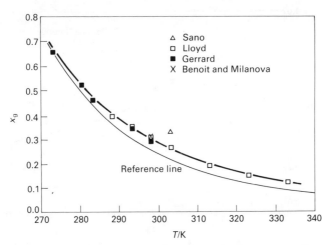

Figure 5.30. Mole fraction solubility in nitrobenzene of sulfur dioxide at a partial pressure of 1.013 bar.

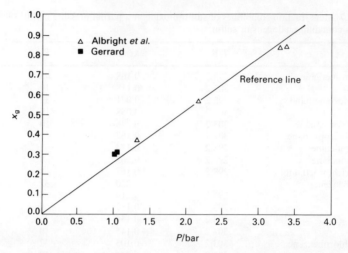

Figure 5.31. Variation with partial pressure of the mole fraction solubility of sulfur dioxide in nitrobenzene at 298.15 K.

this solvent has been measured at a number of temperatures and pressures. It is high compared with that for most other solvents and lies above the reference line. The higher the temperature the more the experimental curve for 1.013 bar deviates from the reference line (Fig. 5.32).

The mole fraction solubility in formamide at a partial pressure of 1.013 bar and 303 K, from measurements by Sano,[11] is 0.222. This is about half the value for DMF under the same conditions.

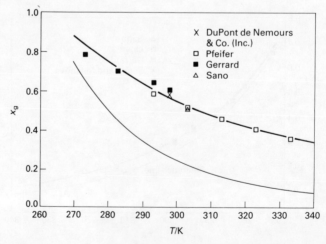

Figure 5.32. Mole fraction solubility in N,N-dimethylformamide of sulfur dioxide at a partial pressure of 1.013 bar.

The solubility in pyridine at several temperatures and pressures has been measured by Gerrard.[8] The value at 303.2 K and a partial pressure of 1.013 bar, as given by Sano and Nakamoto,[11] is close to an extrapolation of Gerrard's values. The mole fraction solubility in this solvent is very high in relation to the reference line. Gerrard's measurements[8,44] of solubility at pressures from 0.136 bar to 1.013 bar show that there is a non-linear variation of mole fraction solubility with pressure in this range. There is a steep increase in solubility with pressure at very low pressures, but, at pressures close to 1.013 bar, the solubility is almost insensitive to changes in pressure (see Fig. 5.33). This behaviour is associated with compound formation in solution between the amine and the gas.

Measurements by Sano using methylpyridine and also 1-methyl-2-pyrrolidinone, at 303 K and a partial pressure of 1.013 bar, indicate mole fraction solubilities close to that in pyridine.[10,11] The mole fraction solubility in quinoline under these conditions is about 15% lower.

Sano and Nakamoto[11] measured the solubility in benzenamine at 303 K and a partial pressure of 1.013 bar. Foote and Fleischer[2] measured the solubility in N,N-diethylbenzenamine at 273 K over almost the complete pressure range (Fig. 5.34). Hill and Fitzgerald[1] made measurements of the solubility in this solvent over a wide pressure range at 298 K. The two sets of measurements are consistent with each other.

Hill and Fitzgerald[1] also measured the solubility in N,N-dimethylbenzenamine at 298 K over a pressure range. These measurements are compatible with the measurement of the solubility at 303 K and a partial pressure of 1.013 bar, reported by Sano and Nakamoto.[11] Hill and Fitzgerald[1] also measured the solubility in N-methyl- and in N-ethylbenzenamine over a range of pressures at 298 K. The mole fraction solubilities in aromatic amines for a partial pressure of 1.013 bar all fall in the shaded area in Fig. 5.35.

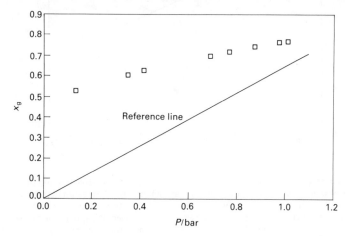

Figure 5.33. Variation with partial pressure of the mole fraction solubility of sulfur dioxide in pyridine at 273.15 K.

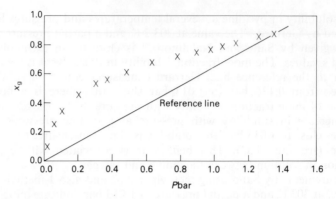

Figure 5.34. Variation with partial pressure of the mole fraction solubility of sulfur dioxide in N,N-diethylbenzenamine at 273 K.

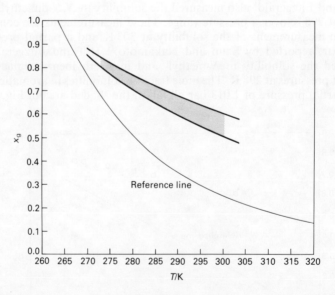

Figure 5.35. General pattern of solubilities of SO_2 in aromatic amines. Mole fraction solubilities, x_g, of SO_2 in aromatic amines for a partial pressure of SO_2 of 1.013 bar, estimated from data in the literature, fall in the shaded area of the graph. Mole fraction solubilities at this partial pressure for other temperatures and other amines may lie between extrapolations of the upper and lower boundaries of the shaded area. In all cases the mole fraction solubility for a partial pressure of 1.013 bar will approach 1.0 as the temperature is reduced to 263.15 K, the boiling point of SO_2 at a pressure of 1.013 bar, provided the system does not split into two liquid layers.

Sano and Nakamoto[11] measured solubilities in 1-butanamine, N,N-dibutylbutanamine and 2-aminoethanol at 303 K and a partial pressure of 1.013 bar. Mole fraction solubilities for the first two of these fall above the values reported for aromatic amines at this temperature. The value for 2-aminoethanol falls within the range for aromatic amines.

5.3.5 Solubilities in compounds containing halogens

Lorimer et al.[35] measured the solubility of sulfur dioxide in trichloromethane at 228 to 250 K over the whole of the pressure range. Lindner[45] made measurements at low pressures to 0.29 bar at 273 K, and to 0.65 bar at 298 K. Approximate values of the mole fraction solubility at a partial pressure of 1.013 bar can be found by extrapolating Lindner's measurements and these values compared with measurements at 1.013 bar and 303 K by Sano[10] (Fig. 5.36). Measurements reported by Lorimer are likely to be reliable but close comparison with measurements at much higher temperatures by the other two workers is not possible. However, the three sets of measurements are not incompatible and some reliance can be placed on values reported by Sano and by Lindner (Table 5.7).

Horiuti[32] reported the solubility in tetrachloromethane at temperatures from 298 to 313 K and pressures less than 1 bar. Sano[10] measured the solubility at a partial pressure of 1.013 bar at 303 K. This measurement is approximately the same as the value obtained by appropriate extrapolation and interpolation of Horiuti's values (Fig. 5.37). Sobolev et al.[26] also

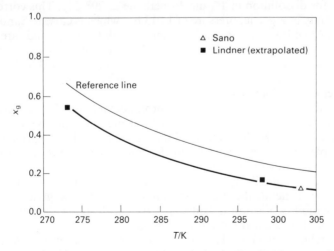

Figure 5.36. Mole fraction solubility in trichloromethane of sulfur dioxide at a partial pressure of 1.013 bar.

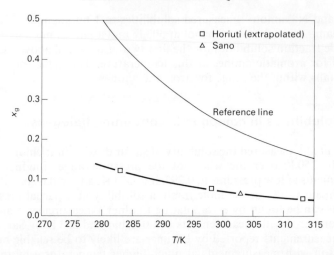

Figure 5.37. Mole fraction solubility in tetrachloromethane of sulfur dioxide at a partial pressure of 1.013 bar.

measured the solubility at atmospheric pressure and temperatures between 283 and 303 K but these values are very low in comparison with measurements by Horiuti and by Sano.

Solubilities at a partial pressure of 1.013 bar in 1,2-dichloroethane and in 1,2-dichloropropane at 303 K were measured by Sano.[10] The mole fraction solubility in 1,2-dichloroethane is higher than that in 1,2-dichloropropane under these conditions. Benoit and Milanova[30] published the Henry's law constant for dissolution in 1,2-dichloroethane at 298.2 K. This corresponds to a solubility at a partial pressure of 1.013 bar which is close to Sano's value at 303 K. Mole fraction solubilities in these dichloro compounds are greater than that in trichloromethane under the same conditions. Sano also measured the solubility in trichloroethene at 1.013 bar and 303 K. In this case the mole fraction solubility is less than that in trichloromethane and close to that in tetrachloromethane.

Solubilities in 1-chlorooctane, in 1-bromooctane and in 2,2,2-trichloroethanol were measured by Gerrard[8] at total pressures of about 1 bar and from 273 to 298 K. Mole fraction solubilities, corrected to a partial pressure of 1.013 bar and extrapolated to 303 K, fall between values for trichloromethane and 1,2-dichloropropane.

The apparent solubility pattern in aliphatic chloro-compounds is shown in Fig. 5.38. In general, the higher the dipole moment the greater the mole fraction solubility of sulfur dioxide. However, with so few data available for comparison, the detailed pattern is uncertain.

The solubility in chlorobenzene, at various pressures to 1.4 bar and temperatures from 273 to 353 K, has been measured by Horiuti,[32] by Sano[10] and by Gerrard.[8] Measurements by the different workers are consistent with

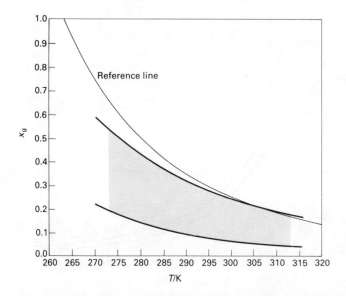

Figure 5.38. General pattern of solubilities of SO_2 in haloalkanes. Mole fraction solubilities, x_g, of SO_2 in haloalkanes for a partial pressure of SO_2 of 1.013 bar, estimated from data in the literature, fall in the shaded area of the graph. Mole fraction solubilities at this partial pressure for other temperatures and other haloalkanes may lie between extrapolations of the upper and lower boundaries of the shaded area. In all cases the mole fraction solubility for a partial pressure of 1.013 bar will approach 1.0 as the temperature is reduced to 263.15 K. the boiling point of SO_2 at a pressure of 1.013 bar, provided the system does not split into two liquid layers.

each other and hence likely to be reliable. They indicate that substituting chlorine into a benzene molecule lowers the mole fraction solubility of sulfur dioxide. Gerrard also measured solubilities in bromobenzene and in iodobenzene at 1.013 bar and temperatures from 273 to 298 K. Mole fraction solubilities, measured under the same conditions, decrease from chloro- to iodobenzene (Fig. 5.39; Table 5.7).

Sano[10] measured the solubility in 1,2-dichlorobenzene at 303 K and 1.013 bar. Substituting a second chlorine atom into a benzene molecule further decreases mole fraction solubility. The mole fraction solubility in 2-chlorophenol, measured by Sano under the same conditions, is greater than the corresponding solubility in benzene.

5.3.6 Solubilities in compounds containing sulfur

The solubility in sulfinylbismethane was measured by Sano[10] at 303 K and a pressure of 1.013 bar. This measurement is consistent with measurements made at this pressure by Smedland[46] over the range 291 to 434 K. Smedland also measured solubilities at low pressures in the range 0.007 bar to 0.095 bar

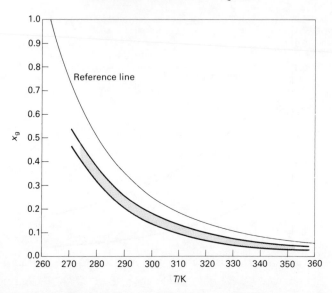

Figure 5.39. General pattern of solubilities of SO_2 in halobenzenes. Mole fraction solubilities, x_g, of SO_2 in chloro-, bromo-, and iodobenzene for a partial pressure of SO_2 of 1.013 bar, estimated from data in the literature, fall in the shaded area of the graph. Mole fraction solubilities at this partial pressure for other temperatures may lie between extrapolations of the upper and lower boundaries of the shaded area. In all cases the mole fraction solubility for a partial pressure of 1.013 bar will approach 1.0 as the temperature is reduced to 263.15 K, the boiling point of SO_2 at a pressure of 1.013 bar, provided the system does not split into two liquid layers.

(5.2 to 71 mmHg) at an unspecified temperature which was probably 293 K. These measurements show that the mole fraction solubility is not proportional to partial pressure in this pressure range and therefore that Henry's law is not a valid approximation in this case. The Henry's law constant of 0.119 bar at 298 K reported by Lenoir et al.[29] and measured by chromatography corresponds to very low partial pressures. The value reported by Benoit and Milanova[30] for the same temperature is 0.110 bar. Neither value can be used to make reliable estimates of solubilities at partial pressures of close to 1.013 bar because of the non-linear variation of mole fraction solubility with pressure. However, the ratio of pressure to mole fraction solubility for the measurement at 0.007 bar, reported by Smedland, is 0.118 bar which is close to the limiting value of 0.119 bar at 298 K given by Lenoir. Measurements at higher pressures reported by Smedland correspond to an increase in the ratio (Table 5.7).

The mole fraction solubility in tetrahydrothiophene 1,1-dioxide at 303.2 K and 1.013 bar from Sano's measurement[10] is 0.482. Benoit and Milanova[30] measured solubility at low pressures at the same temperature and reported a Henry's law constant of 1.01 atm (1.02 bar). These data indicate a very steep rise in mole fraction solubility with increase in pressure at very low pressures with a levelling off as pressures approach 1.013 bar.

Sano[10] also showed that solubility in dimethyl sulfate is fairly high with a mole fraction solubility of 0.325 at 1.013 bar and 303.2 K. In contrast, the mole fraction solubility in carbon disulfide is 0.0263 under these conditions.[10]

Schulze[47] measured the solubility in sulfuryl chloride at 273.2 K. The measurement corresponds to a mole fraction solubility of 0.405 at a partial pressure of 0.967 bar. Very simple apparatus was used and the accuracy cannot be relied upon.

5.3.7 Solubilities in compounds containing phosphorus

Cooper and Smith[41] measured solubility in tributyl phosphate at partial pressures of gas from 0.009 bar to 0.902 bar in the range 279.2 to 342.7 K. Mole fraction solubilities are high, relative to appropriate reference lines, and show a marked non-linear variation with pressure. The single value of 0.485 at 303.2 K and 1.013 bar, from a measurement reported by Sano,[10] is low, compared with a value of 0.4865 at 0.760 bar and 307.0 K from Cooper's data.

The mole fraction solubility in tris(methylphenyl) phosphate, from Sano's measurement[10] at 303.2 K and 1.013 bar, is 0.441. Lenoir et al.[29] measured Henry's law constants for dissolution in triethyl phosphate, tripropyl phosphate and tris(2-methylpropyl) phosphate. These correspond to limiting values at zero pressure. They cannot be used to estimate solubilities at pressure above a few mmHg because of the likelihood of non-linear variation of mole fraction solubility with change of partial pressure.

5.4 TABLES SUMMARISING THE SOLUBILITY OF SULFUR DIOXIDE

Data available from the literature have been used to prepare the following tables which summarise the solubility behaviour of sulfur dioxide.

Table 5.8 Mole fraction solubilities for a partial pressure of 1.013 bar at or near to 298 K with solvents arranged in alphabetical order

Table 5.9 Mole fraction solubilities for a partial pressure of 1.013 bar at or near to 298 K arranged in order.

Table 5.10 Weight ratio solubilities for a partial pressure of 1.013 bar at or near to 298 K arranged in order.

In many cases experimental values correspond to a given total pressure of sulfur dioxide and solvent. Approximations discussed earlier have been made to convert to solubilities corresponding to a partial pressure of gas of 1.013 bar. Details of original measurements on the solubility of sulfur dioxide are given in Solubility Data Series Volume 12.[16]

Table 5.8 Mole fraction solubilities of sulfur dioxide in various solvents for a partial pressure of sulfur dioxide of 1.013 bar

Solvent	A	B	Temp. range/K Low	Temp. range/K High	M_r of solvent	x_g at 298 K[a]
Acetic acid	−4.163	1055	273	303	60.052	0.238
Acetic anhydride	−3.099	816	273	303	102.074	0.436
Acetone	−2.932	778	283	303	58.075	0.477
2-Aminoethanol	bb	b	303	303	61.079	0.520
Aniline	b	b	303	303	93.129	0.534
Anisole	b	b	303	303	108.135	0.298
Benzenamine	b	b	303	303	93.129	0.534
Benzene	−3.841	944	298	333	78.11	0.212
1,2-Benzenediol	b	b	303	303	110.102	0.240
Benzenemethanol	b	b	303	303	108.135	0.219
Benzonitrile	−3.863	1011	273	298	103.124	0.339
Benzyl acetate	b	b	293.15	293.15	150.167	0.481
Benzyl alcohol	b	b	303	303	108.135	0.219
Benzyl benzoate	b	b	293.15	293.15	212.238	0.420
Benzyl ether	b	b	303	303	198.26	0.354
Bromobenzene	−5.584	1434	273	298	157.01	0.169
1-Bromooctane	−5.020	1261	273	298	193.128	0.163
1-Butanamine	b	b	303	303	73.139	0.615
1-Butanol	b	b	303	303	74.123	0.169
2-Butoxyethanol	b	b	303	303	118.166	0.377
Butyl acetate	b	b	303	303	116.15	0.333
Caproic acid	−4.895	1244	273	303	116.15	0.190
Carbon tetrachloride	−5.238	1221	283	313	153.82	0.072
Chlorobenzene	−5.128	1317	273	303	112.56	0.196
Chloroform	−6.741	1770	273	303	119.38	0.158
1-Chlorooctane	−5.068	1285	273	298	148.677	0.175
2-Chlorophenol	b	b	303	303	128.553	0.244
1,2-Cresol	b	b	303	303	110.102	0.240
Cyclohexane	b	b	298	298	84.162	0.063
Cyclohexanol	−6.468	1701	293	303	100.156	0.174
Cyclohexanone	−3.698	992	283	303	98.14	0.427
Decahydronaphthalene	−2.809	450	293	323	138.254	0.050
Decalin	−2.809	450	293	323	138.254	0.050
Decane	−4.967	1136	273	313	142.286	0.070
Dibutyl ether	b	b	293.15	293.15	130.226	0.315
N,N-Dibutylbutanamine	b	b	303	303	185.355	0.669
1,2-Dichlorobenzene	b	b	303	303	147.004	0.105
1,2-Dichloroethane	−1.562	278.9	298	303	98.96	0.237
1,2-Dichloropropane	b	b	303	303	112.987	0.167
Diethyl ether	−2.588	662	273	303	74.118	0.430
Diethylaniline	−2.133	553.8	273	298	149.237	0.531
N,N-Diethylbenzenamine	−2.133	553.8	273	298	149.237	0.531
Dimethyl maleate	b	b	303	303	144.106	0.337
Dimethyl phthalate	b	b	303	303	194.166	0.391
N,N-Dimethylacetamide	−2.190	590	298	366	87.117	0.616
N,N-Dimethylaniline	−1.048	248.4	298	303	121.183	0.610
N,N-Dimethylbenzenamine	−1.048	248.4	298	303	121.183	0.610
1,3-Dimethylbenzene	−4.628	1218	273	298	106.17	0.286
N,N-Dimethylformamide	−2.048	537	273	333	73.09	0.568
Dimethylsulfoxide	−2.982	824	291	434	78.124	0.607
1,4-Dioxane	−2.630	702	288	303	88.096	0.532

Table 5.8 (*continued*)

Solvent	A	B	Temp. range/K Low	High	M_r of solvent	x_g at 298 K[a]
DMF	−2.048	537	273	333	73.09	0.568
DMSO	−2.982	824	291	434	78.124	0.607
Dodecane	−1.851	151	283	313	170.34	0.045
1,2-Ethanediol	−4.603	1162	273	298	62.058	0.197
1,2-Ethanediol diacetate	b	b	303	303	114.134	0.447
Ethanol	−4.560	1160	273	303	46.064	0.215
2-Ethoxyethanol acetate	b	b	303	303	132.144	0.440
Ethyl acetate	b	b	303	303	88.096	0.347
Ethyl laurate	−3.144	798	298	366	228.366	0.342
Ethylaniline	b	b	298	298	121.183	0.560
N-Ethylbenzenamine	b	b	298	298	121.183	0.560
Formamide	b	b	303	303	45.036	0.222
2-Furanmethanol	b	b	303	303	98.091	0.251
Furfuryl alcohol	b	b	303	303	98.091	0.251
Guaiacol	b	b	303	303	124.129	0.328
Heptadecane	−2.729	401	303	333	240.475	0.041
Heptane	−4.032	830	298	313	100.205	0.057
1-Heptanol	−4.274	1057	298	366	116.2	0.187
Hexadecane	−3.225	617	298	475	226.448	0.070
Hexane	−3.944	797	298	313	86.178	0.054
Hexanoic acid	−4.895	1244	273	303	116.15	0.190
1-Hexanol	b	b	303	303	102.17	0.175
Iodobenzene	−5.935	1520	273	298	204.01	0.146
Mesitylene	−3.466	893	273	293	120.195	0.339
Methanol	−4.943	1287	273	303	32.037	0.238
Methoxybenzene	b	b	303	303	108.135	0.298
2-Methoxyethanol	b	b	303	303	76.085	0.395
1,1'-(Methoxymethylene)bisbenzene	b	b	303	303	198.26	0.354
2-Methoxyphenol	b	b	303	303	124.129	0.328
Methyl acetate	−3.307	878	283	313	74.08	0.436
Methyl salicylate	−3.718	970	273	303	152.134	0.344
1-Methyl-2-nitrobenzene	−4.316	1132	288	333	137.128	0.304
2-Methyl-2-propanol	b	b	303	303	74.118	0.160
Methylaniline	b	b	298.15	298.15	107.156	0.561
N-Methylbenzenamine	b	b	298.15	298.15	107.156	0.561
Methylbenzene	−4.512	1174	273	333	92.13	0.268
Nitrobenzene	−4.213	1103	273	303	123.101	0.308
Nitromethane	b	b	303	303	61.03	0.347
Nonane	−4.041	841	298	313	128.259	0.060
Octane	−1.722	95	283	313	114.232	0.040
1-Octanol	−4.468	1116	273	303	130.23	0.189
2-Octanone	−3.269	843	298	366	128.21	0.363
1,1'-Oxybis(2-methoxyethane)	b	b	303	303	134.16	0.257
1,1'-Oxybisbutane	b	b	293.15	293.15	130.226	0.315
1,1'-Oxybisethane	−2.588	662	273	303	74.118	0.430
2,2'-Oxybisethanol	b	b	303	303	106.106	0.333
Pentadecane	−3.927	830	298	313	212.421	0.072
Pentane	−3.990	812.6	298	313	72.151	0.055
2,5,8,11,14-Pentaoxapentadecane	−2.541	711	298	366	222.256	0.700

Table 5.8 (*continued*)

Solvent	A	B	Temp. range/K Low	Temp. range/K High	M_r of solvent	x_g at 298 K[a]
Phenol	−3.617	893.6	303	323	94.108	0.241
2-Phenoxyethanol	b	b	303	303	138.156	0.329
1,2-Propanediol	b	b	303	303	76.085	0.169
1,2,3-Propanetriol	−3.614	816	273	303	92.09	0.133
2-Propanol	b	b	303	303	60.091	0.155
1-Propanol	−4.519	1137	283	313	60.091	0.198
2-Propanone	−2.932	778	283	303	58.075	0.477
Pyridine	−1.220	302	273	303	79.102	0.621
Sulfinylbismethane	−2.982	824	291	434	78.124	0.607
Tetrachloromethane	−5.238	1221	283	313	153.82	0.072
Tetradecane	−3.979	842	298	313	198.394	0.070
Tetrahydro-2-furanmethanol	b	b	303	303	102.123	0.420
Tetrahydrofuran	b	b	303	303	88.096	0.355
Tetrahydrofurfuryl alcohol	b	b	303	303	102.123	0.420
1,2,3,4-Tetrahydronaphthalene	−2.259	474.1	293	303	132.206	0.215
Tetralin	−2.259	474.1	293	303	132.21	0.215
Toluene	−4.512	1174	273	333	92.13	0.268
Tributyl phosphate	−1.777	456	279	343	266.298	0.567
2,2,2-Trichloroethanol	−5.060	1278	273	298	149.399	0.169
Trichloroethene	b	b	303	303	131.39	0.083
Trichloromethane	−6.741	1770	273	303	119.38	0.158
Tridecane	−3.980	839	298	313	184.367	0.068
1,3,5-Trimethylbenzene	−3.466	893	273	293	120.195	0.339
Undecane	−4.067	858	298	313	156.313	0.065
Water	−5.573	1182	278	383	18.02	0.025
m-Xylene	−4.628	1218	273	298	106.17	0.286

[a] Mole fraction solubilities estimated for 298 K except where indicated.
[b] Solubility measured at one temperature only. The value for this temperature is given.

Experimental measurements have been fitted to equations of the form:

$$\log_{10} x_g = A + B/(T/K)$$

Values of A and B are given above.

SOLUBILITY OF SULFUR DIOXIDE

Table 5.9 Mole fraction solubilities of sulfur dioxide in various solvents for a partial pressure of sulfur dioxide of 1.013 bar, in increasing order

Solvent	A	B	Temp. range Low	Temp. range High	M_r of solvent	x_g at 298K[a]
Water	−5.573	1182	278	383	18.02	0.025
Octane	−1.722	95	283	313	114.232	0.040
Heptadecane	−2.729	401	303	333	240.475	0.041
Dodecane	−1.851	151	283	313	170.34	0.045
Decahydronaphthalene (decalin)	−2.809	450	293	323	138.254	0.050
Hexane	−3.944	797	298	313	86.178	0.054
Pentane	−3.990	812.6	298	313	72.151	0.055
Heptane	−4.032	830	298	313	100.205	0.057
Nonane	−4.041	841	298	313	128.259	0.060
Cyclohexane	b	b	298	298	84.162	0.063
Undecane	−4.067	858	298	313	156.313	0.065
Tridecane	−3.980	839	298	313	184.367	0.068
Decane	−4.967	1136	273	313	142.286	0.070
Hexadecane	−3.225	617	298	475	226.448	0.070
Tetradecane	−3.979	842	298	313	198.394	0.070
Pentadecane	−3.927	830	298	313	212.421	0.072
Tetrachloromethane	−5.238	1221	283	313	153.82	0.072
Trichloroethene	b	b	303	303	131.39	0.083
1,2-Dichlorobenzene	b	b	303	303	147.004	0.105
1,2,3-Propanetriol	−3.614	816	273	303	92.09	0.133
Iodobenzene	−5.935	1520	273	298	204.01	0.146
2-Propanol	b	b	303	303	60.091	0.155
Trichloromethane	−6.741	1770	273	303	119.38	0.158
2-Methyl-2-propanol	b	b	303	303	74.118	0.160
1-Bromooctane	−5.020	1261	273	298	193.128	0.163
1,2-Dichloropropane	b	b	303	303	112.987	0.167
1,2-Propanediol	b	b	303	303	76.085	0.169
Bromobenzene	−5.584	1434	273	298	157.01	0.169
2,2,2-Trichloroethanol	−5.060	1278	273	298	149.399	0.169
1-Butanol	b	b	303	303	74.123	0.169
Cyclohexanol	−6.468	1701	293	303	100.156	0.174
1-Hexanol	b	b	303	303	102.17	0.175
1-Chlorooctane	−5.068	1285	273	298	148.677	0.175
1-Heptanol	−4.274	1057	298	366	116.2	0.187
1-Octanol	−4.468	1116	273	303	130.23	0.189
Hexanoicacid	−4.895	1244	273	303	116.15	0.190
Chlorobenzene	−5.128	1317	273	303	112.56	0.196
1,2-Ethanediol	−4.603	1162	273	303	62.068	0.197
1-Propanol	−4.519	1137	283	313	60.091	0.198
Benzene	−3.841	944	298	333	78.11	0.212
1,2,3,4-Tetrahydronaphthalene (tetralin)	−2.259	474.1	293	303	132.206	0.215
Ethanol	−4.560	1160	273	303	46.064	0.215
Benzenemethanol (benzyl alcohol)	b	b	303	303	108.135	0.219
Formamide	b	b	303	303	45.036	0.222
1,2-Dichloroethane	−1.562	278.9	298	303	98.96	0.237
Methanol	−4.943	1287	273	303	32.037	0.238
Acetic acid	−4.163	1055	273	303	60.052	0.238
1,2-Benzenediol (1,2-cresol)	b	b	303	303	110.102	0.240

Table 5.9 (continued)

Solvent	A	B	Temp. range Low	Temp. range High	M_r of solvent	x_g at 298K[a]
Phenol	−3.617	893.6	303	323	94.108	0.241
2-Chlorophenol	b	b	303	303	128.553	0.244
2-Furanmethanol (furfuryl alcohol)	b	b	303	303	98.091	0.251
1,1′-Oxybis(2-methoxyethane)	b	b	303	303	134.16	0.257
Methylbenzene (toluene)	−4.512	1174	273	333	92.13	0.268
1,3-Dimethylbenzene (m-xylene)	−4.628	1218	273	298	106.17	0.286
Methoxybenzene (anisole)	b	b	303	303	108.135	0.298
1-Methyl-2-nitrobenzene	−4.316	1132	288	333	137.128	0.304
Nitrobenzene	−4.213	1103	273	303	123.101	0.308
1,1′-Oxybisbutane (dibutyl ether)	b	b	293.15	293.15	130.226	0.315
2-Methoxyphenol (guaiacol)	b	b	303	303	124.129	0.328
2-Phenoxyethanol	b	b	303	303	138.156	0.329
Butyl acetate	b	b	303	303	116.15	0.333
2,2′-Oxybisethanol	b	b	303	303	106.106	0.333
Dimethyl maleate	b	b	303	303	144.106	0.337
Benzonitrile	−3.863	1011	273	298	103.124	0.339
1,3,5-Trimethylbenzene (mesitylene)	−3.466	893	273	293	120.195	0.339
Ethyl laurate	−3.144	798	298	366	228.366	0.342
Methyl salicylate	−3.718	970	273	303	152.134	0.344
Nitromethane	b	b	303	303	61.03	0.347
Ethyl acetate	b	b	303	303	88.096	0.347
1,1′-(Methoxymethylene)bisbenzene (benzyl ether)	b	b	303	303	198.26	0.354
Tetrahydrofuran	b	b	303	303	88.096	0.355
2-Octanone	−3.269	843	298	366	128.21	0.363
2-Butoxyethanol	b	b	303	303	118.166	0.377
Dimethyl phthalate	b	b	303	303	194.166	0.391
2-Methoxyethanol	b	b	303	303	76.085	0.395
Tetrahydro-2-furanmethanol (tetrahydrofurfuryl alcohol)	b	b	303	303	102.123	0.420
Benzyl benzoate	b	b	293.15	293.15	212.238	0.420
Cyclohexanone	−3.698	992	283	303	98.14	0.427
1,1′-Oxybisethane (diethyl ether)	−2.588	662	273	303	74.118	0.430
Acetic anhydride	−3.099	816	273	303	102.074	0.436
Methyl acetate	−3.307	878	283	313	74.08	0.436
2-Ethoxyethanol acetate	b	b	303	303	132.144	0.440
1,2-Ethanediol diacetate	b	b	303	303	114.134	0.447
2-Propanone (acetone)	−2.932	778	283	303	58.075	0.477
Benzyl acetate	b	b	293.15	293.15	150.167	0.481
2-Aminoethanol	b	b	303	303	61.079	0.520
N,N-Diethylbenzenamine (diethylaniline)	−2.133	553.8	273	298	149.237	0.531
1,4-Dioxane	−2.630	702	288	303	88.096	0.532
Benzenamine (aniline)	b	b	303	303	93.129	0.534
N-Ethylbenzenamine (ethylaniline)	b	b	298	298	121.183	0.560

Table 5.9 (*continued*)

Solvent	A	B	Temp. range Low	Temp. range High	M_r of solvent	x_g at 298K[a]
N-Methylbenzenamine (methylaniline)	[b]	[b]	298.15	298.15	107.156	0.561
Tributyl phosphate	−1.777	456	279	343	266.298	0.567
N,N-Dimethylformamide (DMF)	−2.048	537	273	333	73.09	0.568
1,1-Sulfinylbismethane (dimethylsulfoxide, DMSO)	−2.982	824	291	434	78.124	0.607
N,N-Dimethylbenzenamine (N,N-dimethylaniline)	−1.048	248.4	298	303	121.183	0.610
1-Butanamine	[b]	[b]	303	303	73.139	0.615
N,N-Dimethylacetamide	−2.190	590	298	366	87.117	0.616
Pyridine	−1.220	302	273	303	79.102	0.621
N,N-Dibutylbutanamine	[b]	[b]	303	303	185.355	0.669
2,5,8,11,14-Pentaoxapentadecane	−2.541	711	298	366	222.256	0.700

[a]Mole fraction solubilities, x_g, estimated for 298 K except where indicated.
[b]Solubility measured at one temperature only. The value for this temperature is given.
Experimental measurements have been fitted to equations of the form:

$$\log_{10} x_g = A + B/(T/K)$$

Values of A and B are given.

Table 5.10 Weight ratio solubilities of sulfur dioxide in various solvents for a partial pressure of sulfur dioxide of 1.013, in increasing order of magnitude

Solvent	M_r of solvent	Temp. range of measurements/K Low	High	Wt. of SO_2 at 298 K[a] in 100 g of solvent
Heptadecane	240.475	303	333	1.15
Dodecane	170.34	283	313	1.78
Hexadecane	226.448	298	475	2.13
Octane	114.232	283	313	2.31
Pentadecane	212.421	298	313	2.35
Tetradecane	198.394	298	313	2.44
Decahydronaphthalene (decalin)	138.254	293	323	2.45
Tridecane	184.367	298	313	2.55
Undecane	156.313	298	313	2.84
Nonane	128.259	298	313	3.21
Tetrachloromethane	153.82	283	313	3.25
Decane	142.286	273	313	3.39
Heptane	100.205	298	313	3.84
Hexane	86.178	298	313	4.22
Trichloroethene[b]	131.39	303	303	4.44
Cyclohexane[b]	84.162	298	298	5.07
1,2-Dichlorobenzene[b]	147.004	303	303	5.11
Pentane	72.151	298	313	5.12
Iodobenzene	204.01	273	298	5.39
1-Bromooctane	193.128	273	298	6.45
Bromobenzene	157.01	273	298	8.30
2,2,2-Trichloroethanol	149.399	273	298	8.74
Water	18.02	278	383	9.02
1-Chlorooctane	148.677	273	298	9.17
Trichloromethane	119.38	273	303	10.07
1,2,3-Propanetriol	92.09	273	303	10.68
1,2-Dichloropropane[b]	112.987	303	303	11.37
1-Octanol	130.23	273	303	11.48
1-Heptanol	116.20	298	366	12.72
Hexanoic acid	116.15	273	303	12.96
1,2,3,4-Tetrahydronaphthalene (tetralin)	132.206	293	303	13.25
1-Hexanol[b]	102.17	303	303	13.30
Cyclohexanol	100.156	293	303	13.45
Chlorobenzene	112.56	273	303	13.84
Ethyl laurate	228.366	298	366	14.57
2-Chlorophenol[b]	128.553	303	303	16.08
2-Methyl-2-propanol[b]	74.118	303	303	16.46
1,1'-Oxybis(2-methoxyethane)[b]	134.16	303	303	16.52
Benzenemethanol (benzyl alcohol)[b]	108.135	303	303	16.61
1,2-Propanediol[b]	76.085	303	303	17.12
1-Butanol[b]	74.123	303	303	17.58
1,1'-(Methoxymethylene)bisbenzene (benzyl ether)[b]	198.26	303	303	17.71
1,2-Benzenediol (1,2-cresol)[b]	110.102	303	303	18.37
2-Propanol[b]	60.091	303	303	19.55
1,2-Dichloroethane	98.96	298	303	20.06
1-Methyl-2-nitrobenzene	137.128	288	333	20.39
Dimethyl phthalate[b]	194.166	303	303	21.18
Phenol	94.108	303	323	21.59

Table 5.10 (*continued*)

Solvent	M_r of solvent	Temp. range of measurements/K Low	High	Wt. of SO_2 at 298 K[a] in 100 g of solvent
Benzyl benzoate[b]	212.238	293.15	293.15	21.86
2-Furanmethanol (furfuryl alcohol)[b]	98.091	303	303	21.89
Benzene	78.11	298	333	22.09
Methyl salicylate	152.134	273	303	22.12
Dimethyl maleate[b]	144.106	303	303	22.60
1,1'-Oxybisbutane (dibutyl ether)[b]	130.226	293.15	293.15	22.62
2-Phenoxyethanol[b]	138.156	303	303	22.73
Nitrobenzene	123.101	273	303	23.15
1,3-Dimethylbenzene (*m*-xylene)	106.17	273	298	24.17
Methoxybenzene (anisole)[b]	108.135	303	303	25.15
2-Methoxyphenol (guaiacol)[b]	124.129	303	303	25.19
1,2-Ethanediol	62.068	273	303	25.32
Methylbenzene (toluene)	92.13	273	333	25.41
1-Propanol	60.091	283	313	26.30
1,3,5-Trimethylbenzene (mesitylene)	120.195	273	293	27.38
Butyl acetate[b]	116.15	303	303	27.54
2-Octanone	128.21	298	366	28.47
2,2'-Oxybisethanol[b]	106.106	303	303	30.14
Tributyl phosphate	266.298	279	343	31.44
Benzonitrile	103.124	273	298	31.79
2-Butoxyethanol[b]	118.166	303	303	32.81
Acetic acid	60.052	273	303	33.39
2-Ethoxyethanol acetate[b]	132.144	303	303	38.09
Ethanol	46.064	273	303	38.11
Ethyl acetate[b]	88.096	303	303	38.64
Benzyl acetate[b]	150.167	293.15	293.15	39.54
Tetrahydrofuran[b]	88.096	303	303	40.02
Formamide[b]	45.036	303	303	40.59
1,2-Ethanediol diacetate	114.134	303	303	45.37
Tetrahydro-2-furanmethanol[b] (tetrahydrofurfuryl alcohol)	102.123	303	303	45.42
Acetic anhydride	102.074	273	303	48.47
N,N-Diethylbenzenamine (diethylaniline)	149.237	273	298	48.67
Cyclohexanone	98.14	283	303	48.73
2-Methoxyethanol[b]	76.085	303	303	54.97
Nitromethane[b]	61.03	303	303	55.78
Methanol	32.037	273	303	62.31
1,1'-Oxybisethane (diethyl ether)	74.118	273	303	65.20
Methyl acetate	74.08	283	313	66.80
2,5,8,11,14-Pentaoxapentadecane	222.256	298	366	67.15
N-Ethylbenzenamine (ethylaniline)[b]	121.183	298	298	67.28
N,N-Dibutylbutanamine[b]	185.355	303	303	69.85
N-Methylbenzenamine (methylaniline)[b]	107.156	298.15	298.15	76.40
Benzenamine (aniline)[b]	93.129	303	303	78.82
N,N-Dimethylbenzenamine (N,N-dimethylaniline)	121.183	298	303	82.14
1,4-Dioxane	88.096	288	303	82.58

Table 5.10 (*continued*)

Solvent	M_r of solvent	Temp. range of measurements/K Low	High	Wt. of SO_2 at 298 K[a] in 100 g of solvent
2-Propanone (acetone)	58.075	283	303	100.70
2-Aminoethanol[b]	61.079	303	303	113.62
N,N-Dimethylformamide (DMF)	73.09	273	333	115.03
N,N-Dimethylacetamide	87.117	298	333	118.16
1,1-Sulfinylbismethane (dimethylsulfoxide, DMSO)	78.124	291	434	126.58
Pyridine	79.102	273	303	132.96
1-Butanamine[b]	73.139	303	303	139.91

[a]Solubilities expressed as the weight of sulphur dioxide dissolved in 100 g of solvent have been estimated for 298 K by fitting experimental data to equations of the form:

$$\log_{10} x_g = A + B/(T/K)$$

where x_g is the mole fraction solubility, except where indicated.
[b]Solubility measured at one temperature only. The value for this temperature is given.

Where possible, values of the mole fraction solubility for a partial pressure of 1.013 bar, x_g, have been fitted to equations of the type:

$$\log_{10} x_g = A + B/(T/K)$$

Values for 298 K have then been calculated from such equations. Estimated values of A and B are given in the tables but the corresponding equations may not be reliable outside the temperature range of the original experimental measurements.

Estimations of precision and reliability have not been included as the intention is to present a general overall behaviour of the gas. The more volatile the solvent the less reliable the estimations of the effect of partial pressure of the solvent on the experimental measurements of solubility.

REFERENCES

1. Hill, A.E.; Fitzgerald, T.B. *J. Amer. Chem. Soc.* 1935, 57, 250.
2. Foote, H.W.; Fleischer, J. *J. Amer. Chem. Soc.* 1934, 56, 870.
3. Seyer, W.F.; King, E.G. *J. Amer. Chem. Soc.* 1933, 55, 3140.
4. Albertson, N.F.; Fernelius, W.C. *J. Amer. Chem. Soc.* 1943, 65, 1629.
5. Glavis, F.J.; Ryden, L.L.; Marvel, C.S. *J. Amer. Chem. Soc.* 1937, 59, 707.
6. Locket, G.H. *J. Chem. Soc.* 1932, 1501.
7. Baume, G.; Pamfil, G.-P. *J. Chim. Phys.* 1914, 12, 256.
8. Gerrard, W. *J. Appl. Chem. Biotechnol.* 1972, 22, 623.
9. Gerrard, W. *Solubility of Gases and Liquids*, Plenum Press, New York, 1976.
10. Sano, H. *Nippon Kagaku Zasshi*, 1968, 89, 362.

11. Sano, H.; Nakamoto, Y. *ib.* 369.
12. Makranczy, J.; Megyery-Balog, K.; Rusz, L.; Patyi, L. *Hung. J. Ind. Chem.* 1976, 4(2), 269.
13. Patyi, L.; Furmer, I.E.; Makranczy, J.; Sadilenko, A.S.; Stepanova, Z.G.; Berengarten, M.G. *Zh. Prikl. Khim.* 1978, 51, 1296.
14. Bodor, E.; Bor. G.J.; Mohai, B.; Sipos, G. *Veszpremi. Vegyip. Egy. Kozl.* 1957, 1, 55.
15. Schay, G.; Szekely, G.; Racz, Gy. *Periodica Polytechnica Ser. Chem. Eng. (Budapest)* 1958, 2, 1.
16. *Solubility Data Series Vol 12, Sulfur Dioxide, Chlorine, Fluorine and Chlorine Dioxides*, ed. C.L. Young, Pergamon, Oxford 1983.
17. Ref. 16, p. 3.
18. Hudson, J.C. *J. Chem. Soc.* 1925, 1332.
19. Beuschlein, W.L.; Simenson, L.O. *J. Amer. Chem. Soc.* 1940, 62, 610.
20. Rabe, A.E.; Harris, J.F. *J. Chem. Eng. Data* 1963, 8, 333
21. Lavrova, E.M.; Tudorovskaya, L.L. *Zhur. Prikl. Khim.* 1977, 50, 1146.; *J. Appl. Chem. USSR* 1977, 50, 1102.
22. Fox, C.J.J. *Z. Phys. Chem.* 1902, 41, 458.
23. Jager, L. *Chem. Prumysl.* 1957, 7, 601.
24. Miles, F.D.; Fenton, J. *J. Chem. Soc.* 1920, 117, 59.
25. Cupr, V. *Rec. Trav. Chim.* 1928, 47, 55.
26. Sobolev, I.A.; Kukarin, V.A.; Dzhagatspanyan, R.V.; Kosorotov, V.I.; Zogorets, P.A.; Popov, A.I. *Khim. Prom.* 1970, 46, 668.
27. Nitta, T.; Kido, O.; Katayama, T. *J. Chem. Engng. Japan* 1976, 9, 317.
28. Tremper, K.K.; Prausnitz, J.M. *J. Chem. Engng. Data* 1976, 21, 295.
29. Lenoir, J.-Y.; Renault, P.; Renon, H. *J. Chem. Engng. Data* 1971, 16, 340.
30. Benoit, R.L.; Milanova, E. *Can. J. Chem.* 1979, 57, 1319.
31. Weissenberger, G.; Hadwiger, H. *Z. Angew. Chem.* 1927, 40, 734.
32. Horiuti, J. *Sci. Pap. Inst. Phys. Chem. Res. (Japan)* 1931/32, 17, 125.
33. Ipatieff, V.N.; Monroe, G.S. *Ind. Eng. Chem. Anal. Edn.* 1942, 14, 166.
34. Lloyd, S.J. *J. Phys. Chem.* 1918, 22, 300.
35. Lorimer, J.W.; Smith, B.C.; Smith, G.H. *J.C.S. Faraday I*, 1975, 71, 2232.
36. Albright, L.F.; Shannon, P.T.; Yu, S.-N.; Chueh, P.L. *Chem. Engng. Prog. Symp. Ser.* 1963, 59 (44), 66.
37. Nitta, T.; Itami, J.; Katayami, T. *J. Chem. Engng. Japan*, 1973, 6, 303.
38. Yuferev, R.F.; Maluigin, P.V. *Z. Chim. Prom.* 1930, 7, 553.
39. Foote, H.W.; Fleisher, J. *J. Amer. Chem. Soc.* 1934, 56, 870.
40. Bekarek, V.; Hala, E. *Coll. Czech. Chem. Comm.* 1968, 33, 2598.
41. Cooper, D.F.; Smith, J.W. *J. Chem. Engng. Data*, 1974, 19, 133.
42. DuPont de Nemours and Co. *Chem. Eng. News* 1955, 33, 2366.
43. Pfeifer, G. *Magy. Kem. Folyoirat.* 1963, 69 (3), 138.
44. See additional data published by W. Gerrard in Ref. 16.
45. Lindner, J. *Monatsh.* 1912, 33, 613.
46. Smedslund, T.H. *Finska Kemistsamfundets Medd.* 1950, 59, 40.
47. Schulze, H. *J. Prakt. Chem.* 1881, 24, 168.
48. Johnstone, H.F.; Leppla, P.W. *J. Amer. Chem. Soc.* 1934, 56, 2233.
49. Lobry de Bruyn, C.A. *Rec. Trav. Chim.* 1892, 11, 112; *Zeit. Phys. Chem.* 1892, 10, 782.

Chapter 6
SOLUBILITY OF AMMONIA AND THE AMINES

6.1 PHYSICAL PROPERTIES OF AMMONIA AND AMINES GASEOUS AT 298 K AND 1.013 BAR

Physical properties of ammonia and amines gasous at 298 K and 1.013 bar

Gas	M.pt/K	B.pt (1.013 bar)	Critical temp./K	Critical press/bar	Critical vol./dm^3	M_r	Density of gas at 273.15 K /g dm^{-3}
Ammonia NH$_3$	195.5	239.8	405.4	113.0	0.0725	17.031	0.77141 (1.013 bar)
Methanamine CH$_3$NH$_2$	180.7	266.7	430.1	74.6	0.144	31.058	1.93 (1.37 bar)
N-Methylmethanamine (CH$_3$)$_2$NH	177.2	280.2	437.6	53.1	0.176	45.085	—
N,N-Dimethylmethanamine (CH$_3$)$_3$N	156	276.02	433.3	40.7	0.254	59.112	—
Ethanamine C$_2$H$_5$NH$_2$	192.6	289.7	456.4	56.2	0.1852	45.085	—

Antoine type equations for the variation of vapour pressure with temperature, based upon published vapour pressure data, are as follows:

Ammonia

$$\ln(P_g/\text{bar}) = 11.2489 - 2584.9/[(T/\text{K}) - 9.49]$$

This is based on data for 240 to 371 K.[1]

Methanamine

$$\ln(P_g/\text{bar}) = 10.1442 - 2266.4/[(T/K) - 43.14]$$

This is based on data for 267 to 395 K.[1]

N-Methylmethanamine

$$\ln(P_g/\text{bar}) = 9.8821 - 2354.4/[(T/K) - 41.99]$$

This is based upon data for 281 to 423 K.[1]

N,N-Dimethylmethanamine

$$\ln(P_g/\text{bar}) = 9.1699 - 2201.1/[(T/K) - 35.63]$$

This equation is based upon one given in *Lange's Handbook of Chemistry*[2] which was derived from data for the temperature range 193 to 276 K. The equation is used to construct reference lines for higher temperatures, but it may not be reliable above 276 K.

Ethanamine

$$\ln(P_g/\text{bar}) = 10.1442 - 2590.2/[(T/K) - 34.08]$$

This is based upon data for 290 to 436 K.[1]
Reference lines for the five gases are shown in Fig. 6.1

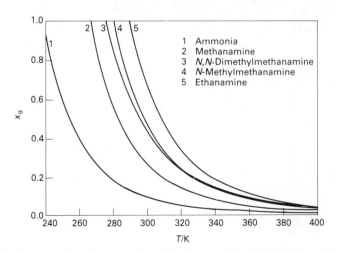

Figure 6.1. Reference lines for ammonia and alkyl amines at a partial pressure of 1.013 bar.

6.2 SOLUBILITY OF AMMONIA AND AMINES IN WATER

Ammonia and the amines are weak bases and react with water according to the equation:

$$R_3N + H_2O \rightleftharpoons R_3NH^+ + OH^-$$

where R may be hydrogen or an alkyl group.

Values of pK_b at 298.15 K are as follows:

NH_3	4.78
CH_3NH_2	3.37
$(CH_3)_2NH$	3.22
$(CH_3)_3N$	4.20
$C_2H_5NH_2$	3.22

The ammonia + water system has been studied over many years because of its practical importance, for example as a refrigerant medium. Data over a temperature range to high pressures have been published.

Values of the mole fraction solubility of ammonia at a partial pressure of 1.013 bar from measurements by Sims[3] are plotted in Fig. 6.2.

The ammonia + water system was studied by Neuhausen and Patrick,[4] Perman[5] and others. Sherwood[6] made use of much of the earlier work in his paper published in 1925.

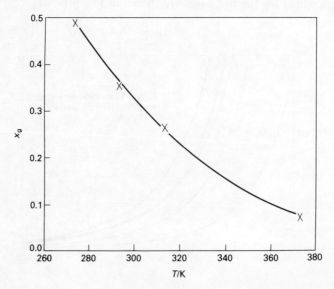

Figure 6.2. Mole fraction solubility in water of ammonia at a partial pressure of 1.013 bar.

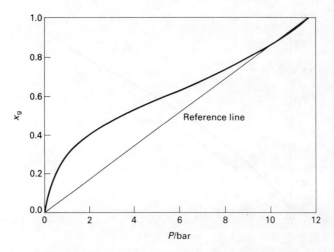

Figure 6.3. Variation with partial pressure of the mole fraction solubility of ammonia in water at 303.15 K.

The system was also investigated by Clifford and Hunter[7] at temperatures from 333 to 420 K and total pressures from about 1 bar to 16 bar. They also made use of earlier published work to present smoothed data for lower temperatures and pressures. A plot of mole fraction solubility against pressure at 303.2 K, from data in this paper, is shown in Fig. 6.3.

The solubilities of methanamine, N-methylmethanamine and N,N-dimethylmethanamine have been measured by Gerrard[8] at one or more temperatures and total pressures to 1.013 bar (Figs 6.4 to 6.6). Other measurements

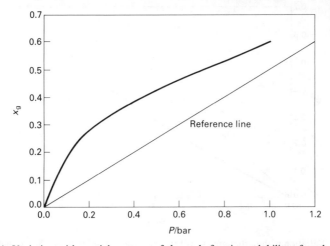

Figure 6.4. Variation with partial pressure of the mole fraction solubility of methanamine in water at 283.15 K.

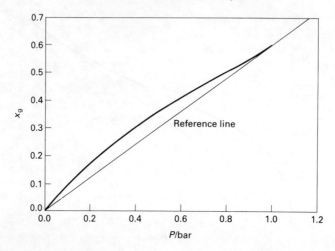

Figure 6.5. Variation with partial pressure of the mole fraction solubility of N-methylmethanamine in water at 293.15 K.

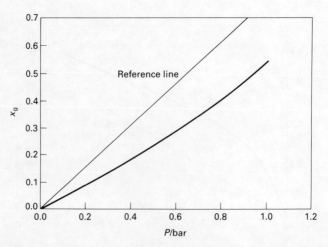

Figure 6.6. Variation with partial pressure of the mole fraction solubility of N,N-dimethylmethanamine in water at 283.15 K.

of solubilities of methanamine,[9,10] N-methylmethanamine[9,10] and ethanamine[9,11] have been published.

6.3 SOLUBILITY OF AMMONIA IN ORGANIC SOLVENTS

6.3.1 Solubility of ammonia in hydrocarbons

Published solubility data for ammonia in organic solvents have been compiled and evaluated in Solubility Data Series, Volume 21.[12]

It is possible to discern a general pattern of solubilities of ammonia in hydrocarbons even though there is a shortage of reliable, modern data.

Mole fraction solubilities in hydrocarbons are low compared with reference line values, and mole fraction solubilities in alkanols, measured under the same conditions. Mole fraction solubilities in aromatic hydrocarbons are greater than in non-aromatic hydrocarbons. A selection of mole fraction solubilities at a partial pressure of 1.013 bar in hydrocarbons and other solvents is given in Table 6.1. The data for hexane should be used with caution. If both solubility measurements were correct the variation of mole fraction solubility with temperature would be unusually large. The variation of mole fraction solubility in hexadecane, at a partial pressure of 1.013 bar, estimated from the Henry's law constants reported by Tremper and Prausnitz[13] is shown in Fig. 6.7. The variation of mole fraction solubility in hexane with change in temperature is likely to follow a similar curve.

Variations of mole fraction solubility in 1,1'-bicyclohexyl, estimated from measurements by Tremper and Prausnitz,[13] and in cyclohexene, from measurements by Noda et al.,[14] are shown in Figs 6.8 and 6.9.

Solubility in benzene was measured by Bell[15] and by Patyi et al.,[16] at a single temperature and barometric pressure, and by Noda et al.[14] at 273.2 and 298.2 K over a wide pressure range. Noda's measurements seem to be the most reliable.

6.3.2 Solubility of ammonia in alcohols and other compounds containing hydrogen and oxygen

Hatem[17] measured the solubility of ammonia in various alcohols at barometric total pressure (unspecified) and various temperatures. Mole fraction solubilities for a partial pressure of 1.013 bar may be estimated on the assumption that the total pressure was approximately 1.013 bar. Values of the mole fraction solubility in methanol are within about 4% of values estimated from measurements by de Bruyn.[18] The single value at 291.2 K, estimated from measurements by Kuznetsov et al.,[19] differs by about 6% from an interpolated value from Hatem's measurements. Measurements by

Table 6.1 Mole fraction solubilities of ammonia in organic solvents at a partial pressure of 1.013 bar

Solvent	T/K	x_g	Ref.
Hexane	293.2	0.0223	15
	298.2	0.0146	16
Hexadecane	300	0.0218	13
1,1'-Bicyclohexyl	300	0.0099	13
Cyclohexene	293.2	0.014	14[a]
Benzene	293.2	0.0359	14[a]
Methylbenzene	273.2	0.078	23
	293.2	0.0313	15
Methylnaphthalene	300	0.0328	13
Methanol	298.2	0.304	17
Ethanol	298.2	0.179	17
1-Propanol	298.2	0.237	17
2-Propanol	298.2	0.210	17
1-Butanol	298.2	0.217	20
	273.2	0.378	20
2-Butanol	273.2	0.312	21
2-Methyl-2-propanol	273.2	0.264	21
2-Methyl-1-propanol	273.2	0.375	21
	291.2	0.28	19
3-Methyl-1-butanol	288.2	0.28	19
1-Octanol	273.2	0.359	23
Cyclohexanol	299.2	0.108	39
1,2-Ethanediol	298.2	0.406	20
2,2'-Oxybisethanol	298.2	0.397	22
1,2,3-Propanetriol	291.2	0.51	19
2-Chloroethanol	273.1	0.507	23
2,2-Dichloroethanol	273.2	0.567	23
2,2,2-Trichloroethanol	273.2	0.571	23
2-Propanone	298.2	0.0842	20
1-Methyl-2-pyrolidinone	298.2	0.096	24
1,1'-Oxybisoctane	273.2	0.09	23
Acetic acid, octyl ester	273.2	0.222	23
Phosphorous acid, triethyl ester	273.2	0.132	21
Phosphorous acid, dibutyl ester	273.2	0.207	21
Phosphorous acid, triphenyl ester	293.2	0.260	21
Phosphoric acid, tributyl ester	273.2	0.17	25
	298.2	0.08	25
1,2,3-Propanetriol triacetate	298.2	0.165	40
Triethoxysilane	298.2	0.0606	41
Trichloromethane	293.2	0.193	15
	298.2	0.125	42[a]
Tetrachloromethane	293.2	0.0281	15
1,2-Dichloroethane	293.2	0.0797	15
1-Chlorooctane	273.2	0.197	23
Chlorobenzene	298.2	0.0409	20
Hydrazine	298.2	0.0787	26
Methylhydrazine	293.2	0.1183	26
1,1-Dimethylhydrazine	293.2	0.1086	26
Benzenamine	298.2	0.1199	27
2-Aminoethanol	290.2	0.19	19
N,N-Diethylethanamine	294.2	0.08	19
Quinoline	291.2	0.06	19
Hexanedinitrile	298.2	0.115	28

[a] Extrapolated.

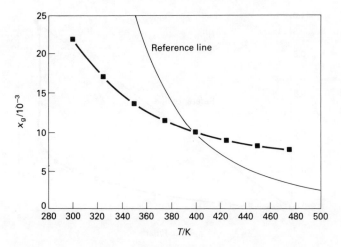

Figure 6.7. Mole fraction solubility in hexadecane of ammonia at a partial pressure of 1.013 bar.

Hatem appear to be more reliable than those of either de Bruyn or Kuznetsov.

Solubility in ethanol has been measured by various workers. Again the measurements by Hatem,[17] at barometric pressure over a temperature range, appear to be the most reliable. Hatem also measured the solubility in 1-propanol. The interpolated value for the mole fraction solubility at 291.2 K, corrected to a partial pressure of 1.013 bar, differs by about 8% from the value from measurements by Kuznetsov et al.[19] The corresponding value for

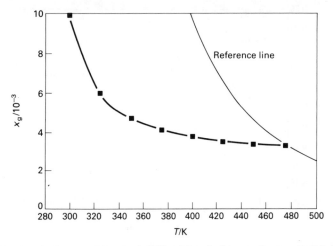

Figure 6.8. Mole fraction solubility in 1,1'-bicyclohexyl of ammonia at a partial pressure of 1.013 bar.

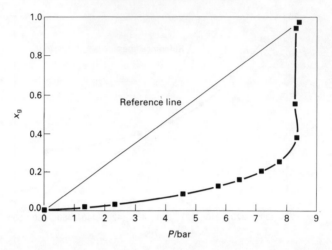

Figure 6.9. Mole fraction solubility in cyclohexene of ammonia at 293.15 K.

2-propanol, based upon Hatem's measurements, is close to the value from Kuznetsov's data.

Short et al.[20] published mole fraction solubilities in 1-butanol for the range 263.2 to 333.2 K at a partial pressure of 1.013 bar. The interpolated value for 273.2 K is within 3% of a value calculated from measurements by Maladkar.[21] The interpolated value for 291.2 K is within 8% of the value from measurements by Kuznetsov.[19]

Mole fraction solubilities of ammonia in monohydric alcohols are shown in Table 6.1 and Fig. 6.10.

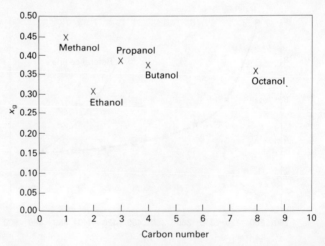

Figure 6.10. Mole fraction solubility in primary alkanols of ammonia at a partial pressure of 1.013 bar at 273.15 K.

Solubility in 1,2-ethanediol (ethylene glycol) at a partial pressure of ammonia of 1.013 bar and temperatures of 263.2 K, 298.2 K and 333.2 K have been reported by Short et al.[20] Mole fraction solubilities in this solvent are about double the values for monohydric alcohols of similar relative molecular mass at the same temperature and under a partial pressure of ammonia of 1.013 bar. Mole fraction solubilities in 2,2'-oxybisethanol (diethylene glycol) calculated from data published by Timonin et al.[22] are close to those available for 1,2-ethanediol under the same conditions. Timonin made measurements over a pressure range of 0.025 to 32.7 bar. The variation of mole fraction solubility with pressure at 298.2 K in this solvent is shown in Fig. 6.11.

The mole fraction solubility in 1,2,3-propanetriol (glycerol) at 291.2 K and a partial pressure of 1.013 bar, calculated from data published by Kuznetsov et al.,[19] is close to that for 1,2-ethanediol and 2,2'-oxybisethanol under the same conditions.

Solubilities in 2-chloroethanol, 2,2-dichloroethanol and 2,2,2-trichloroethanol at a pressure of 1.013 bar and 273.2 K were measured by Gerrard and Maladkar.[23] Mole fraction solubility increases with chlorine content with values for the three solvents of 0.507, 0.567 and 0.571, respectively.

Measurements of the solubility in 2-propanone by Short et al.[20] at 263.2 K and 298.2 K, corrected to a partial pressure of ammonia of 1.013 bar, are likely to be reliable. Kuznetsov et al.[19] reported the solubility at 292.2 K and a partial pressure of 1.013 bar, but their value is about three times the value by interpolation of measurements by Short et al. Chemical reaction between ammonia and 2-propanone may account for this high value.

Solubility in 1-methyl-2-pyrrolidinone was measured over the ranges 263.2 K to 328.2 K and 0.1 to 19.4 bar and reported in graphical form by

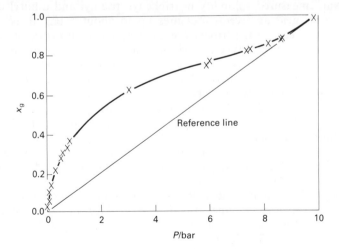

Figure 6.11. Variation with partial pressure of the mole fraction solubility of ammonia in 2,2'-oxybisethanol at 298.15 K.

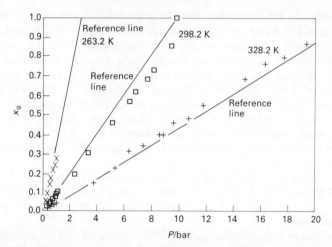

Figure 6.12. Dependence on temperature and partial pressure of the solubility of ammonia in 1-methyl-2-pyrrolidinone.

Freidson et al.[24] The variation of mole fraction solubility at several temperatures is shown in Fig. 6.12.

Solubilities in other compounds of carbon, oxygen and hydrogen are shown in Table 6.1.

6.3.3 Solubilities in solvents containing phosphorus, halogens or nitrogen

Maladkar[21] measured solubility in triphenyl, triethyl and dibutyl esters of phosphorous acid at several pressures up to about 1 bar at 273.2 K, or 293.2 K in the case of the triphenyl ester. At a partial pressure of 1.013 bar the solubility in the dibutyl ester, which contains a hydroxyl group, is nearly double that in the triethyl ester, which contains no hydroxyl group.

Hala and Tuck[25] measured solubility in the tributyl ester of phosphoric acid at 1.013 bar in the temperature range 243.2 K to 293.2 K. The mole fraction solubility at 273.2 K and a pressure of 1.013 bar is close to that in the dibutyl ester of phosphorous acid under the same conditions (Table 6.1).

Variations in solubility in different chlorinated hydrocarbons may reflect differing tendencies towards hydrogen bonding between ammonia and solvent. Bell's value[15] for the mole fraction solubility in tetrachloromethane, at 293.2 K and a partial pressure of 1.013 bar, is 0.0281 and close to that in aliphatic hydrocarbons under the same conditions. The mole fraction solubility in trichloromethane, also given by Bell for the same conditions, is 0.193. The corresponding value in 1,2-dichloroethane is 0.0797.

The solubility at 273.2 K and 1.013 bar in 1-chlorooctane was measured by Gerrard and Maladkar.[23] Solubilities in this and other halogenated compounds are given in Table 6.1.

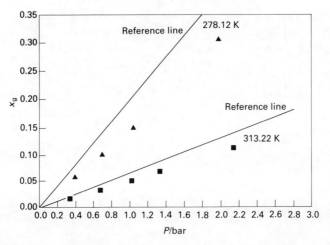

Figure 6.13. Variation with partial pressure of the mole fraction solubility of ammonia in hydrazine at 278.12 K and 313.22 K.

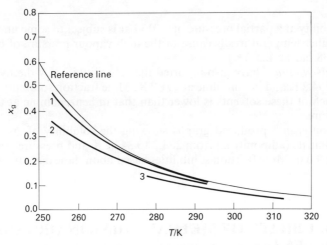

Figure 6.14. Mole fraction solubilities in methylhydrazine (1), dimethylhydrazine (2) and hydrazine (3) of ammonia at a partial pressure of 1.013 bar.

Solubilities in hydrazine, methylhydrazine and 1,1′-dimethylhydrazine were measured by Chang et al.[26] over pressure and temperature ranges. These data are self consistent and appear to be reliable. Some of these data have been plotted in Figs 6.13 and 6.14.

The solubility in benzenamine, at a total pressure equal to barometric pressure, was measured by Stoica et al.[27] from 298.2 to 423.2 K. The extrapolated mole fraction solubility for 291.2 K is 0.148 which is close to the value of 0.13 from measurements by Kuznetsov.[19] Correction of these data to

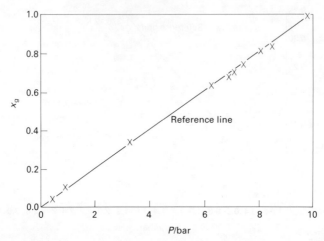

Figure 6.15. Variation with partial pressure of the mole fraction solubility of ammonia in hexanedinitrile at 298.15 K.

give solubility at a partial pressure of 1.013 bar is subject to some uncertainty at the higher temperatures because of the high vapour pressure of benzenamine (0.38 bar at 423.2 K).

Kuznetsov et al.[19] have also reported the solubility in N,N-diethylethenamine at 294.2 K and in quinoline at 291.2 K. Mole fraction solubility at 1.013 bar in each of these solvents is lower than that in benzenamine at the same temperature.

Freidson et al.[28] produced graphs showing the solubility of ammonia in hexanedinitrile (adiponitrile) from 283.2 to 343.2 K and pressures from 0.09 bar to 15.9 bar. Mole fraction solubilities based upon these data are shown in Fig. 6.15.

6.4 SOLUBILITY OF METHANAMINE IN ORGANIC SOLVENTS

Solubilities of methanamine in non-aqueous solvents have been reviewed and assessed by Gerrard,[8] Counsell et al.,[29] and Fogg,[12] and are collected in *Solubility Data Series, Volume 21*.[12] As is the case for ammonia, mole fraction solubilities at a partial pressure of 1.013 bar are higher in solvents containing oxygen or nitrogen than in hydrocarbons.

Wolf et al.[30,31] have carried out detailed studies of equilibria with butane, hexane and nonane. Phase diagrams based upon measurements with butane and hexane are shown in Figs 6.16 and 6.17. The solubility in decane has been measured by Gerrard.[8]

Measurements by Gerrard[8] indicate that the mole fraction solubility in

SOLUBILITY OF AMMONIA AND THE AMINES

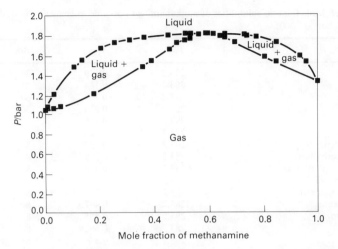

Figure 6.16. Phase equilibria in the methanamine + butane system at 273.15 K.

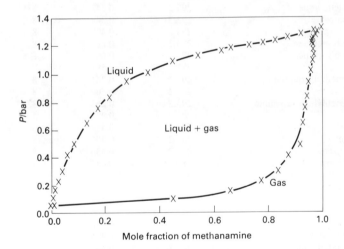

Figure 6.17. Phase equilibria in the methanamine + hexane system at 273.15 K.

benzene at a partial pressure of 1.013 bar and 283.2 K is about twice that in straight chain alkanes under the same conditions. Gerrard also showed that the substitution of methyl groups into benzene lowers the mole fraction solubility at this temperature and partial pressure (Table 6.2).

Mole fraction solubilities in alcohols which have been measured are high compared with corresponding mole fraction solubilities in hydrocarbons. Values lie above the appropriate reference lines. Mole fraction solubilities in dihydric and trihydric alcohols are higher than in monohydric alcohols. The published mole fraction solubilities in ethers and other solvents containing

Table 6.2 Mole fraction solubilities of methanamine in organic solvents at a partial pressure of 1.013 bar

Solvent	T/K	x_g	Ref.
Hexane	293.2	0.119	31
	283.2	0.184	31
Nonane	293.2	0.129	30
	283.2	0.210	30[a]
Decane	293.2	0.156	8
Benzene	283.2	0.423	8
Methylbenzene	283.2	0.397	8
1,3-Dimethylbenzene	283.2	0.359	8
1,3,5-Trimethylbenzene	283.2	0.326	8
1,2-Ethanediol	283.2	0.709	8
1,2,3-Propanetriol	283.2	0.736	8
1,4-Dioxane	283.2	0.487	8
1-Butanol	293.2	0.555	8
	283.2	0.662	8
1-Octanol	293.2	0.554	8
	283.2	0.663	8
Benzenemethanol	293.2	0.605	8
	283.2	0.690	8
1-Phenylethanone	293.2	0.342	8
Ethyl benzoate	293.2	0.348	8
	283.2	0.490	8
1,1'-Oxybispentane	273.2	0.680	8
1,1'-Oxybisoctane	273.2	0.604	8
	283.2	0.388	8
N,N-Dimethylformamide	293.2	0.341	8
	283.2	0.500	8
N,N-Dimethylmethanamine	293.2	0.263	32
	283.2	0.428	32
Pyridine	283.2	0.488	8
Quinoline	283.2	0.456	8
Nitrobenzene	283.2	0.436	8
Benzenamine	283.2	0.634	8
N-Methylbenzenamine	283.2	0.596	8
N,N-Dimethylbenzenamine	283.2	0.404	8
N-Ethylbenzenamine	283.2	0.592	8
N,N-Diethylbenzenamine	283.2	0.360	8
Benzonitrile	283.2	0.516	8
1-Methyl-2-nitrobenzene	283.2	0.408	8
Benzylmethanamine	283.2	0.520	8
Octylamine	283.2	0.493	8
Trichloromethane	283.2	0.674	8
Tetrachloromethane	293.2	0.262	33
	283.2	0.429	33
Bromobenzene	283.2	0.454	8
2,2,2-Trichloro-2-hydroxyethane	293.2	0.653	8

[a]Interpolated.

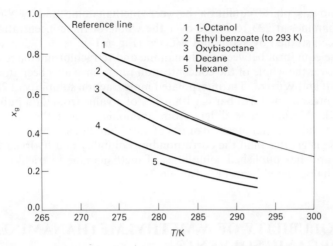

Figure 6.18. Mole fraction solubilities in various solvents of methanamine at a partial pressure of 1.013 bar.

oxygen fall below solubilities in alcohols but above those in hydrocarbons (Table 6.2 and Fig. 6.18).

Gerrard[8] has published solubilities in a number of solvents containing nitrogen (Table 6.2). In the case of benzenamine and the N-substituted benzenamines which have been investigated at 283 K and 1.013 bar the mole fraction solubility of methanamine decreases with the number and size of substituents. At 283.2 K and 1.013 bar, the mole fraction solubilities in quinoline and in pyridine are greater than in benzene. The behaviour of ammonia is similar.[19]

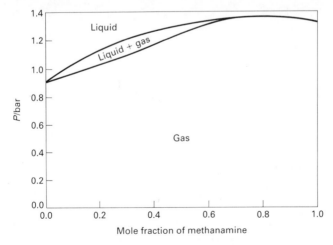

Figure 6.19. Phase equilibria in the methanamine + N,N-dimethylmethanamine system at 273.15 K.

Liquid-gas phase equilibria between methanamine and N,N-dimethylmethanamine, at 223 to 293 K over the whole of the concentration range, have been studied by Wolff and Würtz[32] (Fig. 6.19).

Phase equilibria between methanamine and tetrachloromethane at 253 to 293 K over the whole of the concentration range have also been investigated by Wolff and Würtz.[33] The interpolated mole fraction solubility at 283 K and a total pressure of 1.013 bar is within 1% of a value from data published by Gerrard.[8] Methanamine differs from ammonia in that the mole fraction solubility, at a partial pressure of gas of 1.013 bar and a temperature close to 298.2 K, is greater than the corresponding solubility in a hydrocarbon.

Gerrard[8] has published solubilities of methanamine in trichloromethane and in halogenated benzenes (Table 6.2).

6.5 SOLUBILITY OF N-METHYLMETHANAMINE IN ORGANIC SOLVENTS

N-Methylmethanamine has a higher boiling point than methanamine and therefore, under similar conditions, N-methylmethanamine has the higher solubility. The pattern of relative solubilities of N-methylmethanamine in different solvents is similar to the pattern of relative solubilities of methanamine.

Wolff *et al.*[34,35] studied the hexane + N-methylmethanamine system over the whole of the concentration range at 223 to 293 K. A phase diagram for the system based upon these measurements is shown in Fig. 6.20. The

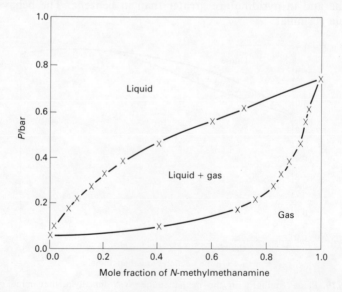

Figure 6.20. Phase equilibria in the N-methylmethanamine + hexane system at 273.15 K.

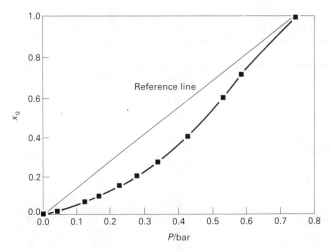

Figure 6.21. Variation with partial pressure of the mole fraction solubility of N-methylmethanamine in hexane at 273.15 K.

corresponding variation of mole fraction solubility of N-methylmethanamine with partial pressure for this system is shown in Fig. 6.21.

Gerrard[8] measured solubilities in decane, benzene and methylbenzenes. At a partial pressure of gas of 1.013 bar at 293.2 K, the mole fraction solubilities in benzene and methylbenzenes are greater than those in the alkanes. Methanamine behaves in a similar way.

At 293 to 313 K, mole fraction solubilities at a partial pressure of 1.013 bar are higher in alcohols than in hydrocarbons (Table 6.3). Solubilities in methanol, ethanol and 1-propanol were measured by Niepel et al.[36] and in 1-butanol and 1-octanol by Gerrard.[8] Mole fraction solubilities at a partial pressure of gas of 1.013 bar tend to increase with chain length of the alcohol, although the solubility in methanol is anomalous. Mole fraction solubilities at a partial pressure of 1.013 bar also increase with increase in the number of hydroxyl groups in a molecule of solvent. Solubilities in other solvents containing oxygen are shown in Table 6.3.

Gerrard[8] has measured solubilities in a wide range of solvents containing nitrogen (Table 6.3).

The N-methylmethanamine + tetrachloromethane system was studied by Wolff and Höppel over the whole of the concentration range, at temperatures from 253 to 293 K. A phase diagram based upon these measurements is shown in Fig. 6.22. The data are consistent with the solubility measurements at 293.2 K reported by Gerrard.[8] Mole fraction solubility in this solvent is higher than in hydrocarbons but lower than in alcohols under the same conditions.

Gerrard[8] measured the mole fraction solubility in trichloromethane at 293.2 K and pressures close to 1.013 bar. The solubility under these conditions is higher than in tetrachloromethane. Methanamine shows a similar difference in solubilities in these two solvents.[8]

Table 6.3 Mole fraction solubility of N-methylmethanamine in organic solvents at a partial pressure of gas of 1.013 bar

Solvent	T/K	x_g	Ref.
Hexane	293.2	0.478	35
Decane	293.2	0.501	8
	298.2	0.390	8
Benzene	293.2	0.567	8
Methylbenzene	293.2	0.587	8
1,3-Dimethylbenzene	293.2	0.566	8
1,3,5-Trimethylbenzene	293.2	0.550	8
Methanol	293.2	0.684	36[a]
	298.2	0.622	36[a]
Ethanol	293.2	0.671	36[a]
	298.2	0.597	36[a]
1-Propanol	293.2	0.694	36[a]
	298.2	0.616	36[a]
1-Butanol	293.2	0.710	8
1-Octanol	293.2	0.719	8
1,2-Ethanediol	293.2	0.719	8
1,2,3-Propanetriol	293.2	0.759	8
1,4-Dioxane	293.2	0.559	8
Benzenemethanol	293.2	0.705	8
	283.2	0.875	8
1-Phenylethanone	293.2	0.559	8
Ethyl benzoate	293.2	0.572	8
1,1'-Oxybispentane	293.2	0.596	8
1,1'-Oxybisoctane	293.2	0.605	8
N,N-Dimethylformamide	298.2	0.413	8
	293.2	0.512	8
	283.2	0.848	8
Pyridine	293.2	0.564	8
Quinoline	293.2	0.522	8
Nitrobenzene	293.2	0.506	8
Benzenamine	293.2	0.687	8
N-Methylbenzenamine	293.2	0.650	8
N,N-Dimethylbenzenamine	283.2	0.848	8
	293.2	0.538	8
N-Ethylbenzenamine	293.2	0.650	8
N,N-Diethylbenzenamine	293.2	0.557	8
Benzonitrile	283.2	0.848	8
	293.2	0.576	8
1-Methyl-2-nitrobenzene	293.2	0.505	8
Benzenemethanamine	283.2	0.815	8
	293.2	0.580	8
1-Octylamine	283.2	0.888	8
	293.2	0.626	8
Trichloromethane	293.2	0.753	8
Tetrachloromethane	293.2	0.603	34
Chlorobenzene	293.2	0.575	8
Bromobenzene	293.2	0.580	8
Iodobenzene	293.2	0.612	8
2,2,2-Trichloro-2-hydroxyethane	293.2	0.756	8
	298.2	0.706	8

[a]Interpolated.

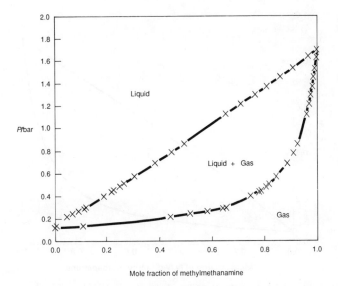

Figure 6.22. Phase equilibria in the methylmethanamine + tetrachloromethane system at 293.15 K.

Gerrard also found that the solubility in chloro-, bromo- and iodobenzene at 293.2 K and a total pressure of 1.013 bar was higher than in benzene. Methanamine again shows similar behaviour.

6.6 SOLUBILITY OF N,N-DIMETHYLMETHANAMINE IN ORGANIC SOLVENTS

The boiling point of N,N-dimethylmethanamine is between that of methanamine and N-methylmethanamine. In the absence of other factors, the mole fraction solubility at temperatures below about 325 K, is likely to be between that of methanamine and N-methylmethanamine. Above about 325 K, the situation is different. N,N-dimethylmethanamine has the lowest vapour pressure of the three amines and the highest reference line values for a fixed pressure (Fig. 6.1). N,N-dimethylmethanamine is, however, a tertiary amine. As such it does not have any hydrogen atoms attached to the nitrogen atom, in contrast to the other two amines. This difference also causes differences in solubility.

Wolff and Würtz[33] investigated phase equilibria between N,N-dimethylmethanamine and hexane over the whole concentration range at temperatures from 223 K to 293 K. Phase equilibria at 273.2 K based upon these data are shown in Fig. 6.23. The corresponding variation of mole fraction

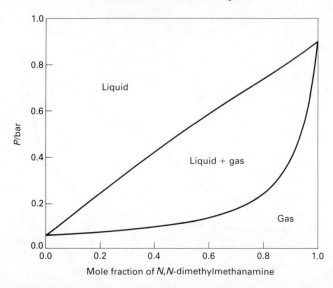

Figure 6.23. Phase equilibria in the N,N-dimethylmethanamine + hexane system at 273.15 K.

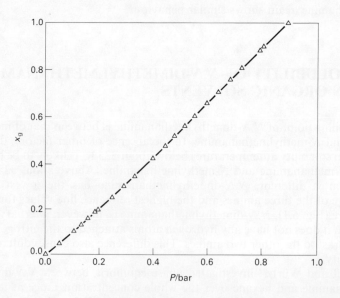

Figure 6.24. Variation with partial pressure of the mole fraction solubility of N,N-dimethylmethanamine in hexane at 273.15 K.

solubility at 273.2 K with variation of partial pressure of N,N-dimethylmethanamine is shown in Fig. 6.24. Solubilities in decane at 283.2 K have been reported by Gerrard.[8]

Gerrard[8] measured the solubility in several alcohols over ranges of pressure up to barometric pressure. Halban[37] made measurements at lower pressures. Linear extrapolation of Halban's measurements to estimate solubilities at a partial pressure of 1.013 bar leads to gross errors. This is clearly seen in Fig. 6.25. This shows the mole fraction solubility in benzenemethanol at 298.2 K and at various pressures, from data published by Gerrard and by Halban.

Gerrard's measurements show that the mole fraction solubilities, at a partial pressure of 1.013 bar, in 1,2-ethanediol and in 1,2,3-propanetriol are less than in the monohydric alcohols which have been studied. This is in contrast to the behaviour of ammonia, methanamine and N-methylmethanamine. This may reflect the reduced tendency of the gas to break up the hydrogen bonded structure of the solvent, because all hydrogen atoms on the nitrogen atom have been replaced by methyl groups.

Solubility data in solvents containing oxygen are given in Table 6.4.

Gerrard[8] also measured the solubility in various solvents containing nitrogen, over a range of pressures up to barometric pressure. In most cases measurements were made at more than one temperature. A selection of solubility data for these compounds, based upon Gerrard's measurements, is given in Table 6.4.

Phase equilibria between N,N-dimethylmethanamine and tetrachloromethane have been studied at 253.2, 273.2 and 293.2 K over the whole of the

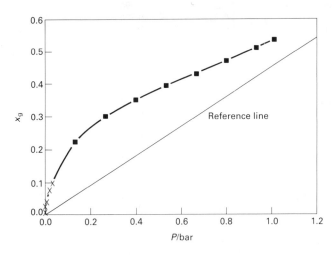

Figure 6.25. Mole fraction solubility in benzenemethanol of N,N- dimethylmethanamine at a partial pressure of 1.013 bar and 298.15 K (X, Halban; □, Gerrard).

Table 6.4 A. Mole fraction solubility of N,N-dimethylmethanamine in organic solvents at a partial pressure of 1.013 bar in organic solvents

Solvent	T/K	x_g	Ref.
Hexane	293.2	0.535	33
Decane	283.2	0.786	8
Benzene	283.2	0.750	8
Methylbenzene	283.2	0.770	8
1,3-Dimethylbenzene	283.2	0.769	8
1,3,5-Trimethylbenzene	283.2	0.756	8
1-Butanol	293.2	0.571	8
	298.2	0.511	8
1-Octanol	283.2	0.800	8
	293.2	0.636	8
	298.2	0.568	8
Benzenemethanol	283.2	0.776	8
	293.2	0.585	8
	298.2	0.541	8
1,2-Ethanediol	283.2	0.672	8
	298.2	0.396	8
1,2,3-Propanetriol	283.2	0.657	8
	298.2	0.408	8
1,4-Dioxane	283.2	0.685	8
1-Phenylethanone	298.2	0.353	8
Ethoxybenzene	283.2	0.750	8
Ethyl benzoate	293.2	0.506	8
	298.2	0.423	8
1,1'-Oxybisoctane	293.2	0.630	8
	298.2	0.567	8
N,N-Dimethylformamide	283.2	0.512	8
	298.2	0.192	8
Pyridine	283.2	0.700	8
Quinoline	283.2	0.648	8
Nitrobenzene	283.2	0.694	8
	298.2	0.326	8
Benzenamine	283.2	0.775	8
N-Methylbenzenamine	283.2	0.775	8
Benzonitrile	283.2	0.720	8
1-Methyl-2-nitrobenzene	283.2	0.676	8
	293.2	0.406	8
	298.2	0.331	8
Benzenemethanamine	283.2	0.732	8
N-Ethylbenzenamine	283.2	0.771	8
1-Octylamine	283.2	0.776	8
	293.2	0.585	8
	298.2	0.500	8
N,N-Dimethylbenzenamine	283.2	0.744	8
N,N-Diethylbenzenamine	283.2	0.750	8
Trichloromethane	298.2	0.625	8
Tetrachloromethane	293.2	0.599	33
2,2,2-Trichloroethanol	283.2	0.820	8
	293.2	0.689	8
	298.2	0.654	8
Bromobenzene	283.2	0.770	8
Iodobenzene	283.2	0.780	8
1-Bromonaphthalene	298.2	0.380	8
1-Chloronaphthalene	298.2	0.365	8

Table 6.4 B. Mole fraction solubilities of N,N-dimethylmethanamine in alcohols at a partial pressure of 0.01333 bar (10 mmHg)

Methanol	298.2	0.0154	37[a]
Ethanol	298.2	0.0146	37[a]
1-Propanol	298.2	0.0187	37[a]
1-Octanol	298.2	0.0220	8
Benzenemethanol	298.2	0.0546	37[a]

[a]Interpolated.

concentration range by Wolff and Würtz.[33] The phase diagram and the variation in mole fraction solubility with pressure at 273.2 K are shown in Figs 6.26 and 6.27. Solubility in trichloromethane, at a total pressure of 1.013 bar and 298.2 K, was measured by Gerrard. The mole fraction solubility corrected to a partial pressure of 1.013 bar is about 0.62. The corresponding value for dissolution in tetrachloromethane, found by extrapolation of Wolff's data, is 0.52. Solubilities in other solvents containing a halogen are given in Table 6.4.

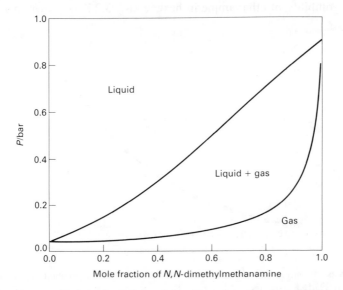

Figure 6.26. Phase equilibria in the N,N-dimethylmethanamine + tetrachloromethane system at 273.15 K.

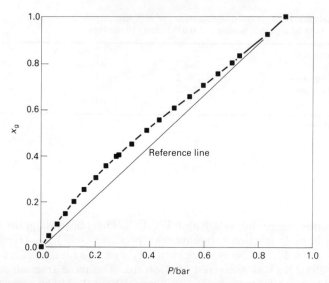

Figure 6.27. Variation with partial pressure of the mole fraction solubility of N,N-dimethylmethanamine in tetrachloromethane at 273.15 K.

6.7 SOLUBILITY OF ETHANAMINE AND HIGHER AMINES

Wolff et al.[30] studied phase equilibria between ethanamine and butane and ethanamine and hexane at 218 to 293 K. The variation with partial pressure of the solubility of ethanamine in hexane at 293.2 K is shown in Fig. 6.28.

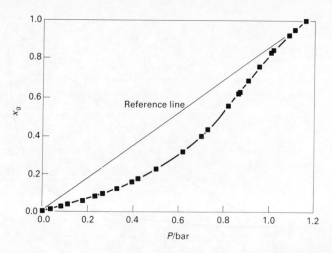

Figure 6.28. Variation with partial pressure of the mole fraction solubility of ethanamine in hexane at 293.15 K.

Phase equilibria between hexane and the propanamines was investigated by Wolff and Shadaikhy.[38] Solubility data for higher amines has also been published by Copley et al.[39] and by Halban.[37]

REFERENCES

1. Stull, D.R. *Ind. Eng. Chem.* 1947, 39, 517
2. *Lange's Handbook of Chemistry, (12th edition)*, McGraw-Hill, New York, 1979.
3. Sims, T.H. *J. Chem. Soc.* 1862, 14, 1.
4. Neuhausen, B.S., Patrick, W.A. *J. Phys. Chem.* 1921, 25, 693.
5. Perman, J. *J. Chem. Soc.* 1903, 83, 1168
6. Sherwood, T.K. *Ind. Eng. Chem.* 1925, 17, 745.
7. Clifford, I.L.; Hunter, E. *J. Phys. Chem.* 1932, 101.
8. Gerrard, W. *Solubility of Gases and Liquids*, Plenum Press, New York, 1976.
9. Doyer, J.W. *Z. Phys. Chem.* 1890, 6, 481.
10. Felsing, W.A.; Phillips, B.A. *J. Amer. Chem. Soc.* 1936, 58, 1973.
11. Dailey, B.P.; Felsing, W.A. *J. Amer. Chem. Soc.* 1939, 61, 2808.
12. *Solubility Data Series, Volume 21, Ammonia, Amines, Phosphine, Arsine, Stibine, Silane and Stannane in Organic Solvents*, ed. Young, C.L.; Fogg, P.G.T., Pergamon Press, New York, 1985.
13. Tremper, K.K.; Prausnitz, J.M. *J. Chem. Eng. Data* 1976, 21, 295.
14. Noda, K.; Morisue, T.; Ishida, K. *J. Chem. Eng. Japan* 1975, 8, 104.
15. Bell, R.P. *J. Chem. Soc.* 1931, 1371.
16. Patyi, L.; Furmer, I.E.; Makranczy, J.; Sadilenko, A.S.; Stepanova, Z.G.; Berengarten, M.G. *Zh. Prikl. Khim.* 1978, 51, 1296.
17. Hatem, S. *Bull. Soc. Chim. Fr.* 1949, 16, 337.
18. de Bruyn, L. *Rec. Trav. Chim. Pays-Bas* 1892, 11, 112.
19. Kuznetsov, A.I.; Panchenkov, G.M.; Gogoleva, T.V. *Zh. Fiz. Khim.* 1968, 42, 982.; *Russ. J. Phys. Chem.* 1968, 42, 510.
20. Short, I.; Sahgal, A.; Hayduk, W. *J. Chem. Eng. Data* 1983, 28, 63.
21. Maladkar, V.K. *Thesis, Univ. of London*, 1970
22. Timonin, V.E.; Timofeeva, E.G.; Marchenkova, T.G.; Marchenkov, V.F. 1980, *VINITI deposited document* 2874-80. (see ref. 12).
23. Gerrard, W.; Maladkar, V.K. *Chem. Ind.* 1970, 925.
24. Freidson, G.S.; Furmer, I.E.; Amelin, A.G. 1974, *VINITI deposited document* 1697-74. (see ref. 12)
25. Hala, J.; Tuck, D.G. *J. Chem. Soc. (A)* 1970, 3242.
26. Chang, E.T.; Gocken, N.A.; Poston, T.M. *J. Chem. Eng. Data* 1971, 16, 404.
27. Stoica, T.; Bota, T.D.; Teusdea, G.M.; Sarbu, L.; Herscovici, J. *Rev. Chim. (Bucharest)* 1981, 32, 1018.
28. Freidson, G.S.; Furmer, I.E.; Amelin, A.G. 1974, *VINITI deposited document* 1543-74. (see ref. 12)
29. Counsell, J.F.; Ellender, J.H.; Hicks, C.P. *Vapour-Liquid Equilibrium Data for Binary Systems Containing Aliphatic Amines*, N.P.L. Report, Chem. 79, 1978.
30. Wolff, A.; Höpfner, A.; Höpfner, H,-M. *Ber. Bunsenges. Phys. Chem.* 1964, 68, 410.
31. Wolff, A.; Höpfner, A. *Z. Elektrochem.* 1962, 66, 149.
32. Wolff, H.; Würtz, R. *Z. Phys. Chem. (Frankfurt am Main)* 1969, 67, 115.
33. Wolff, H.; Würtz. R. *Ber. Bunsenges. Phys. Chem.* 1968, 72, 101.
34. Wolff, H.; Höppel. H.-E. *Ber. Bunsenges. Phys. Chem.* 1966, 70, 874.
35. Wolff, H.; Würtz, R. *J. Phys. Chem.* 1970, 74, 1600.
36. Niepel, W.; Novak, J.P.; Matous, J,; Sobr, J. *Chem. Zvesti* 1972, 26 (1), 44.
37. Halban, H. *Z. Phys. Chem.* 1913, 84, 129.
38. Wolff, H.; Shadiakhy, A. *Fluid Phase Equilibria* 1983, 11, 267.
39. Copley, M.J.; Ginsberg, E.; Zellhoefer, G.F.; Marvel, C.S. *J. Am. Chem. Soc.* 1941, 63, 254.

39. Cauquil, G. *J. Chim. Phys. Phys. Chim. Biol.* 1927, 24, 53.
40. Furmer, I.E.; Amelin, A.G.; Freidson, G.S. *Tr. Mosk. Teknol. Inst.* 1974, 79, 96.
41. Ditsent, V.E.; Zolotareva, M.N. 1973, *VINITI deposited document* 5307-73 (see ref. 12)
42. Seward, R.P. *J. Am. Chem. Soc.* 1932, 54, 4598.

Chapter 7
SOLUBILITIES OF METHANE AND OTHER GASEOUS HYDROCARBONS

7.1 PHYSICAL PROPERTIES OF HYDROCARBONS WITH BOILING POINTS BELOW 298 K AT 1.013 BAR

Mole fraction solubilities of these hydrocarbons, in particular solvents at the same temperature and partial pressure, show a wide variation because of the wide range of volatilities of the liquefied gases. The general trends in solubilities can be judged from Fig. 7.1 which depicts the reference lines at a partial pressure of 1.013 bar for some of these gases.

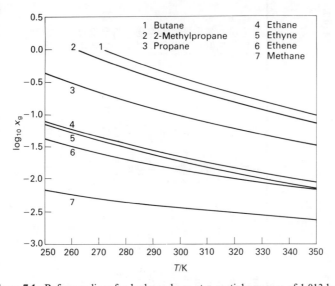

Figure 7.1. Reference lines for hydrocarbons at a partial pressure of 1.013 bar.

Physical properties of hydrocarbons with boiling points below 298 K at 1.013 bar

Gas		M.pt/K	B.pt (1.013 bar)/K	Crit.temp. /K	Crit.press. /bar	Crit.vol. /dm³ mol⁻¹	M_r	Density of gas (273.15 K; 1.013 bar)/g dm⁻³
Methane	CH_4	90.65	111.54	191.05	46.42	0.099	16.043	0.71683
Ethene	C_2H_4	103.65	169.45	282.40	50.68	0.123	28.054	1.26037
Ethane	C_2H_6	89.85	184.55	305.4	48.8	0.148	30.069	1.356
Ethyne (acetylene)	C_2H_2	191.7[a]	198[a]	308.4	61.4	0.113	26.038	1.1747
Propane	C_3H_8	85.44	231.08	369.95	42.57	0.200	44.097	2.02
Propene	C_3H_6	87.90	225.45	364.95	46.22	0.181	42.081	1.9149
Propadiene	$CH_2=C=CH_2$	136.59	238.65	393.85	52.49	0.162	40.065	
Cyclopropane	C_3H_6	145.53	240.29	398.30	55.79	0.163	42.081	1.88
Propyne	$CH_3C\equiv CH$	170.45	249.95	402.39	56.28	0.164	40.065	
2-Methylpropane	$CH(CH_3)_3$	113.45	261.45	408.05	36.49	0.263	58.124	
2-Methylpropene	$CH_2=C(CH_3)_2$	132.80	266.25	417.85	39.99	0.240	56.108	
1-Butene	$C_2H_5CH=CH_2$	87.8	266.92	419.55	40.21	0.240	56.108	
Buta-1,3-diene	$CH_2=CHCH=CH_2$	164.23	268.74	425.15	43.28	0.221	54.092	
(E)-2-Butene	$CH_3CH=CHCH_3$	167.60	274.04	428.15	41.55	0.236	56.108	
Butane	C_4H_{10}	134.85	272.65	425.15	37.95	0.258	58.124	2.7032
(Z)-2-Butene	$CH_3CH=CHCH_3$	134.25	276.87	433.15	41.55	0.236	56.108	
1-Buten-3-yne (vinyl acetylene)	$CH\equiv CCH=CH_2$		278.3				52.076	
1-Butyne	$C_2H_5C\equiv CH$	146.55	281.25	463.65	47.12	0.221	54.092	
2,2-Dimethylpropane	$C(CH_3)_4$	256.55	282.65	433.75	32.03	0.303	72.151	
Cyclobutane	C_4H_8	182.45	285.66	459.95	49.65	0.210	56.108	
3-Methyl-1-butene	$(CH_3)_2CH-CH=CH_2$	104.66	293.21	444.65	32.65	0.312	70.134	

[a] At 1.641 bar

7.2 SOLUBILITY OF METHANE

The critical temperature of methane is 191.05 K. Most measurements of solubility have been made at temperatures well above the critical temperature. The vapour pressure of liquid methane, at temperatures below the critical temperature, may be extrapolated to temperatures above the critical temperature. Reference lines may be drawn which are based upon the vapour pressure of a hypothetical liquid methane above the critical temperature. These reference lines indicate the order of magnitude of the solubility at temperatures above the critical and give a standard against which experimental measurements may be compared. However, the value of the vapour pressure of a hypothetical liquid methane at temperatures above the critical temperature, and hence the positions of the reference lines, depends upon the way in which the extrapolation is carried out.

The following data[1] fit an Antoine equation given below.

T/K	Vapour pressure/bar
148.35	10.1325
164.65	20.265
190.6	46.2

$$\ln(P/\text{bar}) = 8.3077 - 746.61/[(T/\text{K}) - 23.74]$$

This equation can be used to calculate reference lines for methane.

7.2.1 Solubility of methane in water and aqueous solutions

The complete range of experimental measurements of the solubility of methane has been compiled and evaluated in *Solubility Data Series Vol. 27/28*.[2] Solubilities presented below are intended to give a brief summary.

Measurements of the solubility in water have been evaluated by Battino.[2] He obtained a smoothing equation for solubility at a partial pressure of 1.013 bar from 16 solubility values reported by Rettich *et al.*[3] and 9 from the work by Crovetto *et al.*[4] It can be written in the form

$$\ln x_g = -338.217 + 13282.1/(T/\text{K}) + 51.9144 \ln(T/\text{K}) - 0.0425831(T/\text{K})$$

standard deviation in $\ln x_g = 0.015$
standard deviation in $x_g = 1.5\%$ at about 400 K.
temperature range 273 to 523 K.
This equation is plotted in Fig. 7.2.

The solubility measurements by Crovetto *et al.* show that mole fraction

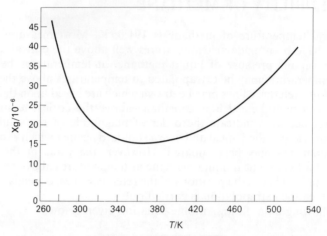

Figure 7.2. The mole fraction solubility in water of methane at a partial pressure of 1.013 bar.

solubility in water for a partial pressure of 1.013 bar passes through a minimum as temperature rises. The value of x_g in the above equation is at a minimum when T, is 365.3 K.

Battino[2] also gave a smoothing equation for solubility in water at pressures from 6 to 2000 bar and temperatures in the range 298 to 627 K. This is based upon 242 solubility values published by various authors and can be written in the form:

$$\ln x_g = -152.777 + 7478.8/(T/K) + 20.6794\ln(T/K) + 0.75316\ln(P_t/\text{bar})$$

In this equation P_t is the *total* pressure, i.e. the sum of the partial pressures of methane and of water. The value of P_t must be greater than the vapour pressure of water at a given temperature. The variation of x_g with variation of both P_t and T is shown in Fig. 7.3.

Clever[5] has evaluated published data on the solubility of methane in electrolyte solutions for about 25 systems and has calculated Sechenov constants. In some cases, concentrations are recorded on a molar scale, and in other cases, on a molal scale. Accurate conversion is not possible unless densities of solutions are available.

The solubility in solutions of sodium chloride has received most attention and has been measured to high temperatures and pressures. Clever referred to twelve papers dealing with this system and noted significant differences between the different sets of data. The variation of mole fraction solubility with concentration of sodium chloride, from some of the data published by Stoessell and Byrne,[6] is shown in Fig. 7.4.

Muccitelli and Wen[7] reported the solubility in hydrochloric acid of various concentrations at temperatures from 278 K to 298 K. Variation of the Ostwald coefficient with concentration of acid at 298 K is shown in Fig. 7.5.

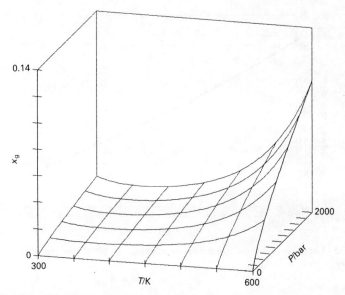

Figure 7.3. The mole fraction solubility of methane in water as a function of temperature and total pressure.

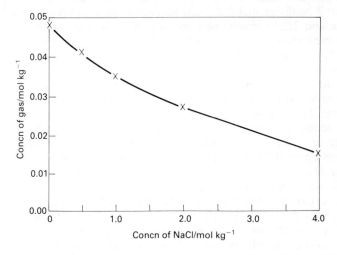

Figure 7.4. Dependence on the concentration of salt of the solubility of methane in aqueous solutions of sodium chloride at 298.15 K and a partial pressure of gas of 37.9 bar.

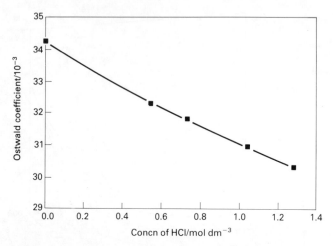

Figure 7.5. Dependence of the solubility of methane at 298 K on the concentration of hydrochloric acid.

7.2.2 Solubility of methane in hydrocarbons

Mole fraction solubility in alkanes, at a partial pressure of 1.013 bar, rises with increase in chain length. A selection of experimental measurements is given in Table 7.1.

High pressure data have been published for solubilities in various hydrocarbons and have been evaluated by Young.[8] The solvents include straight chain alkanes ethane to decane, hexadecane, eicosane and dotriacontane. Data for 2-methylpropane, 2,2-dimethylpropane, 2-methylbutane, 3-methylpentane and 2,2,3-trimethylpentane are also available. High pressure data for the methane + ethane system have also been evaluated by Hiza et al.[9]

Extensive measurements of the solubility of methane in ethane have been reported by Wichterle and Kobayashi[10] for 139 to 200 K, over the whole of the liquid composition range (mole fraction of methane from 0 to 1). A typical solubility curve is shown in Fig. 7.6. A solubility curve for dissolution in hexane, based upon measurements by Lin et al.,[11] is shown in Fig. 7.7.

A selection of measurements of the solubility in cycloalkanes, at a partial pressure of 1.013 bar, is given in Table 7.1. Solubilities in cyclohexane at pressures up to 282 bar, in the range 294 to 444 K, have been published by Reamer et al.[12] The variation of mole fraction solubility with pressure at 344.3 K is shown in Fig. 7.8.

The most extensive data for the solubility of methane in ethene are those of Miller et al.[13] for 150 to 190 K and 1.5 to 37 bar. Variation of mole fraction solubility with pressure at 160 K is shown in Fig. 7.9.

Experimental measurements of the solubilities of methane in benzene and other arenes have been evaluated by Clever.[2] A selection of mole fraction solubilities in these solvents is shown in Table 7.1.

Table 7.1 Solubility of methane in hydrocarbons at a partial pressure of gas of 1.013 bar

Solvent	T/K	x_g	Ref.
Hexane	298.2	0.00502	16
		0.00506	15
Heptane	298.2	0.00506	16
Octane	298.2	0.00511	16
Hexadecane	298.2	0.00602	16
Cyclohexane	298.2	0.00328	15
	298.2	0.00333	29
Methylcyclohexane	298.2	0.003957	25
(Z)-1,2-Dimethylcyclohexane	298.0	0.003963	26
(E)-1,2-Dimethylcyclohexane	298.0	0.004275	26
1,2-Bicyclohexyl	300.0	0.00336	23
Benzene	293.2	0.002100	14
	298.2	0.002078	14[a]
	298.2	0.00209	15
	298.2	0.00206	16
Methylbenzene	297.8	0.00224	25
1,2-Dimethylbenzene	298.2	0.002507	27
1,3-Dimethylbenzene	298.2	0.002713	27
1,4-Dimethylbenzene	298.1	0.002831	27
1-Methylnaphthalene	298.2	0.00155	24[b]
1,1'-Methylenebisbenzene	298.2	0.00181	23

[a]Interpolated.
[b]Extrapolated.

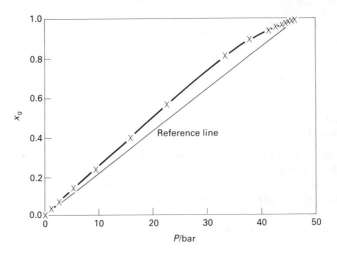

Figure 7.6. Variation with partial pressure of the mole fraction solubility of methane in ethane at 190.94 K.

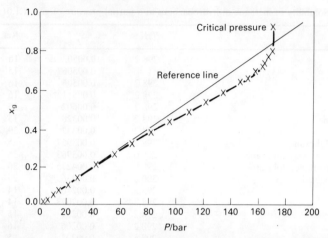

Figure 7.7. Variation with partial pressure of the mole fraction solubility of methane in hexane at 273.16 K. The solubility curve terminates at the critical pressure of the system.

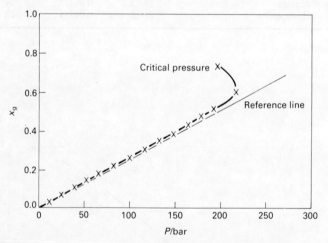

Figure 7.8. Variation with partial pressure of the mole fraction solubility of methane in cyclohexane at 344.3 K. The solubility curve terminates at the critical pressure of the system.

Clever has published a smoothing equation for the mole fraction solubility of methane in benzene at a partial pressure of 1.013 bar. This is based upon data for 286 to 333 K of Horiuti,[14] Lannung and Gjalkbaek[15] and Hayduk and Buckley.[16] It can be written in the form

$$\ln x_g = -6.6679 + 147.79/(T/K)$$

standard deviation in $x_g = 1.7 \times 10^{-5}$
temperature 286 to 333 K.

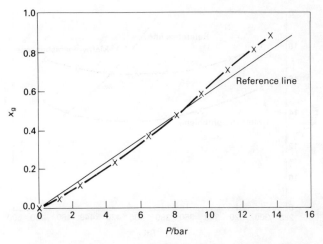

Figure 7.9. Variation with partial pressure of the mole fraction solubility of methane in ethene at 160 K.

The solubility in benzene at elevated pressures has been measured by various workers. These measurements have been discussed by Young[17] who has found some disagreement between different sets of data. At high temperatures and pressures solubilities increase with rise in temperature (Fig. 7.10 which is based upon data published by Lin et al.[18]). Mole fraction solubilities in methylbenzene,[18,19] 1,3-dimethylbenzene[20,21] and 1,3,5-trimethylbenzene[22] change in a similar manner.

Mole fraction solubilities at a partial pressure of 1.013 bar in 1,1'-

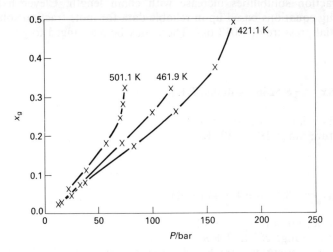

Figure 7.10. Dependence upon pressure and temperature of the solubility of methane in benzene.

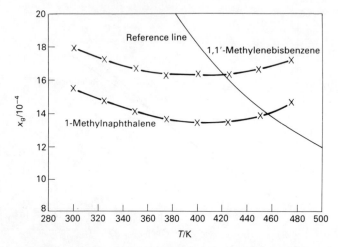

Figure 7.11. Solubilities in 1,1'-methylenebisbenzene and in 1-methylnaphthalene of methane at a partial pressure of 1.013 bar.

methylenebisbenzene[23] and in 1-methylnaphthalene[24] also pass through minimum values with increase in temperature (Fig. 7.11).

7.2.3 Solubility of methane in solvents containing carbon, hydrogen and oxygen

Solubilities of methane in alkanols have been critically reviewed by Clever.[28] Mole fraction solubilities increase with chain length. Clever has given smoothing equations based upon reliable data for mole fraction solubilities at a partial pressure of 1.013 bar. These may be rearranged to give:

Methanol

$$\ln x_g = -8.5246 + 440.00/(T/K)$$

standard deviation in $x_g = 1.35 \times 10^{-5}$
temperature range 283 to 308 K.

Ethanol

$$\ln x_g = -8.1113 + 431.44/(T/K)$$

standard deviation in $x_g = 1.4 \times 10^{-5}$
temperature range 283 to 308 K.

The above equations are based upon measurements by Lannung and Gjaldbaek,[15] Ben-Naim and Yaacobi,[29] and Boyer and Bircher.[30]

1-Propanol

$$\ln x_g = 3.7562 + 323.39/(T/K) - 2.047\ln(T/K)$$

standard deviation in $x_g = 0.0000543$
temperature range 173 to 243 K.
This equation is based upon measurements by Komarenko and Manzhelii.[31]

$$\ln x_g = -8.0574 + 477.19/(T/K)$$

standard deviation in $x_g = 1.3 \times 10^{-5}$
temperature range 283 to 308 K.
This equation is based upon measurements by Boyer and Bircher[30] and by Ben-Naim and Yaacobi.[29]

$$\ln x_g = -32.6469 + 1483.15/(T/K) + 3.7230\ln(T/K)$$

standard deviation in $x_g = 0.00010$
temperature range 273 to 308 K.
This equation is based upon a combination of the three sets of solubility data used above.

Table 7.2 Mole fraction solubilities of methane at a partial pressure of 1.013 bar in solvents containing carbon, oxygen and hydrogen

Solvent	T/K	x_g	Ref.
Methanol	298.2	0.000871	15
	298.2	0.000862	29
Ethanol	298.2	0.00128	15
	298.2	0.00126	29
1-Propanol	298.2	0.00156	29
1-Butanol	298.2	0.00184	29
1-Pentanol	298.2	0.00207	29
1-Hexanol	298.2	0.00228	29
1-Heptanol	298.2	0.00260	30
1-Octanol	298.2	0.00284	30
	298.2	0.002687	34
1-Decanol	298.2	0.003166	34
Cyclohexanol	298.2	0.00127	15
1,1'-Oxybisethane	298.2	0.004544	14[a]
1,2-Epoxyethane	298.2	0.001629	35
1,4-Dioxane	298.2	0.00138	29
	298.2	0.001372	36
Methoxybenzene	298.2	0.001598	37
2-Propanone	298.2	0.00185	15
Methyl acetate	298.2	0.00196	14[b]

[a]Extrapolated.
[b]Interpolated.

A selection of measurements of the mole fraction solubilities in alkanols is shown in Table 7.2.

Francesconi et al.[32] have studied the solubility in methanol over the whole of the composition range at pressures up to 2935 bar and at 303.2 to 502.2 K.

The solubility in 1,1'-oxybisethane (diethyl ether) was measured by Horiuti[14] at 1.013 bar. Mole fraction solubilities at a partial pressure of 1.013 bar fit the equation

$$\ln x_g = -23.796 + 1161.2/(T/K) + 2.5461 \ln(T/K)$$

standard deviation in $x_g = 3.3 \times 10^{-5}$.
temperature range 192 to 293 K.

Solubility in 2-propanone has been measured by Horiuti[14] and by Lannung and Gjaldbaek[15] with close agreement between the two sets of measurements. Mole fraction solubilities from the two sources fit the equation

$$\ln x_g = -27.992 + 1119.4/(T/K) + 3.1515 \ln(T/K)$$

standard deviation in $x_g = 2.4 \times 10^{-5}$
temperature range 197 to 310 K.

Solubility in 2-propanone at 298 and 323 K under total pressures up to 118 bar has been measured by Yokoyama et al.[33] The variation of mole fraction solubility with partial pressure of methane is approximately linear under these conditions (Fig. 7.12).

Horiuti[14] measured solubility in methyl acetate at 1.013 bar. Mole fraction solubilities at a partial pressure of 1.013 bar fit the equation

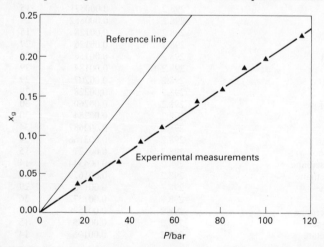

Figure 7.12. Variation with partial pressure of the mole fraction solubility of methane in 2-propanone at 298.2 K.

$$\ln x_g = -20.8475 + 799.69/(T/K) + 2.09415\ln(T/K)$$

standard deviation in $x_g = 8.5 \times 10^{-6}$
temperature range 197 to 313 K.

Solubility in 3-methylphenol has been measured by Simnick et al.[20] from 462 to 663 K and pressures from 20 to 253 bar. Mole fraction solubility increases with temperature (Fig. 7.13).

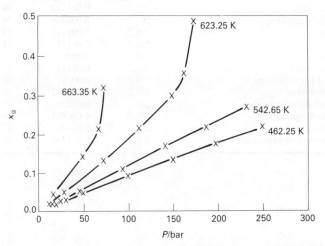

Figure 7.13. Dependence upon pressure and temperature of the solubility of methane in 3-methylphenol.

7.2.4 Solubility of methane in non-aqueous solvents containing halogens, nitrogen, sulfur, silicon or phosphorus

Several workers have measured solubilities in halogenated compounds. The mole fraction solubility in hexadecafluoroheptane at a partial pressure of 1.013 bar in the temperature range 291 to 303 K, as measured by Kobatake and Hildebrand,[38] is high compared with that in other solvents (Table 7.3). It is also high in dichlorodifluoromethane and in chlorodifluoromethane at elevated pressures in the temperature range 263 to 298 K (Yorizane et al.[39]; see Figs 7.14 and 7.15). Measurements of the solubility in tetrachloromethane by Horiuti[14] at a partial pressure of 1.013 bar are consistent with more recent measurements by Tominaga et al.[40] Mole fraction solubilities at a partial pressure of 1.013 bar from the two sources fit the equation

$$\ln x_g = -17.7376 + 819.76/(T/K) + 1.60299\ln(T/K)$$

standard deviation in values of $x_g = 2.4 \times 10^{-5}$
temperature range 253 to 333 K.

Table 7.3 Mole fraction solubilities of methane at a partial pressure of 1.013 bar in solvents containing a halogen, sulfur, nitrogen or silicon

Solvent	T/K	x_g	Ref.
Hexadecafluoroheptane	298.2	0.008262	38
Hexafluorobenzene	297.8	0.003844	67
1,1,2-Trichloro-1,2,2-trifluoroethane	298.2	0.004978	68
Tetrachloromethane	298.1	0.002818	40
Chlorobenzene	298.2	0.001986	14, 41
Carbon disulfide	298.2	0.001272	69
1,1-Sulfinylbismethane	298.2	0.000386	70
Pyrrolidine	298.2	0.00141	71
Pyridine	298.2	0.00112	71
Piperidine	298.2	0.00190	71
Cyclohexylamine	303.2	0.00192	72
1-Methyl-2-pyrrolidinone	298.2	0.0018	44
N-Methylformamide	298.2	0.000492	73
N,N-Dimethylformamide	298.2	0.00094	74
Perfluorotributylamine	298.2	0.006883	69
Nitrobenzene	298.2	0.00106	43
Octamethylcyclotetrasiloxane	298.0	0.009346	48

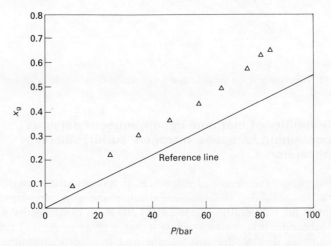

Figure 7.14. Mole fraction solubility of methane in dichlorodifluoromethane at 263.2 K.

The solubility in chlorobenzene was also measured by Horiuti.[14] More recent measurements by Lopez et al.[41] confirm the early work by Horiuti. The mole fraction solubility in chlorobenzene at a partial pressure of 1.013 bar from the two sources fits the equation

$$\ln x_g = -24.0799 + 1053.01/(T/\text{K}) + 2.51448\ln(T/\text{K})$$

standard deviation in values of $x_g = 1.3 \times 10^{-5}$
temperature range 232 to 372 K.

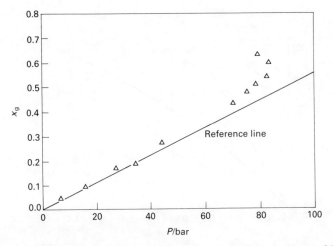

Figure 7.15. Mole fraction solubility of methane in chlorodifluoromethane at 263.2 K.

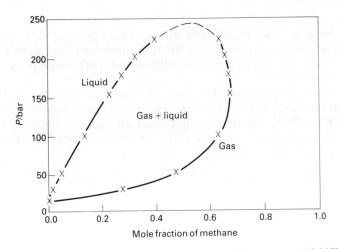

Figure 7.16. Phase equilibria in the methane + quinoline system at 702.85 K.

Solubilities in a variety of halocompounds are shown in Table 7.3.

There has been comparatively little interest in solvents containing nitrogen. A phase diagram for methane + quinoline[47] at high pressures is shown in Fig. 7.16. Solubility in 1-methyl-2-pyrrolidinone has been studied by four groups.[42–45] Young[46] notes discrepancies between the different sets of data. The variation in solubility with pressure at 298.2 K[44] is shown in Fig. 7.17.

Mole fraction solubilities at a partial pressure of 1.013 bar in solvents containing sulfur, nitrogen or silicon are given in Table 7.3.

The solubility in tributyl phosphate to a maximum of 110 bar in the temperature range 298 to 343 K has been reported by Shakhova and Zubchenko.[42] The variation in mole fraction solubility in tributyl phosphate

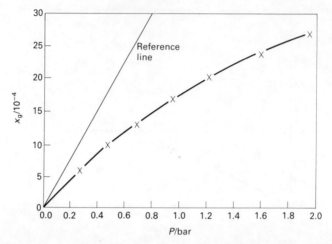

Figure 7.17. Variation with partial pressure of the mole fraction solubility of methane in 1-methyl-2-pyrrolidinone at 298.15 K.

with pressure at 298.2 K is shown in Fig. 7.18. Lenoir et al.[43] have made chromatographic measurements of limiting values of Henry's law constant for dissolution in triethyl, tripropyl, tributyl and tris(2-methylpropyl) esters of phosphoric acid.

Solubility in octamethylcyclotetrasiloxane has been measured by Wilcock et al.[48] in the range 292 to 313 K and by Chappelow and Prausnitz[24] in the range 300 to 425 K. Clever[49] has published an equation for mole fraction solubilities at a partial pressure of 1.013 bar which is based upon these data. This may be written in the form

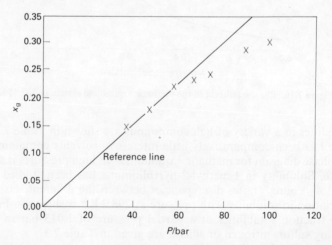

Figure 7.18. Variation with partial pressure of the mole fraction solubility of methane in tributyl phosphate at 298.15 K.

$$\ln x_g = 286.1351 - 7702.1/(T/K) - 50.6322\ln(T/K) + 0.078863(T/K)$$

standard deviation in $x_g = 5.4 \times 10^{-5}$
temperature range 293 to 425 K.

This equation gives a minimum value of x_g at about 393 K. Measurements by Chappelow and Prausnitz indicate that minimum solubility occurs between 375 and 425 K.

Phase equilibria in the methane + ammonia;[50] methane + carbon dioxide;[51-58,61-2] methane + hydrogen sulfide;[58-62] methane + sulfur dioxide;[63] methane + carbonyl sulfide[64] and methane + nitrous oxide[65] systems have also been investigated. Young[66] has compiled all available data. He has evaluated the measurements on systems containing carbon dioxide or hydrogen sulfide. Phase diagrams and curves showing variation of mole fraction solubility with variation of partial pressure of methane, based upon these data, are given in Figs 7.19 to 7.26.

It is interesting to compare the phase diagram for the methane + carbonyl sulfide system at 298.2 K (Fig. 7.19) with Fig. 7.20 which shows the variation of mole fraction solubility of methane with change of its partial pressure at this temperature. The same experimental data[65] have provided the basis for each diagram. The partial pressure of methane for the second diagram has been taken to be equal to the total pressure multiplied by the mole fraction of methane in the vapour phase. Similar diagrams for the methane + hydrogen sulfide system are shown in Figs 7.22 and 7.23 but in this case measurements have been made up to the critical pressure.[59] The corresponding curve showing the variation of mole fraction solubility with partial pressure shows a marked change of slope as the critical conditions are approached. At total pressures above the critical pressure the system cannot exist as a liquid so the concept of dissolving a gas in a liquid cannot apply.

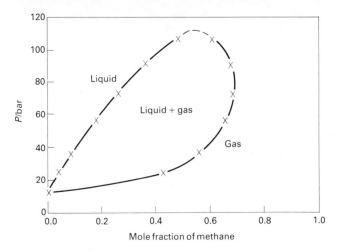

Figure 7.19. Phase equilibria in the methane + carbonyl sulfide system at 298.15 K. (Ref. 64).

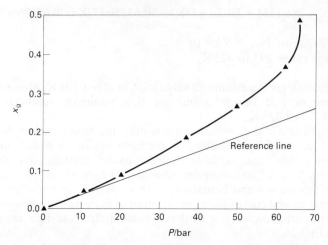

Figure 7.20. Variation with partial pressure of the mole fraction solubility of methane in carbonyl sulfide at 298.15 K.

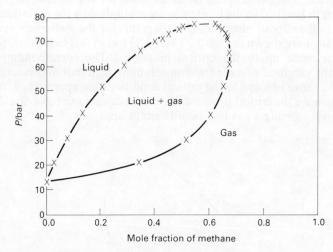

Figure 7.21. Phase equilibria in the methane + carbon dioxide system at 240.0 K (see Ref. 57).

SOLUBILITIES OF METHANE AND OTHER GASEOUS HYDROCARBONS   131

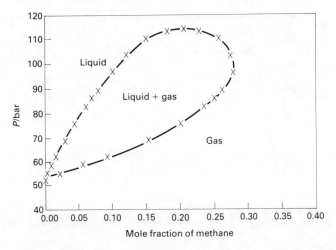

Figure 7.22. Phase equilibria in the methane + hydrogen sulfide system at 344.3 K (See Ref. 59).

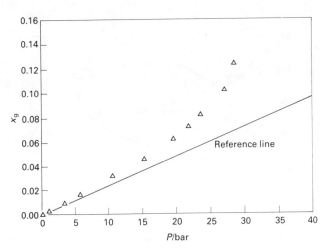

Figure 7.23. Variation with partial pressure of the mole fraction solubility of methane in hydrogen sulfide at 344.3 K (see Ref. 59).

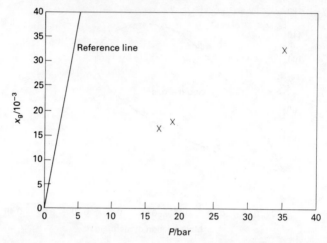

Figure 7.24. Variation with partial pressure of the mole fraction of methane in sulfur dioxide at 241.1 K.

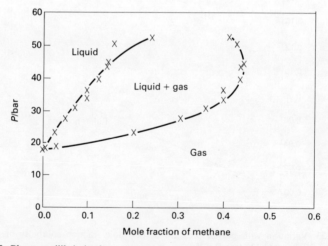

Figure 7.25. Phase equilibria in the system methane + nitrous oxide at 253.15 K (see Ref. 65).

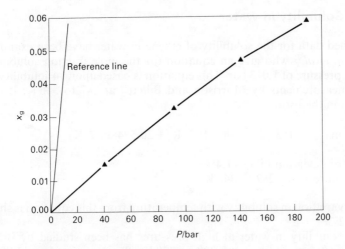

Figure 7.26. Variation of mole fraction solubility in liquid ammonia with change of partial pressure of methane at 298.15 K (see Ref. 50).

7.3 SOLUBILITY OF ETHENE

Ethene has a critical temperature of 282.40 K and a critical pressure of 50.68 bar. Solubility data are available for temperatures above and below the critical temperature. The vapour pressure of liquid ethene at temperatures below the critical temperature may be extrapolated above the critical temperatures to determine reference lines. As in the case of methane, the values of hypothetical vapour pressures above the critical temperature depend upon the method of extrapolation. Nevertheless it is felt that reference lines derived in this way are useful in the comparison and discussion of solubility data. These lines are based upon an Antoine equation which was derived from the following vapour pressure data[1] for the liquefied gas at temperatures below the critical temperature:

T/K	P/bar
244.05	20.265
258.95	30.397
282.4	51.2

These data fit the equation

$$\ln(P/\text{bar}) = 8.0763 - 868.19/[(T/\text{K}) - 72.72].$$

7.3.1 Solubility in water

Published data for the solubility of ethene in water have been examined by Wilhelm et al.[75] who give an equation for the mole fraction solubility at a partial pressure of 1.013 bar. This equation is based upon 14 solubility values from measurements by Morrison and Billett[76] at 287 to 346 K. It may be written in the form:

$$\ln x_g = -153.027 + 7965.2/(T/K) + 20.5248\ln(T/K)$$

Standard deviation in $x_g = 1.4\%$
temperature range 287 to 346 K.

The variation in solubility with temperature from this equation is shown in Fig. 7.27.

The solubility in water at high pressures has been studied by Bradbury et al.[77] at temperatures from 308 to 379 K. The variation in mole fraction solubility with total pressure at 308.2 K is shown in Fig. 7.28.

7.3.2 Solubility in non-aqueous solvents

Waters et al.[78] and McDaniel[79] measured the solubility of ethene in hexane at pressures below 1.013 bar at various temperatures. Leites and Ivanovskii[80] and also McDaniel measured the solubility in heptane. Jadot[81] has published Henry's law constants of dissolution of the gas in alkanes from pentane to decane. These were found by chromatography. Lenoir et al.[43] have determined Henry's law constants for heptadecane and hexadecane also by chromatography.

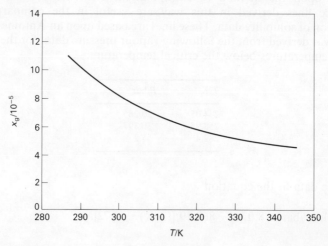

Figure 7.27. Mole fraction solubility in water of ethene at a partial pressure of 1.013 bar.

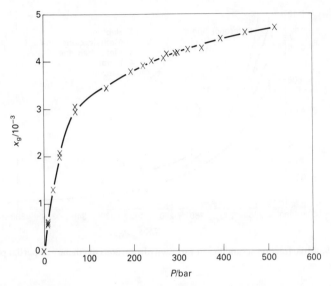

Figure 7.28. Dependence of the mole fraction solubility of ethene in water upon the total pressure at 308.15 K.

Zhuze and Zhurba[82] measured the solubility in hexane at pressures up to about 90 bar in the range 303 to 423 K. They reported linear variations of mole fraction solubility with change of pressure. The variation in mole fraction solubility in other lower alkanes is likely to be nearly linear, in the same temperature range, to pressures above 1.013 bar. It follows that mole fraction solubilities in alkanes at a partial pressure of 1.013 bar may be estimated from limiting values of Henry's law constants. However, Henry's law constants obtained by chromatography are subject to some uncertainty because of the possibility of adsorption of gas at the surface of a liquid phase.

Jadot[81] published a value of Henry's law constant for dissolution in benzene. Horiuti[14] measured the solubility at 1.013 bar from 278 to 323 K. The solubility in methylbenzene at various temperatures has been measured by Waters et al.[78], Hannaert et al.[83] and Leites and Ivanovskii.[80] Mole fraction solubility in aromatic hydrocarbons is less than in alkanes under similar conditions (Fig. 7.29).

Boyer and Bircher[84] have shown that mole fraction solubilities in straight chain primary alkanols at 298.2 K and 1.013 bar are low compared with solubilities in alkanes. Solubility increases with the length of the carbon chain of the alkanol (Fig. 7.30).

Horiuti[14] measured the solubility in methyl acetate and in 2-propanone at 1.013 bar from 273 to 313 K. Hannaert et al.[83] also measured the solubility in 2-propanone at various temperatures. Solubilities in other solvents containing oxygen have been found by chromatography. In most solvents which contain oxygen solubilities are low compared with those in alkanes.

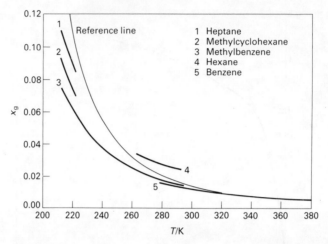

Figure 7.29. Mole fraction solubilities in various hydrocarbons of ethene at a partial pressure of 1.013 bar.

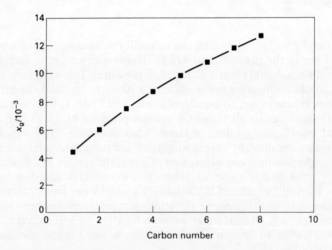

Figure 7.30. Mole fraction solubilities of ethene in straight chain primary alkanols—the dependence on chain length at 298.15 K and a partial pressure of 1.013 bar.

Horiuti also measured solubilities in tetrachloromethane and in chlorobenzene at 1.013 bar at various temperatures (Fig. 7.31).

A selection of solubility data for organic solvents is given in Table 7.4

Table 7.4 Mole fraction solubilities of ethene at a partial pressure of 1.013 bar in organic solvents

Solvent	T/K	x_g	Ref.
Pentane	298.2	0.0153	81[a]
Hexane	298.2	0.0159	81[a]
	298.2	0.0227	78[b]
Heptane	298.2	0.0173	81[a]
	223.2	0.0833	80
Octane	298.2	0.0184	81[a]
Nonane	298.2	0.0196	81[a]
Decane	298.2	0.0211	81[a]
Hexadecane	298.2	0.0234	43
Decahydronaphthalene	298.2	0.0138	43[a]
Benzene	293.2	0.0133	14
	298.2	0.0125	14
	298.2	0.0112	81[a]
Methylbenzene	293.2	0.0158	78
	293.2	0.0148	83
Methanol	293.2	0.0047	83
Benzenemethanol	298.2	0.0062	43[a]
1,2-Ethanediol	303.2	0.00031	83[a]
	298.2	0.0013	43[a]
Oxybispropanol	298.2	0.0049	43[a]
2,2'-[1,2-Ethanediylbis(oxy)] bisethanol (triethylene glycol)	303.2	0.0018	83[a]
2-Propanone	293.2	0.0135	83
	293.2	0.0118	14
	298.2	0.0111	14
Methyl acetate	293.2	0.0123	14
	298.2	0.0116	14
4-Methyl-1,3-dioxolan-2-one (propylene carbonate)	298.2	0.0052	43[a]
Phenol	323.2	0.0039	43[a]
Tetrachloromethane	293.2	0.0157	83
	293.2	0.0158	14
	298.2	0.0148	81[a]
	298.2	0.0146	83[b]
	298.2	0.0148	14
1,2-Dichloroethane	293.2	0.0093	83
	298.2	0.0089	83
Chloroethene	243.2	0.0503	83
Chlorobenzene	293.2	0.0128	14
	298.2	0.0121	14
Benzenamine	298.2	0.0048	43[a]
1-Methyl-2-pyrrolidinone	298.2	0.0079	43[a]
Nitrobenzene	298.2	0.0079	43[a]
Sulfinylbismethane	298.2	0.0032	43[a]
Tributyl phosphate	325.2	0.0190	43[a]
Triethyl phosphate	325.2	0.0115	43[a]
Trimethyl phosphate	325.2	0.0050	43[a]
Tripropyl phosphate	323.2	0.0175	43[a]
Triisobutyl phosphate	325.2	0.0205	43[a]

[a]Obtained by chromatographic methods.
[b]Extrapolated.

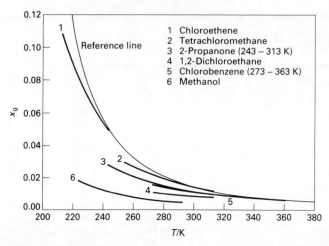

Figure 7.31. Mole fraction solubilities of ethene, at a partial pressure of 1.013 bar, in solvents containing oxygen or chlorine.

7.4 SOLUBILITY OF ETHANE

Ethane has a critical temperature of 305.4 K and critical pressure of 48.8 bar. The following data[1] have been used to derive an equation for vapour pressure of liquid ethane for construction of reference lines:

T/K	P/bar
241.15	10.1325
266.75	20.2650
305.4	48.8

This equation is:

$$\ln(P/\text{bar}) = 12.1279 - 3304.67/[(T/\text{K}) + 95.64]$$

A reference line, based upon this equation, is shown in Fig. 7.32

Hayduk[85] has edited an extensive compilation and evaluation of the solubility of ethane in aqueous and non-aqueous solvents. He has shown the extent to which solubility can be correlated with the Hildebrand solubility parameter of the solvent.

7.4.1 Solubility of ethane in water

Various measurements have been made of the solubility of ethane in water. The most reliable for solubilities at partial pressures close to 1.013 bar are

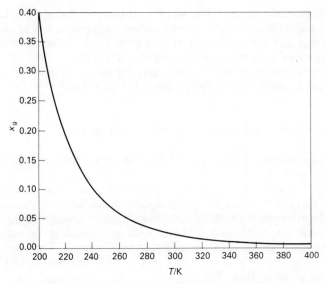

Figure 7.32. Reference line for ethane at a partial pressure of 1.013 bar.

those given by Rettich et al.[86] for the range 275 to 323 K and 0.51 to 1.11 bar. Mole fraction solubilities for a partial pressure of 1.013 bar from these data fit an equation derived by Battino.[85] This can be written in the form:

$$\ln x_g = -250.8120 + 12695.6/(T/K) + 34.7413\ln(T/K).$$

The variation in solubility with temperature from this equation is shown in Fig. 7.33.

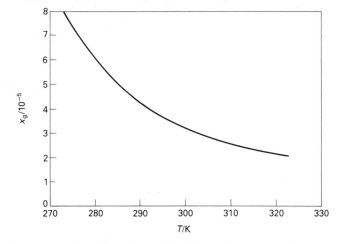

Figure 7.33. Mole fraction solubility in water of ethane at a partial pressure of 1.013 bar.

Solubilities in the range 311 to 444 K and from 4.1 to 68.5 bar have been measured by McKetta et al.[87] Extrapolation of the data for 311 K to give solubility at a partial pressure of 1.013 bar gives a value which is within about 3% of the value from measurements by Rettich et al.

The water + ethane system has been investigated by Danneil et al.[88] in the pressure range 200 to 3700 bar and from 473 to 673 K.

7.4.2 Solubility of ethane in organic solvents

Solubilities at a partial pressure of 1.013 bar in a selection of solvents are shown in Figs 7.34 to 7.36.

A number of workers have measured solubilities in alkanes at pressures at or below 1.013 bar. There are some discrepancies between measurements. The trend is for mole fraction solubilities at a partial pressure of 1.013 bar to increase with length of carbon chain, with solubilities in branched chain alkanes higher than those in the straight chain isomers. The more reliable values of mole fraction solubilities for a partial pressure of 1.013 bar at 298.2 K are given in Table 7.5.

Solubilities of ethane, at pressures appreciably above barometric pressure, have been measured by various authors[85] in propane, butane, 2-methylpropane, pentane, hexane, heptane, octane, decane, dodecane and eicosane. The dependence on pressure of the solubility in hexane, from data published by Ohgaki et al.,[92] is shown in Fig. 7.37.

In solvents having little or no polarity, other than alkanes, mole fraction

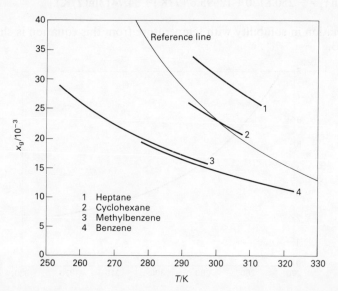

Figure 7.34. Mole fraction solubilities of ethane, at a partial pressure of 1.013 bar, in hydrocarbons (see Refs 14, 78, 89, 94).

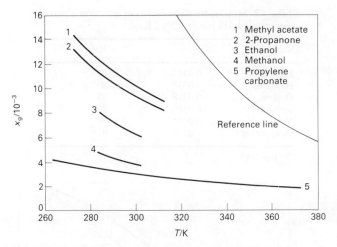

Figure 7.35. Mole fraction solubilities of ethane, at a partial pressure of 1.013 bar, in solvents containing oxygen (see Refs 14, 94, 96, 98).

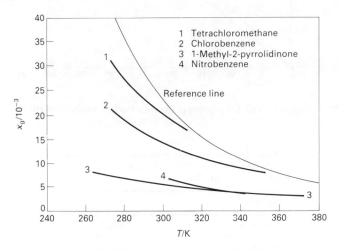

Figure 7.36. Mole fraction solubilities of ethane, at a partial pressure of 1.013 bar, in solvents containing chlorine or nitrogen (see Refs 14, 98, 99).

solubilities at 298.2 K and a partial pressure of 1.013 bar are between 0.01 and 0.03. Hayduk[85] has shown that the maximum is reached when the solubility parameter of the solvent is about 7.5. Mole fraction solubilities recommended by Hayduk[85] and based upon literature data are shown in Table 7.6.

Table 7.5 Mole fraction solubilities of ethane at a partial pressure of 1.013 bar and 298.2 K

Solvent	x_g	Ref.
Hexane	0.0317	89
Heptane	0.0315	89
Octane	0.0318	89
Nonane	0.0336	90
Dodecane	0.0345	89
Hexadecane	0.0366	91[a]

[a]Extrapolated.

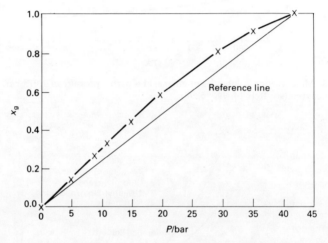

Figure 7.37. Variation with partial pressure of the mole fraction solubility of ethane in hexane at 298.15 K.

Table 7.6 Mole fraction solubilities of ethane in solvents of low polarity at 298.2 K and a partial pressure of 1.013 bar

Solvent	x_g
Cyclohexane	0.0233
Bicyclohexyl	0.0228
Decahydronaphthalene	0.0224
Methylnaphthalene	0.0106
Benzene	0.0149
Methylbenzene	0.0155
Tetrachloromethane	0.0211
Chlorobenzene	0.0146
Perfluorotributylamine	0.0227
Perfluoroheptane	0.0216
1,1,2-Trichloro-1,2,2-trifluoroethane (Freon 113)	0.0269
Carbon disulfide	0.0107

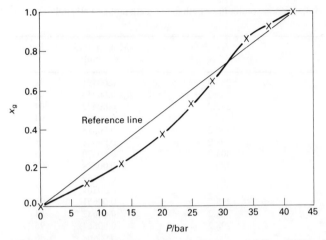

Figure 7.38. Variation with partial pressure of the mole fraction solubility of ethane in benzene at 298.15 K.

The phase equilibrium with benzene in the range 269 to 504 K and 0.7 to 10.0 bar has been investigated by Kay and Nevens.[93] Dew points and bubble points were determined for various mixtures. The equilibrium at 298.2 K to a total pressure of 38 bar has been studied by Ohgaki et al.[92] The curve for the variation of mole fraction solubility with partial pressure from these measurements crosses the reference line (see Fig. 7.38).

Mole fraction solubilities in normal alcohols increase with chain length. Hayduk has shown that mole fraction solubilities, at a partial pressure of 1.013 bar and 298.2 K, from data published by Ben-Naim and Yaacobi,[94] Boyer and Bircher,[84] and Gjaldbaek and Niemann[95] may be fitted to the equation

$$\ln x_g = 0.70796 \ln C_n - 5.5244$$

where C_n is the carbon number up to a value of eight.

A selection of solubilities in methanol and ethanol is given in Table 7.7. Solubility in methanol at 248 to 373 K at pressures to 61 bar have been measured by Ma and Kohn[97] and by Ohgaki et al.[92] at 298 K to 41 bar. Measurements are consistent with solubilities given in Table 7.7. The variation of mole fraction solubility with partial pressure of ethane from Ohgaki's work is shown in Fig. 7.39.

The solubilities in numerous other polar solvents containing oxygen and/or nitrogen have been compiled and evaluated by Hayduk.[85] Mole fraction solubilities at 298.2 K and a partial pressure of 1.013 bar which he considers reliable enough to be accepted on a tentative basis are listed in Table 7.8.

Table 7.7 Mole fraction solubilities of ethane in methanol and ethanol at a partial pressure of 1.013 bar

Solvent	T/K	Mole fraction solubility	Ref.
Methanol	283.2	0.00491	94
	288.2	0.00454	94
	293.2	0.00422	94
	298.2	0.00395	94
	298.2	0.00388	84
	298.2	0.00388	95
	303.2	0.00370	94
Ethanol	283.2	0.00828	96
	288.2	0.00759	96
	293.2	0.00701	96
	298.2	0.00651	96
	298.2	0.00682	84
	298.2	0.00664	95
	303.2	0.00608	96

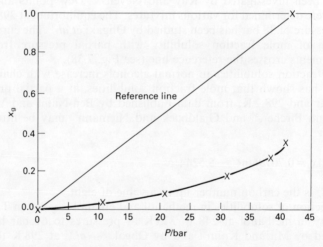

Figure 7.39. Variation with partial pressure of the mole fraction solubility of ethane in methanol at 298.15 K.

Table 7.8 Mole fraction solubilities of ethane at a partial pressure of 1.013 bar bar in miscellaneous polar solvents at 298.2 K

Solvent	x_g
Cyclohexylamine	0.0145
Diphenylmethane	0.0120
Methyl acetate	0.0105
2-Ethoxyethanol	0.0298
2-Propanone	0.00972
1,4-Dioxane	0.00816
1,1'-Oxybis(2-chloroethane)	0.0258
Nitrobenzene	0.00735
1,2-Epoxyethane	0.00917
Methyl acetamide	0.00484
1-Methyl-2-pyrrolidinone	0.00533
2-Furancarboxaldehyde	0.00861
N,N-Dimethylformamide	0.00514
1,1,-Sulfinylbismethane	0.00178
4-Methyl-1,3-dioxolan-2-one	0.00300
2-(2-Aminoethoxy)-ethanol	0.00215
2-Aminoethanol	0.000804
Octamethylcyclotetrasiloxane	0.0509
Tetrahydrothiophene 1,1-dioxide	0.00208

Solubilities at 298.2 K in 2-propanone and in methyl acetate to high pressures have been measured by Ohgaki et al.[92] (Figs 7.40 and 7.41).

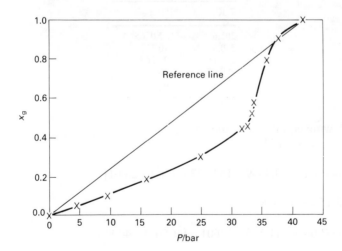

Figure 7.40. Variation with partial pressure of the mole fraction solubility of ethane in methyl acetate at 298.15 K.

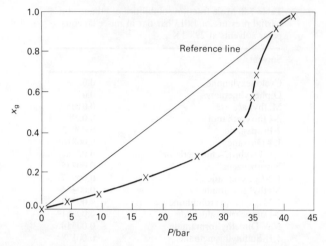

Figure 7.41. Variation with partial pressure of the mole fraction solubility of ethane in 2-propanone at 298.15 K.

7.5 SOLUBILITY OF ETHYNE (ACETYLENE)

The critical temperature of ethyne is 308.4 K and critical pressure 61.40 bar. Vapour pressures of liquid ethyne from data published by Stull[1] are tabulated below:

T/K	P/bar
201.6	2.0265
240.5	10.132
263.15	20.265
289.95	40.530
308.4	61.40

The following equations fit the data for
(a) 201.6 to 263.15 K

$$\ln(P/\text{bar}) = 9.4959 - 1525.37/[(T/\text{K}) - 28.01]$$

(b) 263.15 to 308.4 K

$$\ln(P/\text{bar}) = 11.5657 - 2601.69/[(T/\text{K}) + 40.89]$$

These equations have been used to derive reference lines for ethyne (Fig. 7.42).

Miller[100] has published a review of the properties of ethyne. He has

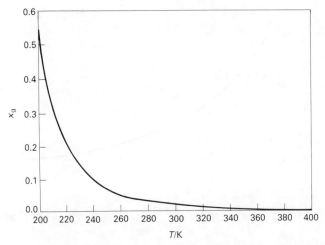

Figure 7.42. Reference line for ethyne at a partial pressure of 1.013 bar.

included a compilation and evaluation of much of the solubility data which have been published.

7.5.1 Solubility of ethyne in water

Solubility in water at partial pressures of gas less than 1.013 bar has been measured by Vitovec[101] at 293 K and by Schoen[102] from 291 K to 316 K. Hiraoka[103] measured solubilities at 274 K to 303 K and pressures from 5 to 39 bar. The solubility is low compared with that in all other solvents. A summary of the mole fraction solubility at a partial pressure of 1.013 bar from these measurements is given below.

T/K	Vitovec	Schoen	Hiraoka
293.2	0.000810	0.000841[a]	0.000802[b]
298.2		0.000763[a]	0.000718[b]

[a]Extrapolated and interpolated.
[b]Extrapolated.

Wilhelm et al.[75] published an equation for the mole fraction solubility at a partial pressure of 1.013 bar. This was based upon data published by Hiraoka[103], Flid and Golynets[104] and Billitzer.[105] It can be written in the form

$$\ln x_g = -156.51 + 8160.2/(T/K) + 21.403\ln(T/K)$$

standard deviation in $x_g = 1.3\%$
temperature range 274 to 343 K.

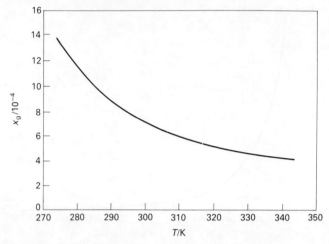

Figure 7.43. Mole fraction solubility in water of ethyne at a partial pressure of 1.013 bar.

The equation is plotted in Fig. 7.43.

Solubilities in various aqueous solutions of electrolytes and non-electrolytes have been evaluated by Miller.[100]

7.5.2 Solubility of ethyne in non-aqueous solvents

Solubilities in selected solvents are shown in Figs 7.44, 7.45 and 7.46.

Solubility in benzene at pressures close to barometric over the range 283.2

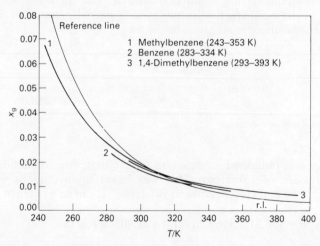

Figure 7.44. Mole fraction solubilities in aromatic hydrocarbons of ethyne at a partial pressure of 1.013 bar.

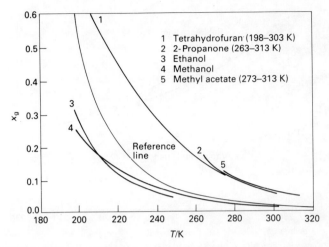

Figure 7.45. Mole fraction solubilities, in solvents containing oxygen, of ethyne at a partial pressure of 1.013 bar.

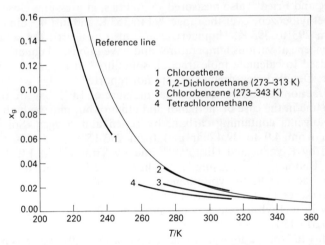

Figure 7.46. Mole fraction solubilities in chloro compounds of ethyne at a partial pressure of 1.013 bar.

to 318.2 K was measured by Horiuti[14] and from 293.35 K to 333.55 K by Vitovec and Fried.[106] The two sets of values for mole fraction solubilities at a partial pressure of 1.013 bar are plotted in Fig. 7.47 and fit the equation

$$\ln x_g = -90.6995 + 5162.50/(T/K) + 12.1700\ln(T/K)$$

standard deviation in $x_g = 1.98 \times 10^{-4}$
temperature range 283 to 334 K.

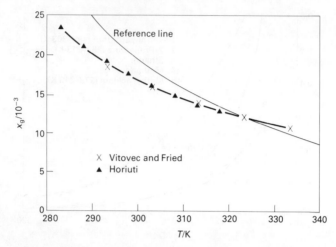

Figure 7.47. Mole fraction solubilities in benzene of ethyne at a partial pressure of 1.013 bar.

Vitovec and Fried[106] also measured solubilities, at pressures close to 1.013 bar, in methylbenzene over the range 293 to 363 K and in 1,4−dimethylbenzene from 293 to 393 K. Hannaert et al.[83] also measured the solubility in methylbenzene at various temperatures and have published equations which may be used to calculate mole fraction solubilities at a partial pressure of 1.013 bar. Detailed measurements were not reported.

Mole fraction solubilities at a partial pressure of 1.013 bar are below the reference line in the case of methanol and ethanol but may be high in other organic solvents containing carbon, hydrogen and oxygen. Solubility in methanol from 4.9 to 29.4 bar and from 273.15 to 303.15 K has been measured by Kiyama and Hiraoka,[107] and by Vitovec[101] at 1.013 bar and 293.15 K. Bodor et al.[108] measured solubilities in methanol, ethanol, acetaldehyde, ethyl acetate, 1,1′-oxybisethane, tetrahydrofuran and 2-methoxyethanol at pressures up to about 1.013 bar in the range 198.15 to 273.15 K. Solubility in methyl acetate was measured at 1.013 bar by Horiuti[14] from 273.15 to 313.15 K.

Solubility in 2-propanone at pressures up to about 1.013 bar was measured by Bodor[108] from 198 to 273 K and by Horiuti from 273 to 313 K. Holemann and Hasselmann[109] measured the solubility at temperatures from 276 to 313 K and total pressures from 1.39 to 30.03 bar. An equation for the solubility from 243 to 293 K has been published by Hannaert et al.[83] At temperatures below about 273 K this equation gives values of the solubility which are very high in comparison with other published data.

Miller[100] has published smoothed values of solubilities of ethyne in 2-propanone at partial pressures of gas from 0.0101 to 1.013 bar and also for total pressures from 1.013 bar to 30.39 bar. The temperature range is 183 to 323 K. These values were based upon unpublished measurements by British Oxygen Co Ltd and were correlated with a modified van Laar equation.

Miller gives an equation for mole fraction solubility at partial pressures of ethyne of 5.05–30.39 bar in the range 273–323 K:

$$x_g = (P/\text{atm})(a - b\log_{10}(P/\text{atm})).$$

Miller stated that this equation fits experimental measurements to within 1% for values of x_g of 0.25 to 0.65.
a and b are given by:

$$\log_{10} a = 703.8/(T/\text{K}) - 3.544$$
$$\log_{10} b = 903.5/(T/\text{K}) - 4.578$$

The variation with temperature of the solubility at a partial pressure of 1.013 bar, from Miller's data, is shown in Fig. 7.48. The variation with total pressure at 293.15 K is shown in Fig. 7.49.

Kiyama and Hiraoka[110] measured the solubility in tetrahydrofuran in the range 273.15 to 303.15 K and from 4.9 to 24.5 bar. Extrapolation of the data to 1.013 bar shows that the measurements are consistent with those made by Bodor et al.[108]

The mole fraction solubility in chlorobenzene at a partial pressure of 1.013 bar in the range 273 to 343 K may be calculated from measurements by Horiuti.[14] Mole fraction solubilities are below the reference line and close to those for benzene at the same temperatures. Measurements by the same author of the solubility in tetrachloromethane over the range 273 K to 313 K

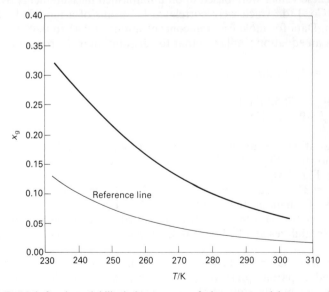

Figure 7.48. Mole fraction solubility in 2-propanone of ethyne at a partial pressure of 1.013 bar.

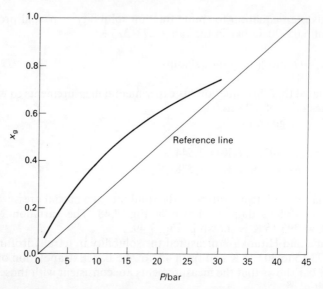

Figure 7.49. Variation with partial pressure of the mole fraction solubility of ethyne in 2-propanone at 293.15 K.

also correspond to mole fraction solubilities below the reference line (see Fig. 7.46).

Miller published smoothed values of the solubility in N,N-dimethylformamide at partial pressures of ethyne from 0.101 to 30.40 bar in the range 213 to 333 K. These values were based upon unpublished measurements by British Oxygen Co. Ltd. which were correlated by means of a modified van Laar equation. Data for mole fraction concentrations of 0.41 to 0.68 at 273 K to 333 K fit an equation similar to that for dissolution in 2-propanone, i.e.

$$x_g = (P/\text{atm})(a - b\log_{10}(P/\text{atm}))$$

where $\log_{10} a = 705/(T/\text{K}) - 3.4305$
and $\log_{10} b = 917/(T/\text{K}) - 4.4427$

The variation of solubility with temperature at a partial pressure of 1.013 bar is shown in Fig. 7.50. The variation with partial pressure at 293 K is shown in Fig. 7.51.

Bodor et al. measured the solubility in liquid ammonia at various partial pressures of ethyne in the range 0.31 to 1.07 bar from 199 to 223 K. The solubility in this solvent was also measured by Hannaert et al.[83] Mole fraction solubilities at 1.013 bar are high, relative to the reference line (Fig. 7.52).

McKinnis[111] has published an extensive list of solubilities of ethyne at 298 K and a partial pressure of 1.013 bar. Some of these are original measurements but the sources of other measurements are not stated. He has

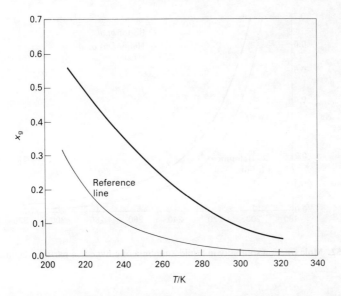

Figure 7.50. Mole fraction solubility in N,N-dimethylformamide of ethyne at a partial pressure of 1.013 bar.

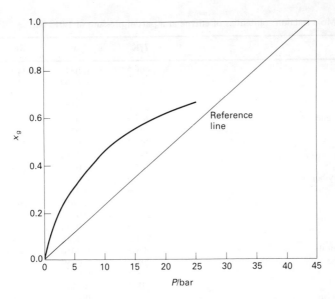

Figure 7.51. Variation with partial pressure of the mole fraction solubility of ethyne in N,N-dimethylformamide at 293.15 K.

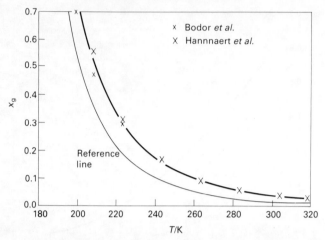

Figure 7.52. Mole fraction solubility in liquid ammonia of ethyne at a partial pressure of 1.013 bar.

derived an empirical formula to correlate solubilities with structures of solvent molecules. A more recent list was published by Miller.

Mole fraction solubilities in various non-aqueous solvents are shown in Table 7.9.

Table 7.9 Mole fraction solubilities of ethyne at a partial pressure of 1.013 bar

Solvent	T/K	x_g	Ref.
Hexane	298.2	0.047	112
Benzene	293.2	0.0193	14
	293.4	0.0186	106
	298.2	0.0177	14
Methylbenzene	243.2	0.0743	80
	293.2	0.0219	83
	293.2	0.0197	106
	298.2	0.0214	80
	303.3	0.0170	106
1,4-Dimethylbenzene	293.4	0.0208	106
	303.4	0.0179	106
Methanol	248.2	0.0617	108
	293.2	0.0202	101
	298.2	0.0190	107[b]
Ethanol	248.2	0.0466	108
1,2-Ethanediol	303.2	0.0050	83
2-Propanone	298.2	0.0659	14
	298.2	0.0628	109
Acetaldehyde	248.2	0.1445	108
Methyl acetate	273.2	0.1257	14
	298.2	0.0684	14
Ethyl acetate	273.2	0.1128	108

Table 7.9 (continued)

Solvent	T/K	x_g	Ref.
Acetic acid, ethenyl ester (vinyl acetate)	293.2	0.0438	113
Tetrahydrofuran	273.2	0.1202	108
	273.2	0.1174	110[a]
	298.2	0.0613	110[b]
1,1'-Oxybisethane	248.2	0.1624	108
2-Methoxyethanol	273.2	0.0790	108
Chlorobenzene	298.2	0.0148	14
Tetrachloromethane	293.2	0.0125	14
	293.2	0.0128	83
	298.2	0.0117	14
Trichloromethane	243.2	0.0479	80
	298.2	0.014	111[c]
1,2-Dichloroethane	298.2	0.0214	83
Chloroethene	298.2	0.0144	83
N,N-Dimethylformamide	298.2	0.1204	80
Ammonia	223.2	0.3008	108
	298.2	0.0428	83
Hexamethylphosphoric triamide	298.2	0.253	111

[a] Extrapolated.
[b] Extrapolated and interpolated.
[c] Quoted in Ref. 111, original source not given.

7.6 SOLUBILITY OF PROPANE, BUTANE AND 2-METHYLPROPANE

The solubilities of these three gases in a wide range of solvents have been compiled and evaluated.[114] Critical temperatures and pressures of the three gases are:

Gas	T_c/K	P_c/bar
Propane	369.95	42.57
Butane	425.15	37.95
2-Methylpropane	408.05	36.49

In the liquid phase the compounds have the following vapour pressures[1]:

Gas	T/K	P/bar
Propane	274.6	5.066
	331.3	20.265
	368.0	40.530
Butane	272.7	1.013
	323.2	5.066
	389.2	20.265
2-Methylpropane	261.5	1.013
	312.2	5.066
	372.7	20.265

These vapour pressures fit the following equations which will be used to calculate reference lines:

Propane

$$\ln(P/\text{bar}) = 10.7611 - 2897.6/[(T/K) + 42.46]$$

Butane

$$\ln(P/\text{bar}) = 8.7991 - 1978.2/[(T/K) - 47.55]$$

2-Methylpropane

$$\ln(P/\text{bar}) = 10.7807 - 3106.2/[(T/K) + 26.96]$$

Reference lines for a partial pressure of 1.013 bar are shown in Fig. 7.53.

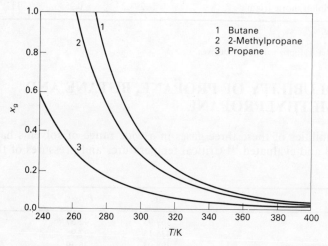

Figure 7.53. Reference lines for propane, 2-methylpropane and butane at partial pressures of 1.013 bar.

7.6.1 Solubilities of propane, butane and 2-methylpropane in water

Battino has published[13] smoothing equations for solubilities at a partial pressure of 1.013 bar, based on the published experimental measurements which he considers the most reliable. The equation for propane may be written in the form:

$$\ln x_g = -283.828 + 14434.5/(T/K) + 39.4740\ln(T/K)$$

standard deviation in $\ln x_g = 0.012$
standard deviation in $x_g = 2\%$

This equation is based upon 30 solubility values in the temperature range 273 to 347 K. It indicates that x_g is likely to be a minimum at about 366 K. The corresponding equation for butane is:

$$\ln x_g = -280.525 + 14604.0/(T/K) + 38.7599\ln(T/K)$$

This is based upon 45 solubility values in the range 273 to 349 K and indicates that x_g is likely to be a minimum at about 377 K.

There is a typographical error in Battino's equation[114] for the solubility of 2-methylpropane. A correct form of the equation is

$$\ln x_g = -371.325 + 18304.4/(T/K) + 52.4651\ln(T/K).$$

This equation is based on only four points in the range 278 to 318 K and must be considered much less reliable than smoothing equations for the other two gases. Variations in solubility with temperature from these three equations are shown in Fig. 7.54. The mole fraction solubility is greatest in propane. This is in contrast to the relative positions of the reference lines (Fig. 7.53) and to the relative solubilities in many other solvents.

Wilhelm, Battino and Wilcock[75] have, in an earlier publication, reported another equation for the solubility of 2-methylpropane in water. This was based upon 14 solubility values from a paper by Nosov and Barlyaev[115] and may be written as:

$$\ln x_g = 96.107 - 2472.35/(T/K) - 17.3665\ln(T/K)$$

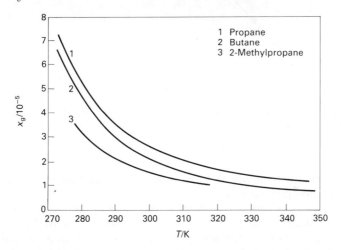

Figure 7.54. Mole fraction solubilities in water of propane, butane and 2-methylpropane at partial pressures of 1.013 bar.

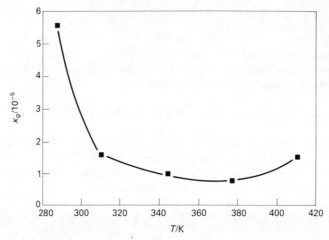

Figure 7.55. Mole fraction solubility in water of propane at a partial pressure of 1.013 bar by extrapolation of data published by Azarnoosh and McKetta.

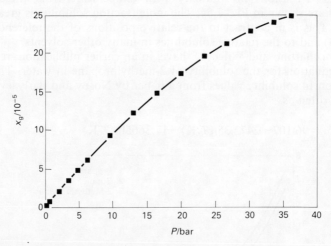

Figure 7.56. Variation with partial pressure of the mole fraction solubility of propane in water at 360.9 K.

This corresponds to solubilities about 20% less than those fitting the more recent equation given above.

The solubility of propane in water in the temperature range 289 to 411 K and partial pressure range 1 to 34 bar was measured by McKetta et al.[116] (Figs 7.55 and 7.56). Measurements are consistent with those carried out by Kobayashi and Katz[117] in the temperature range 311 to 422 K and pressure range 5 to 33 bar.

Similar curves for the variation in solubility of butane in water with change

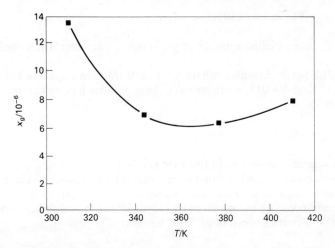

Figure 7.57. Mole fraction solubility in water of butane at a partial pressure of 1.013 bar by extrapolation of data published by Le Breton and McKetta.

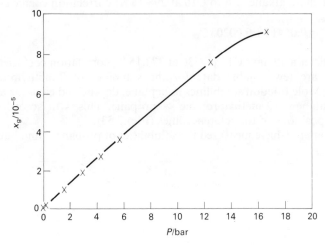

Figure 7.58. Variation with partial pressure of the mole fraction solubility of butane in water at 377.6 K.

in pressure are shown in Figs 7.57 and 7.58. These are based upon data published by Le Breton and McKetta.[118]

7.6.2 The solubilities of propane, butane and 2-methylpropane in hydrocarbons

Hayduk[114] has shown that mole fraction solubilities of propane at 298.15 K and 1.013 bar in straight chain alkanes having carbon number 6 to 16 increase with carbon number. He gives the equation

$$x_g = 0.1036 + 0.00197 C_n$$

where C_n is the carbon number of the solvent. The correlation coefficient is 0.975.

Hayduk[11] gives a similar equation for mole fraction solubilities of propane at 373.15 K and 1.013 bar in straight chain alkanes having carbon numbers of 17 to 36.

$$x_g = 0.02250 + 0.00105 C_n$$

The correlation coefficient in this case is 0.964.

In both carbon number ranges the solubilities in branched-chain alkanes are close to those in straight chain alkanes having the same carbon number.

Corresponding equations given by Hayduk[11] for the solubility of butane in alkanes at a partial pressure of 1.013 bar are:

$$x_g = 0.3975 + 0.00484 C_n$$

(straight chain alkanes C6 to C16 at 298.15 K; correlation coefficient 0.813)

$$x_g = 0.06414 + 0.00203 C_n$$

(straight chain alkanes C16 to C36 at 373.15 K; correlation coefficient 0.980)

There are few reliable data for the solubility of 2-methylpropane in alkanes. Mole fraction solubilities in heptane, decane and eicosane are in the order butane > 2-methylpropane > propane. This is in accord with the relative positions of the reference lines (Fig. 7.53).

Reamer et al. have measured the solubilities of propane[126] and butane[127] in

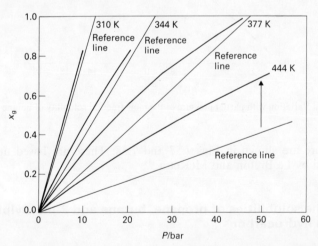

Figure 7.59. Dependence of the mole fraction solubility of propane in decane upon the temperature and partial pressure.

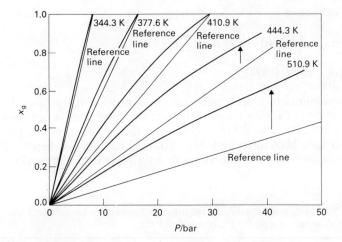

Figure 7.60. Dependence of the mole fraction solubility of butane in decane upon the temperature and partial pressure.

decane to high pressures at 278 to 511 K (Figs 7.59 and 7.60). Below about 311 K the variation of mole fraction solubility with partial pressure of propane or butane is close to the reference line. The higher the temperature the more the experimental values lie above the reference lines.

7.6.3 Solubility of propane, butane and 2-methylpropane in solvents containing oxygen

Various measurements have been made of the solubility of propane, butane and 2-methylpropane in normal alcohols. Hayduk[114] has shown that mole fraction solubilities from methanol to octanol at 298.15 K and a partial pressure of gas of 1.013 bar fit the following equations:

Propane

$$\ln x_g = -4.3982 + 0.7774 \ln C_n$$

(correlation coeff. 0.997)

Butane

$$\ln x_g = -3.1867 + 0.9053 \ln C_n$$

(correlation coeff. 0.996)

2-Methylpropane

$$\ln x_g = -3.6401 + 0.9128 \ln C_n$$

(correlation coeff. 0.997)
C_n = length of carbon chain.

The solubilities in 2-propanol[130] are close to those in 1-propanol.

A selection of data for the solubilities of the three gases in organic solvents is shown in Table 7.10. Where appropriate, the data for a partial pressure of 1.013 bar have been interpolated or extrapolated to give values for 298.2 K. Changes in mole fraction solubilities with temperature are shown in Figs 7.61 to 7.64.

Table 7.10 Mole fraction solubilities in organic solvents of propane, butane and 2-methylpropane at a partial pressure of 1.013 bar

Solvent	T/K	Propane	Butane	2-Methylpropane
Hydrocarbons				
Hexane	298.2	0.116 (119)	0.386 (123)	
	288.2	0.138 (112)	0.441 (112)	0.313 (112)
2-Methylpentane	298.2	0.118 (119)		
2,2-Dimethylbutane	298.2	0.116 (119)	0.290 (112)	
Heptane	298.2	0.117 (120)	0.423 (123)	
Octane	298.2	0.119 (121)	0.432 (123)	
Decane	298.2	0.125 (122)	0.493 (122)	
	273.2	0.244 (124)		0.645 (125)
Eicosane	375.0	0.0391 (24)	0.1012 (24)	0.0752 (24)
Cyclohexane	298.2	0.0910 (119,128)		
Decahydronaphthalene	298.2	0.0826 (43)	0.280 (43)	
Ethylbenzene	298.2	0.0724 (119)		
1,3-Dimethylbenzene	298.2	0.0731 (119)	0.353 (124)	
Methylbenzene	298.2	0.071 (83)	0.370 (83)	
Benzene	298.2	0.0573 (119,121)		
1,1'-Bicyclohexyl	298.2	0.0889 (24)	0.226 (24)	0.164 (24)
Diphenylmethane	298.2	0.0395 (24)		
1-Methylnaphthalene	298.2	0.0380 (24)		0.0869 (24)
Solvents containing oxygen				
Methanol	298.2	0.0115 (130)	0.0398 (130)	0.0256 (130)
Ethanol	298.2	0.0228 (131)	0.0815 (131)	0.0512 (131)
2-Propanol	298.2	0.0295 (130)	0.1179 (130)	0.0741 (130)
1-Butanol	298.2	0.0367 (123)	0.139 (129)	0.0889 (129)
Benzenemethanol	273.2	0.038 (124)		0.107 (125)
	293.2	0.0237 (124)	0.0680 (124)	
	298.2	0.0219 (124)		
2-Propanone	298.2	0.0302 (123)		
1-Phenylethanone	293.2	0.0276 (124)	0.141 (124)	
Acetic acid anhydride	298.2	0.011 (124)	0.042 (124)	
Ethoxybenzene	298.2	0.038 (124)	0.21 (124)	
1,4-Dioxane	298.2	0.0284 (121)	0.145 (124)	

Table 7.10 (*continued*)

Solvent	T/K	Propane	Butane	2-Methylpropane
Acetic acid	294.7	0.0151 (132)		
Hexanoic acid	273.2	0.125 (124)		0.416 (125)
	278.2	0.113 (124)	0.643 (124)	
	298.2	0.083 (124)	0.281 (124)	
2-Hydroxybenzoic acid, methyl ester	298.2	0.030 (124)	0.149 (124)	
1,1'-Oxybispentane	268.2	0.291 (124)		
	278.2		0.831 (124)	
	298.2		0.449 (124)	
1,1'-Oxybisoctane	273.2	0.274 (124)		0.664 (125)
	278.2	0.244 (124)	0.850 (124)	
	298.2	0.138 (124)	0.512 (124)	
Halogenated solvents				
Tetrachloromethane	298.2	0.0813 (119)	0.339 (129)	0.231 (129)
Chlorobenzene	298.2	0.0570 (123)	0.269 (129)	0.162 (129)
Bromobenzene	275.2			0.435 (125)
Iodobenzene	298.2	0.0958 (124)	0.142 (124)	
Tetradecaflurohexane	298.2	0.0480 (123)		
Hexadecafluoroheptane	298.2	0.0505 (121)		
2,2,2-Trichloroethanol	298.2		0.0909 (124)	
Dichloroacetic acid	298.2		0.0625 (124)	
1,2-Dibromoethane	278.2		0.644 (124)	
1-Bromo-3-methylbenzene	298.2	0.070 (124)	0.240 (124)	
1-Bromooctane	273.2			0.580 (125)
1-Iodooctane	298.2	0.102 (124)	0.349 (124)	
Solvents containing nitrogen				
1-Octanamine	298.2	0.100 (124)	0.313 (124)	
Cyclohexanamine	298.4	0.0537 (133)		
Benzenamine	298.2	0.0124 (124)		
	278.2		0.0689 (124)	
N-Methylbenzenamine	298.2	0.0267 (124)	0.0934 (124)	
N,N-Dimethylbenzenamine	298.2	0.055 (124)	0.156 (124)	
Benzenemethanamine	298.2	0.0291 (124)	0.094 (124)	
Benzonitrile	298.2	0.0375 (124)	0.111 (124)	
Quinoline	293.2	0.0236 (124)		
	278.2		0.250 (124)	
Nitrobenzene	298.2	0.0247 (124)	0.0711 (124)	0.058 (125, 99)
1-Methyl-2-nitrobenzene	298.2		0.128 (124)	
1-Methyl-2-pyrrolidinone	298.2	0.0175 (43)	0.0543 (43)	0.0300 (43)
N,N-Dimethylformamide	298.15	0.0119 (124)	0.0380 (124)	

(Literature references are given in parentheses)

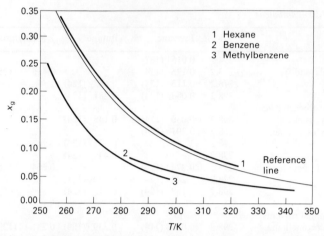

Figure 7.61. Mole fraction solubilities, at a partial pressure of 1.013 bar, of propane in hydrocarbons.

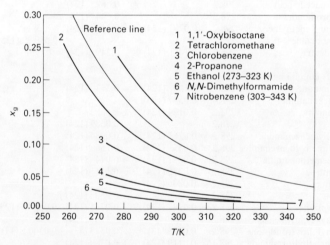

Figure 7.62. Mole fraction solubilities, at a partial pressure of 1.013 bar, of propane in solvents containing oxygen, nitrogen or chlorine.

Details of measurements of solubilities have been given in the volume of The Solubility Data Series devoted to these gases.[114]

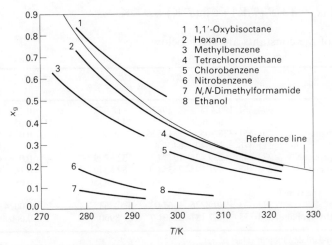

Figure 7.63. Mole fraction solubilities, at a partial pressure of 1.013 bar, of butane in various solvents.

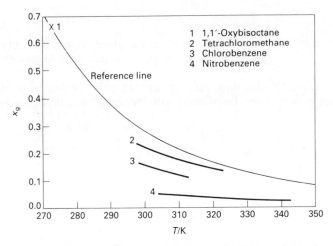

Figure 7.64. Mole fraction solubilities, at a partial pressure of 1.013 bar, of 2-methylpropane in various solvents.

7.7 SOLUBILITIES OF OTHER GASEOUS HYDROCARBONS

Equations for the mole fraction solubilities of numerous gaseous hydrocarbons in water for a partial pressure of 1.013 bar have been published by Wilhelm et al.[75] These are based upon the published data by other workers which were judged to be reliable. These equations may be written in the form:

SOLUBILITY OF GASES IN LIQUIDS

$$\ln x_g = A + B/(T/\text{K}) + C\ln(T/\text{K}) + DT/\text{K}$$

Values of the constants are given below:

Gas		A	B	C	D	d^a	x_g at 298.2 K
Propene	294-361	100.473	−1983.17	−18.0325		0.27	0.0001336
Propene[b]	273-361	−8476.26	228911	1477.41	−2.40111	0.054	0.0001239
Cyclopropane	298-361	326.905	−13526.9	−50.9015		0.074	0.0002074
1-Butene	311-378	−29.840	6406.4			[c]	0.0002357
2-Methylpropene	273-343	−239.426	12774.4	32.8909		0.076	0.0001029
Buta-1,3-diene	298-363	−491.195	25354.0	69.8370		0.16	0.0002601
1-Butyne	273-333	86.5215	−2558.71	−14.8356		0.082	0.001377
1-Butene-3-yne (vinyl acetylene)	273-333	−49.8226	3367.03	5.48363		0.0024	0.0006851
2,2-Dimethylpropane	288-353	−437.187	21801.7	61.8900		0.060	0.00001077

[a] d is the standard deviation in values of x_g.
[b] The value of the constant B for propyne, given in the paper, appears to be incorrect. A corrected value is given here.
[c] The equation for 1-butene is based upon only three measurements.

Solubilities in water as a function of temperature, at a partial pressure of 1.013 bar, for all gaseous hydrocarbons for which data are available, are shown in Fig. 7.65.

Hannaert et al.[83] have published equations for the solubilities of hydrocarbon gases in petroleum fractions, methylbenzene, dimethylbenzenes, 'Car-

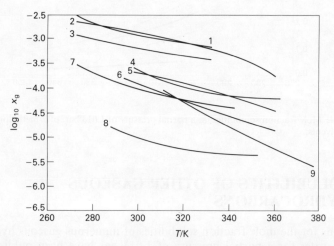

Figure 7.65. Mole fraction solubilities of various hydrocarbon gases in water at a partial pressure of gas of 1.013 bar: 1, propyne: 2, 1-butyne; 3, 1-buten-3-yne (vinylacetylene); 4, buta-1,3-diene; 5, cyclopropane; 6, propene: 7, 2-methylpropene; 8, 2,2-dimethylpropane; 9, 1-butene.

bowaxes', solvents containing oxygen, chlorinated solvents and liquid ammonia. Solubility curves for gases, not discussed earlier, in some of these solvents are shown in Figs 7.66 to 7.70.

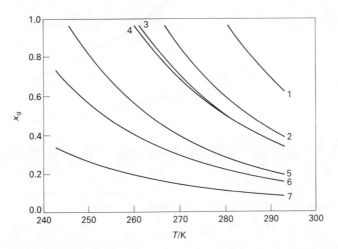

Figure 7.66. Solubilities of various hydrocarbon gases in methylbenzene at a partial pressure of gas of 1.013 bar: 1, 1-buten-3-yne (vinylacetylene); 2, 1,3-butadiene; 3, 1-butene; 4, butane; 5, methylethyne (methylacetylene); 6, propadiene; 7, propene.

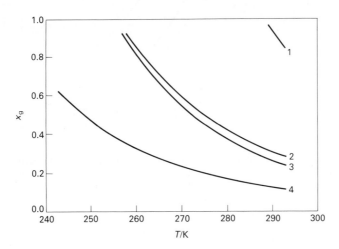

Figure 7.67. Solubilities of hydrocarbon gases in 2-propanone at a partial pressure of gas of 1.013 bar: 1, 1-buten-3-yne (vinylacetylene); 2, 1,3-butadiene; 3, methylethyne (methylacetylene); 4, propadiene.

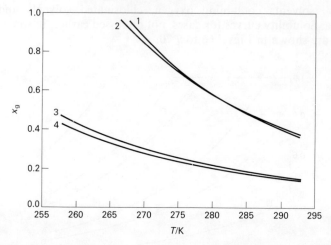

Figure 7.68. Mole fraction solubilities in tetrachloromethane at a partial pressure of gas of 1.013 bar: 1, 1-buten-3-yne (vinylacetylene); 2, 1,3-butadiene; 3, methylethyne (methylacetylene); 4, propadiene.

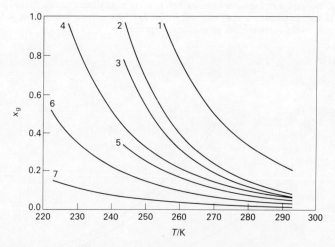

Figure 7.69. Mole fraction solubilities of hydrocarbon gases in methanol at a partial pressure of gas of 1.013 bar: 1, 1-buten-3-yne (vinylacetylene); 2, 1,3-butadiene; 3, 2-butene $(E)+(Z)$; 4, methylethyne (methylacetylene); 5, 1-butene; 6, propadiene; 7, propene.

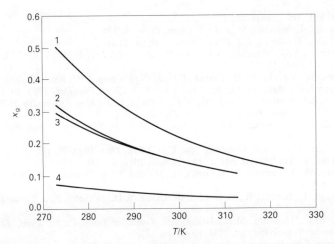

Figure 7.70. Mole fraction solubilities of hydrocarbon gases in 1,1-dichloroethane at a partial pressure of gas of 1.013 bar: 1, 1,3-butadiene; 2, methylethyne (methylacetylene); 3, 2-butene $(E)+(Z)$; 4, propene.

REFERENCES

1. Stull, D.R. *Ind. Eng. Chem.* 1947, 39, 517.
2. *Solubility Data Series, Vol. 27/28, Methane*, ed. Clever, H.L.; Young, C.L., Pergamon Press, Oxford, 1987.
3. Rettich, T.R.; Handa, Y.P.; Battino, R.; Wilhelm, E. *J. Phys. Chem.* 1981, 85, 3230.
4. Crovetto, R.; Fernandez-Prini, R.; Japas, M.L. *J. Chem. Phys.* 1982, 76, 1077.
5. Clever, H.L. ref. 2, page 56.
6. Stoessel, R.K.; Byrne, P.A. *Geochim. Cosmochim. Acta* 1982, 46, 1327.
7. Muccitelli, J.A.; Wen, W.-Y. *J. Solution Chem.* 1980, 9, 141.
8. Young, C.L. ref. 2, pages 248, 266, 281, 297, 303, 314, 320, 338, 349, 360, 370.
9. Hiza, M.J.; Miller, R.C.; Kidnay, A.J. *J. Phys. Chem. Ref. Data* 1979, 8, 799.
10. Wichterle, I.; Kobayashi, R. *J. Chem. Engng. Data* 1972, 17, 9.
11. Lin, Y.-N.; Chen, R.J.J.; Chappelear, P.S.; Kobayashi, R. *J. Chem. Engng. Data* 1977, 22, 402.
12. Reamer, H.H.; Sage, B.H.; Lacey, W.N. *Ind. Eng. Chem.* 1958, 3, 240.
13. Miller, R.C.; Kidnay, A.J.; Hiza, M.J. *J. Chem. Thermodynamics* 1977, 9, 167.
14. Horiuti, J. *Sci. Pap. Inst. Phys. Chem. Res. (Japan)* 1931/32, 17, 125.
15. Lannung, A.; Gjaldbaek, J.C. *Acta Chem. Scand.* 1960, 14, 1124.
16. Hayduk, W.; Buckley, W.D. *Can. J. Chem. Eng.* 1971, 49, 667.
17. Young, C.L. ref. 2, page 493.
18. Lin. H.-M., Sebastian, H.M.; Simnick, J.J.; Chao, K.-C. *J. Chem. Engng. Data* 1979, 24, 146.
19. Lin, Y.-N.; Hwang, S.-C.; Kobayashi, R. *J. Chem. Engng. Data* 1978, 23, 231.
20. Simnick, J.J.; Sebastian, H.M.; Lin, H.-M.; Chao, K.-C. *Fluid Phase Equilibria* 1979, 3, 145.
21. Ng, H.-J.; Hwang, S.S.-S.; Robinson, D.B. *J. Chem. Eng. Data* 1982, 27, 119.
22. Huang, S.S.-S.; Robinson, D.B. *Can. J. Chem. Eng.* 1985, 63, 126.
23. Cukor, P.M.; Prausnitz, J.M. *J. Phys. Chem.* 1972, 76, 598.
24. Chappelow, C.C.; Prausnitz, J.M. *Am. Inst. Engnrs. J.* 1974, 20, 1097.
25. Field, L.R.; Wilhelm, E.; Battino, R. *J. Chem. Thermodyn.* 1974, 6, 237.
26. Geller, E.B.; Battino, R.; Wilhelm, E. *J. Chem. Thermodyn.* 1976, 8, 197.
27. Byrne, J.E.; Battino, R.; Wilhelm, E. *J. Chem. Thermodyn.* 1975, 7, 515.

28. Clever, H.L. ref. 2, page 582.
29. Ben-Naim, A.; Yaacobi, M. *J. Phys. Chem.* 1974, 78, 175.
30. Boyer, F.L.; Bircher, L.J. *J. Phys. Chem.* 1960, 64, 1330.
31. Komarenko, V.G.; Manzhelii, V.G. *Ukr. Fiz. Zh. (Ukr. Ed.)* 1968, 13, 387.; *Ukr. Fiz. Zh. (Engl. Ed.)* 1968, 13, 273.
32. Francesconi, A.Z.; Lentz, H.; Franck, E.U. *J. Phys. Chem.* 1981, 85, 3303.
33. Yokoyama, C.; Masuoka, H.; Aral, K.; Saito, S. *J. Chem. Eng. Data* 1985, 30, 177.
34. Wilcock, R.J.; Battino, R.; Danforth, W.F.; Wilhelm, E. *J. Chem. Thermodyn.* 1978, 10, 817.
35. Olsen, J.D. *J. Chem. Engng. Data* 1977, 22, 326.
36. Gallardo, M.A.; Urieta,J.S.; Gutierrez Losa, C. *J. Chim. Phys. Phys. Chim. Biol.* 1983, 80, 621.
37. Gibanel, F.; Urieta, J.S.; Gutierrez Losa C. *J. Chim. Phys.* 1980, 78, 171.
38. Kobatake, Y.; Hildebrand, J.H. *J. Phys. Chem.* 1961, 65, 331.
39. Yorizane, M.; Yoshimura, S.; Masuoka, H.; Miyano, Y.; Kakimoto, Y. *J. Chem. Eng. Data* 1985, 30, 174.
40. Tominaga, T.; Battino, R.; Gorowara, B.; Dixon, R.D.; Wilhelm, E. *J. Chem. Eng. Data* 1986, 31, 175.
41. Lopez, M.C.; Gallardo, M.A.; Urieta, J.S.; Gutierrez Losa, C. *Int. Conf. Thermodyn. Solutions of Non-electrolytes*, 1984, Paper no. 127.
42. Shakhova, S.F.; Zubchenko, Yu. P. *Khim Prom.* 1973, 49, 595.
43. Lenoir, J.-Y.; Renault, P.; Renon, H. *J. Chem. Eng. Data* 1971, 16, 340.
44. Murrieta-Guevara, F.; Rodriguez, A.T. *J. Chem. Eng. Data* 1984, 29, 456.
45. Wu, Z.; Zeck, S.; Knapp, H. *Ber. Bunsenges. Phys. Chem.* 1985, 89, 1009.
46. Young, C.L. ref. 2, page 712.
47. Simnick, J.J.; Sebastian, H.M.; Lin, H.-M.; Chao, K.-C. *J. Chem. Eng. Data* 1979, 24, 239.
48. Wilcock, R.J.; McHale, J.L.; Battino, R.; Wilhelm, E. *Fluid Phase Equilib.* 1978, 2, 225.
49. Clever, H.L. ref. 2, page 724.
50. Kaminishi, G. *Kogyo Kogaku Zaashi* 1965, 68, 419.
51. Donnelly, H.G.; Katz, D.L. *Ind. Eng. Chem.* 1954, 46, 511.
52. Kaminishi, G.; Arai, Y,; Saito, S.; Maeda, A. *J. Chem. Eng. Japan* 1968, 1, 109.
53. Arai, Y.; Kaminishi, G.; Saito, S. *J. Chem. Eng. Japan* 1971, 4, 131.
54. Davalos, J.; Anderson, W.R.; Phelps, R.E.; Kidnay, A.J. *J. Chem. Eng. Data* 1976, 21, 81.
55. Mraw, S.C.; Hwang, S.-C.; Kobayashi, R. *J. Chem. Eng. Data* 1978, 23, 135.
56. Somait, F.A.; Kidnay, A.J. *J. Chem. Eng. Data* 1978, 23, 301.
57. Al-Sahhaf, T.A.; Kidnay, A.J.; Sloan, E.D. *Ind. Eng. Chem. Fundam.* 1983, 22, 372.
58. Robinson, D.B.; Bailey, J.A. *Can. J. Chem. Engng.* 1957, 35, 151
59. Reamer, H.H.; Sage, B.H.; Lacey, W.N. *Ind. Eng. Chem.* 1951, 43, 976.
60. Kohn, J.P.; Kurata, F. *Am. Inst. Chem. Engnrs. J.* 1958, 4, 211.
61. Robinson, D.B.; Lorenzo, A.P.; Macrygeorgos, C.A. *Can. J. Chem. Engng.* 1959, 37, 212.
62. Hensel, W.E. Jr.; Massoth, F.E. *J. Chem. Eng. Data* 1964, 9, 352.
63. Dean, M.R.; Walls, W.S. *Ind. Eng. Chem.* 1947, 39, 1049.
64. Senturk, N.H.; Kalra, H.; Robinson, D.B. *J. Chem. Engng. Data* 1979, 24, 311.
65. Zeininger, H. *Chemi-Ing.-Techn.* 1972, 44, 607.
66. Young, C.L. ref. 2, page 736.
67. Evans, D.F.; Battino, R. *J. Chem. Thermodyn.* 1971, 3, 753.
68. Hiraoka, H.; Hildebrand, J.H. *J. Phys. Chem.* 1964, 68, 213.
69. Powell, R.J. *J. Chem. Eng. Data* 1972, 17, 302.
70. Dymond, J.H. *J. Phys. Chem.* 1967, 71, 1829.
71. Guerry, D. Jr. *Ph.D Thesis, Vanderbilt University, Nashville, TN* (quoted in ref. 1, page 707)
72. Keevil, T.A.; Taylor, D.R. Streitwieser, A. *J. Chem. Engng. Data* 1978, 23, 237.
73. de Ligny, C.; Denessen, H.J.M.; Alfenaar, M. *Recl. Trav. Chim. Pays-Bas* 1971, 90, 1265.
74. Haideggar, E.; Szebenyi, I.; Szekely, A. *Magy. Kem. Foly.* 1958, 64, 365.
75. Wilhelm, E.; Battino, R.; Wilcock, R.J. *Chem. Rev.* 1977, 77, 219.
76. Morrison, T.J.; Billett, F. *J. Chem. Soc.* 1952, 3819.
77. Bradbury, E.J.; McNulty, D.; Savage, R. L.; McSweeney, E. E. *Ind. Eng. Chem.* 1952, 44, 211
78. Waters, J. A.; Mortimer, G. A.; Clements, H. E. *J. Chem. Eng. Data* 1970, 15, 174.

79. McDaniel, A. S. *J. Phys. Chem.* 1911, 15, 587.
80. Leites, I. L.; Ivanovskii, F. P. *Khim. Prom.* 1960, (9), 653.
81. Jadot, R. *J. Chim. Phys. Physicochim. Biol.* 1972, 69, 1036.
82. Zhuze, T. P.; Zhurba, A. S. *Bull. Akad. Sci. USSR, Div. Chem. Sci.* 1960 No.2, 335.
83. Hannaert, H.; Haccuria, M.; Mathieu, M. P. *Ind. Chim. Belg.* 1967, 32 (2), 156.
84. Boyer, F. L.; Bircher, L. J. *J. Phys. Chem.* 1960, 64, 1330.
85. Hayduk, W. *Solubility Data Series, Vol. 9, Ethane*, Pergamon Press, Oxford, 1982.
86. Rettich, T. R.; Handa, Y. P.; Battino, R.; Wilhelm, E. *J. Phys. Chem.* 1981, 85, 3230.
87. Culbertson, O. L.; McKetta, J. J. *J. Petr. Technol. Trans. AIME Petr. Div.* 1950, 189, 319.; Culbertson, O. L.; Horn, A. B.; McKetta, J. J. *loc. cit.* page 1; Anthony, R. G.; McKetta, J.J. *J. Chem. Eng. Data* 1967, 12, 17.
88. Danneil, A.; Todheide, K.; Franck, E. U. *Chem. Ing-Tech.* 1967, 13, 816.
89. Hayduk, W.; Cheng, S.C. *Can. J. Chem. Eng.* 1970, 48, 93.
90. Thomsen, E.S.; Gjaldbaek, J.C. *Acta Chem. Scand.* 1963, 17, 127.
91. Cukor, P.M.; Prausnitz, J.M. *J. Phys. Chem.* 1972, 76, 598.
92. Ohgaki, K.; Sano, R.; Katayama, T. *J. Chem. Eng. Data* 1976, 21, 55.
93. Kay, W.B.; Nevens, T.D. *Chem. Eng. Prog. Symp. Ser.* No. 3 1952, 48, 108.
94. Ben-Naim, A.; Yaacobi, M. *J. Phys. Chem.* 1974, 78, 175.
95. Gjaldbaek, J. C.; Niemann, H. *Acta Chem. Scand.* 1958, 12, 1015.
96. Yaacobi, M.; Ben-Naim, A. *J. Solution Chem.* 1973, 2, 425.
97. Ma, Y. H.; Kohn, J. P. *J. Chem. Eng. Data* 1964, 9,3.
98. Rivas, O. R.; Prausnitz, J. M. *Am. Inst. Chem. Eng. J.* 1979, 25, 975.
99. Ezheleva, A. E.; Zorin, A. D. *Tr. Khim. Khim. Tech. (Gorkii)* 1961, 1, 37.
100. Miller, S.A. *Acetylene, Its Properties, Manufacture and Uses, Vol. I*, Academic Press, New York, 1965.
101. Vitovec, J. *Collect. Czech. Chem. Comm.* 1968, 33, 1203.
102. Schoen, R. *Z. Physiol. Chem.* 1923, 127, 243.
103. Hiraoka, H. *Rev. Phys. Chem. Japan* 1954, 24, 13.
104. Flid, R. M.; Golynets, Yu. F. *Izv. Vyssh. Uchebn. Zaved., Khim. Khim. Tekhnol.* 1959, 2, 173.
105. Billitzer, J. *Zeit. Phys. Chem.* 1902, 40, 535.
106. Vitovec, J.; Fried, V. *Coll. Czech. Chem. Comm.* 1960, 25, 1552.
107. Kiyama, R.; Hiraoka, H. *Rev. Phys. Chem. Japan* 1955, 25, 16.
108. Bodor, E.; Bor, G.; Szeness, M. M.; Mesko, G.; Mohai, B.; Siposs, G. *Veszpremi. Vegyip. Egy. Kozl.* 1957, 1, 63; Bodor, E.; Pfeifer, G. *ib.* 1957, 1, 109; Bodor, E.; Mohai, B.; Pfeifer, G. *ib.* 1959, 3, 205; Mohai, B.; Szeness, M. M. *ib.* 1959, 3, 211.
109. Holemann, P.; Hasselmann, R. *Chem. Ing. Tech.* 1953, 25, (8/9), 466.
110. Kiyama, R.; Hiraoka, H. *Rev. Phys. Chem. Japan* 1956, 26, 1.
111. McKinnis, A. C. *Ind. Eng. Chem.* 1955, 47, 850.
112. Tilquin, B.; Decanniere, L.; Fontaine, R.; Claes, P. *Ann. Soc. Sci. Brux.* 1967, 81, (2), 191.
113. Haspra, J.; Paulech, J. *Chem. Prumysl.* 1957, 7, 569.
114. Hayduk, W. *Solubility Data Series Volume 24, Propane, Butane and 2-Methylpropane*, Pergamon Press, Oxford, 1986.
115. Nosov, E. F.; Barlyaev, E. V. *Zh. Obsshch. Khim.* 1968, 38, 211.
116. Azarnoosh, A.; McKetta, J. J. *Pet. Ref.* 1958, 37, 275; Wehe, A. H.; McKetta, J. J. *Anal. Chem.* 1961, 33, 291.
117. Kobayashi, R.; Katz, D. L. *Ind. Eng. Chem.* 1953, 45, 440.
118. Le Breton, J. G.; McKetta, J. J. *Hydroc. Prog. Pet. Ref.* 1964, 43, 136.
119. Fleury, D.; Hayduk, W. *Can. J. Chem. Eng.* 1975, 53, 195.
120. Hayduk, W.; Walter, E. B.; Simpson, P. *J. Chem. Eng. Data* 1972, 17, 59.
121. Thomsen, E. S.; Gjaldbaek J. C., *Acta Chem. Scad.* 1963, 17, 134.
122. Monfort, J. P.; Arriaga, J. L. *Chem. Eng. Commun.* 1980, 7, 17.
123. Hayduk, W.; Castaneda, R. *Can. J. Chem. Eng.* 1973, 51, 353.
124. Gerrard, W. *J. Appl. Chem. Biotechnol.* 1973, 23, 1.
125. Gerrard, W. *Solubility of Gases and Liquids*, Plenum Press, New York, 1976.
126. Reamer, H. H.; Sage, B. H. *J. Chem. Eng. Data* 1966, 11, 17.
127. Reamer, H. H.; Sage, B. H.; Lacey, W. N. *Ind. Eng. Chem.* 1946, 38, 986.

128. Miller, K. W. *J. Phys. Chem.* 1968, 72, 2248.
129. Blais, C.; Hayduk, W. *J. Chem. Eng. Data* 1983, 28, 181.
130. Kretschmer, C. B.; Wiebe, R. *J. Am. Chem Soc.* 1952, 74, 1276.
131. Kretschmer, C. B.; Wiebe, R. *J. Am. Chem Soc.* 1951, 73, 3778.
132. Barton, J. R.; Hsu, C. C. *Chem. Eng. Sci.* 1972, 27, 1315.
133. Keevil, T. A.; Taylor, D. R.; Streitwieser, A. *J. Chem. Eng. Data* 1978, 23, 237.

Chapter 8
SOLUBILITY OF CHLORINE

8.1 GENERAL BEHAVIOUR

Chlorine has the following physical properties:
Melting point	=	172.11 K
Boiling point (1.013 bar)	=	239.04 K
Critical temperature	=	417.15 K
Critical pressure	=	77 bar
Critical volume	=	0.124 dm^3 mol^{-1}
Relative molecular mass	=	70.906
Density of gas at 273.15 K; 1.013 bar	=	3.212 g dm^{-3}

The vapour pressure of the pure liquid follows the equation

$$\ln(P/\text{bar}) = A - B/[(T/\text{K}) + C]$$

Values of A, B, C, calculated from vapour pressure data given by Stull,[1] are as follows:

Temp. range/K	A	B	C
189–239	9.7669	2132.63	−20.70
239–309	9.2495	1930.26	−30.36
309–400	9.4933	1974.14	−33.71

The reference line for solubility of chlorine at a partial pressure of 1.013 bar is shown in Fig. 8.1.

Compiled and evaluated solubility data for chlorine have been published in the Solubility Data Series.[2]

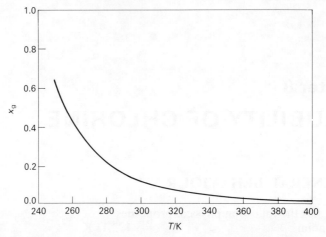

Figure 8.1. Reference line for chlorine at a partial pressure of 1.013 bar.

8.2 SOLUBILITY OF CHLORINE IN WATER AND AQUEOUS SOLUTIONS

When chlorine dissolves in water it exists in equilibrium with chloride ions and hypochlorous acid i.e.

$$Cl_2 + H_2O \rightleftharpoons H^+ + Cl^- + HClO.$$

The equilibrium constant defined by

$$K = \frac{[H^+][Cl^-][HClO]}{[Cl_2]}$$

is 1.56×10^{-4} mol dm^{-3} at 273.15 K and 10.0×10^{-4} mol dm^{-3} at 343.15 K.[3] The interpolated value for 298.15 K is about 3.0×10^{-4} mol dm^{-3}. Water, saturated at 298.15 K with chlorine at a partial pressure of 1.013 bar, contains about half of the chlorine as chloride ions or hypochlorous acid. The bulk solubility of chlorine corresponds to the quantity of Cl_2 which is equivalent to the total Cl_2, Cl^- and HClO in solution.

Chlorine forms a solid hydrate which may separate from aqueous solution. The formula of this hydrate is $Cl_2.8H_2O$ and is in equilibrium with chlorine at a partial pressure of 1.013 bar when the temperature is 282.8 K.[4]

Solubility of chlorine in water has been evaluated by Battino,[5] who has recommended values of the mole fraction bulk solubility for a partial pressure of 1.013 bar in the range 283.15 to 383.15 K. (Table 8.1 and Fig. 8.2.)

SOLUBILITY OF CHLORINE

Table 8.1 Bulk mole fraction solubility of chlorine in water at a partial pressure of 1.013 bar recommended by Battino[5]

T/K	x_g
283.15	0.00248
293.15	0.00188
303.15	0.00150
313.15	0.00123
323.15	0.00106
333.15	0.000939
343.15	0.000849
353.15	0.000784
363.15	0.000737
373.15	0.000697
383.15	0.000668

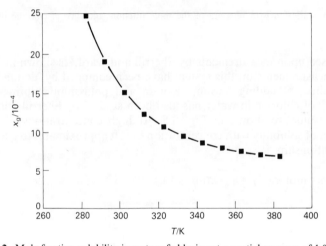

Figure 8.2. Mole fraction solubility in water of chlorine at a partial pressure of 1.013 bar.

Mole fraction solubilities fall well below the reference line. The variation of mole fraction solubility with pressure for 298.2 K from measurements by Whitney and Vivian[6] is shown in Fig. 8.3. The curve has a reverse sigmoid shape. The lower the pressure the greater, in proportion, is the contribution of hydrolysis to the overall solubility. Linear extrapolations of mole fraction solubilities from low pressures to 1.013 bar, based upon Henry's law, cannot give reliable estimates of solubility.

Adding hydrochloric acid to water lowers the solubility of chlorine at low concentrations of acid. This is probably due to the reduction of the degree of hydrolysis of the chlorine. High concentrations of hydrochloric acid enhance the solubility of chlorine. The variation of solubility with concentration of

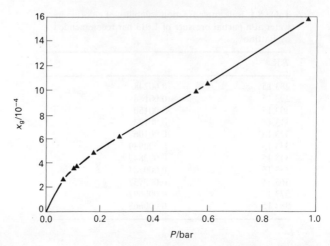

Figure 8.3. Variation with pressure of the mole fraction solubility of chlorine in water at 298.2 K.

acid, based upon measurements by Sherrill and Izard,[7] is shown in Fig. 8.4. Other measurements on this system have been compiled by Young.[2]

The effect of adding barium, sodium and potassium chlorides to the solubility of chlorine in water has also been studied by Sherrill and Izard,[7] whose results are shown in Fig. 8.5. At high concentrations of salt the variation of solubility with concentration of salt approximates to a Sechenov type relationship

$$\ln(s/\text{mol kg}^{-1}) = A - B(m/\text{mol kg}^{-1})$$

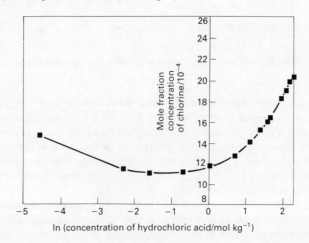

Figure 8.4. Effect of the concentration of hydrochloric acid on the solubility of chlorine at 298.15 K and a partial pressure of 1.013 bar.

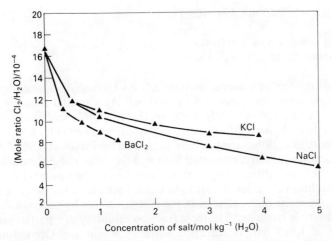

Figure 8.5. The effect of electrolytes on the solubility of chlorine in water under barometric pressure at 298.2 K.

where s is solubility of the gas, m concentration of salt and A and B are constants. Adding chlorides reduces the degree of hydrolysis of chlorine. This is reflected in the marked change in solubility with addition of chlorides at low concentration of chlorides.

8.3 SOLUBILITY OF CHLORINE IN NON-AQUEOUS SOLVENTS

Solubilities of chlorine in non-aqueous solvents have been compiled for the Solubility Data Series and have been evaluated by Fogg[8] and Gerrard.[21]

8.3.1 Solubilities in hydrocarbons

Chlorine reacts with hydrocarbons, either by addition or by substitution of hydrogen by chlorine. The extent of reaction can often be reduced by adding an inhibitor such as phenol or by excluding light.

The solubility in heptane at total or partial pressures of about 1.013 bar has been measured by Taylor and Hildebrand,[9] by Semb[10] and by Zetkin et al.[11] The measurements reported by Semb may be unreliable because of chlorination of the hydrocarbon during the absorption of chlorine. Solubility at 273.2 K for a partial pressure of 1.013 given by Taylor and Hildebrand is consistent with that at 283 to 323 K given by Zetkin et al. Solubility from these two sources fit the equation:

$$\ln x_g = -8.6398 + 1992.01/(T/K)$$

standard deviation in $x_g = 0.005$
temperature range 283 to 323 K.

The solubilities of chlorine in octane, 2,2,4-trimethylpentane, decane and tetradecane were also measured by Zetkin.[11] At a partial pressure of 1.013 bar there is an approximately linear increase of mole fraction solubility in straight chain alkanes with increase in carbon number (Fig. 8.6).

Solubility in cyclohexane has been measured by Tsiklis and Svetlova.[12] The mole fraction solubility, estimated for 1.013 bar from these measurements, is close to the value obtained by extrapolation to a carbon number of six of the corresponding values for the straight chain hydrocarbons (see Fig 8.6).

There are some discrepancies in the measurements by different authors of the solubility in benzene.[8] The mole fraction solubility at a partial pressure of chlorine of 1.013 bar, based upon work by Pozin and Opykhtina,[13] Cervinka,[14] Zetkin et al.,[15] and Lohse and Deckwer,[16] fits the equation:

$$\ln x_g = -9.811 + 2374/(T/K)$$

standard deviation in $x_g = 0.018$
temperature range 283 to 341 K.

Experimental measurements are shown in Fig. 8.7.

In the temperature range of 288 to 348 K there is little difference between mole fraction solubilities at a partial pressure of 1.013 bar in the solvents benzene, methylbenzene, ethylbenzene, 1,2-dimethylbenzene, 1,3-dimethylbenzene and 1,4-dimethylbenzene (Table 8.2).

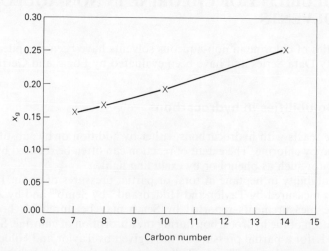

Figure 8.6. Variation in the mole fraction solubility of chlorine with the carbon number of straight chain alkanes ($T/K = 293.15$; $P/\text{bar} = 1.013$).

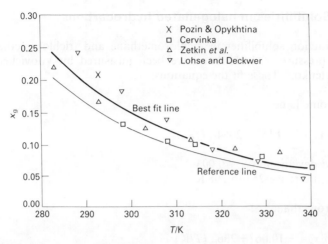

Figure 8.7. Mole fraction solubility in benzene of chlorine at a partial pressure of 1.013 bar.

Table 8.2

Solvent	x_g at 298.15 K ($P(Cl_2) = 1.013$ bar)	Smoothing equation	Temp. range/K	Ref.
Benzene	0.157 ± 0.018	$\ln x_g = -9.811 + 2374/(T/K)$	283–341	13–16
Methylbenzene	0.167 ± 0.006	$\ln x_g = -10.030 + 2457/(T/K)$	288–353	16,17
Ethylbenzene	0.167 ± 0.002	$\ln x_g = -8.425 + 1978/(T/K)$	288–348	16
1,2-Dimethylbenzene	0.144 ± 0.001	$\ln x_g = -7.615 + 1692/(T/K)$	298–338	16
1,3-Dimethylbenzene	0.148	$\ln x_g = -7.403 + 1637/(T/K)$	308–328	16
1,4-Dimethylbenzene	0.147 ± 0.004	$\ln x_g = -7.820 + 1759/(T/K)$	288–348	16

8.3.2 Solubilities in compounds containing nitrogen

Solubility in nitrobenzene at a total pressure equal to barometric (unspecified) was measured by Zetkin et al.[18] Solubility at a partial pressure of 1.013 bar, estimated from these data on the assumption that the barometric pressure was approximately equal to 1.013 bar, fits the equation

$$\ln x_g = -9.058 + 2052/(T/K)$$

standard deviation in $x_g = 0.002$
temperature range 313 to 433 K.

Solubilities lie close to the reference line.

The solubility in N,N-dimethylformamide at 273.2 K has been published[19] as an Ostwald coefficient. The corresponding mole fraction solubility for a partial pressure of 1.013 bar is 0.569, a value which is double the reference line value and very high compared with values for other solvents.

8.3.3 Solubilities in halogenated hydrocarbons

Mole fraction solubilities in dichloromethane and trichloromethane at a partial pressure of 1.013 bar have been measured by Vdovichenko and Kondratenko.[20] These fit the equations:

Dichloromethane

$$\ln x_g = -9.153 + 2054/(T/K)$$

standard deviation in $x_g = 0.003$
temperature range 254 to 298 K.

Trichloromethane

$$\ln x_g = -10.067 + 2362/(T/K)$$

standard deviation in $x_g = 0.013$
temperature range 257 to 298 K.

There have been many measurements of the solubility in tetrachloromethane. These have been discussed by Gerrard,[21] who recommended numerical values for the mole fraction solubility of chlorine in the temperature range 253 to 353 K on the basis of the more reliable experimental measurements (Table 8.3).

Table 8.3 Recommended values for the solubility of chlorine in tetrachloromethane[21]. Partial pressure of chlorine = 1.013 bar

T/K	x_g
253.15	0.600
263.15	0.434
273.15	0.305
283.15	0.262
293.15	0.181
303.15	0.142
313.15	0.112
323.15	0.088
333.15	0.072
343.15	0.058
353.15	0.050

These fit the equation

$$\ln x_g = 2.898 + 1702/(T/K) - 1.8309\ln(T/K)$$

standard deviation in $x_g = 0.008$
temperature range 253 to 353 K.

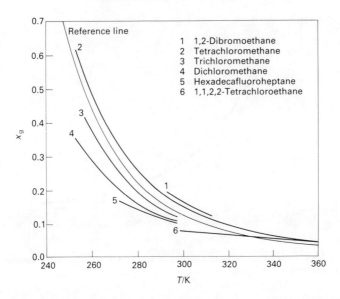

Figure 8.8. Mole fraction solubilities in halo-alkanes of chlorine at a partial pressure of 1.013 bar.

At the same partial pressure of chlorine and temperature mole fraction solubilities are in the order $CCl_4 > CHCl_3 > CH_2Cl_2$ (Fig 8.8).

Taylor and Hildebrand[9] measured the solubility in 1,2-dibromoethane by bubbling the gas through the liquid under a total pressure equal to barometric. These authors made allowances for the vapour pressure of the solvent in calculating the mole fraction solubility under a partial pressure of 1.013 bar.

Solubility in 1,1,2,2-tetrachloroethane was measured by Kalz and Naumann,[23] at a total pressure equal to an unspecified barometric pressure. The solubility in 1,1,2,3,4,4-hexachloro-1,3-butadiene has been reported by Kogan et al.[24] at a partial pressure of chlorine of 1.013 bar from 253 to 365 K and over the pressure range 0.54 to 1.30 bar at 303.2 K (see Fig. 8.9 and Fig. 8.10).

Solubility in hexadecafluoroheptane has been reported by Gjaldbaek and Hildebrand[25] for a partial pressure of 1.013 bar. Measurements were made at a total pressure equal to barometric. Appropriate allowance was made for the vapour pressure of the solvent and the exact value of the barometric pressure was taken into account. The general pattern of solubilities in halogenated alkanes is shown in Fig. 8.8.

Mole fraction solubility in tetrachloroethene under a total pressure equal to barometric (unspecified) has been reported by Curda and Holas.[22]

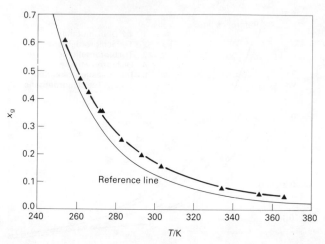

Figure 8.9. Mole fraction solubility of chlorine in 1,1,2,3,4,4-hexachloro-1,3-butadiene (partial pressure of chlorine = 1.013 bar).

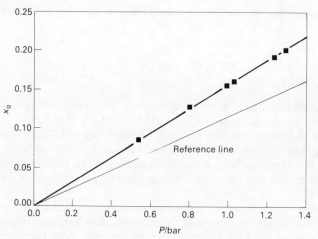

Figure 8.10. Variation with partial pressure of the mole fraction solubility of chlorine in 1,1,2,3,4,4-hexachloro-1,3-butadiene at 303.2 K.

8.4 TABLES SUMMARIZING THE SOLUBILITY OF CHLORINE

Data from the literature have been used to prepare tables of the solubilities of chlorine in various solvents. Details of the original measurements are given in Solubility Data Series Volume 12.[2]

SOLUBILITY OF CHLORINE

Table 8.4 Mole fraction solubilities of chlorine in various solvents at a partial pressure of chlorine of 1.013 bar in alphabetical order

Solvent	A	B	Temp. range/K Low	High	M_r of solvent	x_g at 298 K[a]
Benzene	−4.261	1031	283	341	78.114	0.158
Carbon disulphide	−3.096	699	275	298	76.131	0.178
Carbon tetrachloride	−4.07	976	253	353	153.823	0.160
Chlorobenzene	−4.24	1026	283	343	112.559	0.160
Chloroform	−4.372	1026	257	298	119.378	0.118
(Chloromethyl)benzene	−3.927	927	303	383	126.586	0.153
1-Chloro-2-methylbenzene	−3.711	876	288	348	126.586	0.169
1-Chloro-3-methylbenzene	−3.667	845	298	348	126.586	0.147
1-Chloro-4-methylbenzene	−3.805	883	298	348	126.586	0.144
Cyclohexane	−4.034	927	283	313	84.162	0.119
Decane	−3.689	869	283	323	142.286	0.169
1,2-Dibromoethane	−3.969	954	293	313	187.862	0.171
1,2-Dichlorobenzene	−4.241	1049	288	333	147.004	0.190
Dichloromethane	−3.975	892	254	298	84.933	0.104
(Dichloromethyl)benzene	−4.043	966	303	403	161.031	0.158
1,2-Dimethylbenzene	−3.307	735	298	338	106.168	0.144
1,4-Dimethylbenzene	−3.396	764	288	348	106.168	0.147
N,N-Dimethylformamide	[b]	[b]	273	273	73.095	
Ethylbenzene	−3.659	859	288	348	106.168	0.167
Heptane	−3.368	759	254	334	100.205	0.151
1,1,2,3,4,4-Hexachloro-1,3-butadiene	−3.771	901	253	365	260.762	0.179
Hexadecafluoroheptane	−3.416	719	273	298	388.051	0.099
Methylbenzene	−4.356	1067	288	353	92.141	0.168
Nitrobenzene	−3.994	891	313	433	123.111	0.114
Octane	−3.439	782	283	323	114.232	0.153
1,1,2,2-Tetrachloroethane	−2.552	433	298	383	167.85	0.080
Tetrachloroethene	−4.066	971	273	363	165.834	0.156
Tetrachloromethane	−4.07	976	253	353	153.823	0.160
Tetradecane	−3.609	881	283	323	198.394	0.223
Toluene	−4.356	1067	288	353	92.141	0.168
1,2,4-Trichlorobenzene	−4.681	1175	288	333	181.449	0.183
Trichloromethane	−4.37	1026	257	298	119.378	0.118
(Trichloromethyl)benzene	−4.176	1006	303	423	195.476	0.158
(Trifluoromethyl)benzene	−4.883	1179	279	353	146.112	0.118
2,2,4-Trimethylpentane	−2.92	622	283	323	114.232	0.147
Water	−4.803	604	283	383	18.015	0.002
o-Xylene	−3.307	735	298	338	106.168	0.144
p-Xylene	−3.396	764	288	348	106.168	0.147

[a] Mole fraction solubilities, x_g, estimated for 298.15 K except where indicated by an asterisk.
[b] Solubilities measured at one temperature only. The value for this temperature is given.

Experimental measurements have been fitted to equations of the form:

$$\log_{10} x_g = A + B/(T/\text{K})$$

Values of A and B are given.

Table 8.5 Mole fraction solubilities of chlorine in various solvents at a partial pressure of chlorine of 1.013 bar in increasing magnitude at 298.15 K

Solvent	A	B	Temp. range/K Low	Temp. range/K High	M_r of solvent	x_g at 298 K[a]
Water	−4.803	604	283	383	18.015	0.002
1,1,2,2-Tetrachloroethane	−2.552	433	298	383	167.85	0.080
Hexadecafluoroheptane	−3.416	719	273	298	388.051	0.099
Dichloromethane	−3.975	892	254	298	84.933	0.104
Nitrobenzene	−3.934	891	313	433	123.111	0.114
(Trifluoromethyl)benzene	−4.883	1179	279	353	146.112	0.118
Trichloromethane (chloroform)	−4.372	1026	257	298	119.378	0.118
Cyclohexane	−4.034	927	283	313	84.162	0.119
1-Chloro-4-methylbenzene	−3.805	883	298	348	126.586	0.144
1,2-Dimethylbenzene (o-xylene)	−3.307	735	298	338	106.168	0.144
2,2,4-Trimethylpentane	−2.920	622	283	323	114.232	0.147
1,4-Dimethylbenzene (p-xylene)	−3.396	764	288	348	106.168	0.147
1-Chloro-3-methylbenzene	−3.667	845	298	348	126.586	0.147
Heptane	−3.368	759	254	334	100.205	0.151
(Chloromethyl)benzene	−3.927	927	303	383	126.586	0.153
Octane	−3.439	782	283	323	114.232	0.153
Tetrachloroethene	−4.066	971	273	363	165.834	0.156
(Dichloromethyl)benzene	−4.043	966	303	403	161.031	0.158
Benzene	−4.261	1031	283	341	78.114	0.158
(Trichloromethyl)benzene	−4.176	1006	303	423	195.476	0.158
Chlorobenzene	−4.240	1026	283	343	112.559	0.160
Tetrachloromethane (carbon tetrachloride)	−4.070	976	253	353	153.823	0.160
Ethylbenzene	−3.659	859	288	348	106.168	0.167
Methylbenzene (toluene)	−4.356	1067	288	353	92.141	0.168
Decane	−3.689	869	283	323	142.286	0.169
1-Chloro-2-methylbenzene	−3.711	876	288	348	126.586	0.169
1,2-Dibromoethane	−3.969	954	293	313	187.862	0.171
Carbon disulfide	−3.096	699	275	298	76.131	0.178
1,1,2,3,4,4-Hexachloro-1,3-butadiene	−3.771	901	253	365	260.762	0.179
1,2,4-Trichlorobenzene	−4.681	1175	288	333	181.449	0.183
1,2-Dichlorobenzene	−4.241	1049	288	333	147.004	0.190
Tetradecane	−3.609	881	283	323	198.394	0.223

Experimental measurements have been fitted to equations of the form:

$$\log_{10} x_g = A + B/(T/K)$$

Table 8.6 Weight ratio solubilities of chlorine in various solvents at a partial pressure of chlorine of 1.013 bar in increasing magnitude

Solvent	M_r of solvent	Temp. range of measurements/K Low	Temp. range of measurements/K High	Wt of Cl_2/g in 100 g solvent at 298.15 K[a]
Water	18.015	283	383	0.66
Hexadecafluoroheptane	388.051	273	298	2.01
1,1,2,2-Tetrachloroethane	167.850	298	383	3.65
1,1,2,3,4,4-Hexachloro-1,3-butadiene	260.762	253	365	5.90
(Trifluoromethyl)benzene	146.112	279	353	6.48
(Trichloromethyl)benzene	195.476	303	423	6.80
1,2-Dibromoethane	187.862	293	313	7.74
Tetrachloroethene	165.834	273	363	7.85
Trichloromethane (chloroform)	119.378	257	298	7.95
Nitrobenzene	123.111	313	433	7.37
(Dichloromethyl)benzene	161.031	303	403	8.22
1,2,4-Trichlorobenzene	181.449	288	333	8.69
Tetrachloromethane (carbon tetrachloride)	153.823	253	353	8.77
1-Chloro-4-methylbenzene	126.586	298	348	9.38
1-Chloro-3-methylbenzene	126.586	298	348	9.65
Dichloromethane	84.933	254	298	9.68
(Chloromethyl)benzene	126.586	303	383	10.05
Decane	142.286	283	323	10.07
Tetradecane	198.394	283	323	10.18
2,2,4-Trimethylpentane	114.232	283	323	10.66
Octane	114.232	283	323	11.19
1,2-Dimethylbenzene (o-xylene)	106.168	298	338	11.23
1,2-Dichlorobenzene	147.004	288	333	11.27
1-Chloro-2-methylbenzene	126.586	288	348	11.37
Cyclohexane	84.162	283	313	11.37
1,4-Dimethylbenzene (p-xylene)	106.168	288	348	11.48
Chlorobenzene	112.559	283	343	11.90
Heptane	100.205	254	334	12.54
Ethylbenzene	106.168	288	348	13.37
Methylbenzene (toluene)	92.141	288	353	15.43
Benzene	78.114	283	341	16.96
Carbon disulphide	76.131	275	298	20.06

[a] solubilities expressed as the weight of chlorine dissolved in 100 g of solvent have been estimated for 298.15 K by fitting experimental data to equations of the form:

$$\log_{10} x_g = A + B/(T/K)$$

where x_g is the mole fraction solubility.

REFERENCES

1. Stull, D.R. *Ind. Eng. Chem.* 1947, 39, 517.
2. *Solubility Data Series, Vol. 12, Sulfur Dioxide, Chlorine, Fluorine, and Chlorine Oxides*, ed. C.L. Young, Pergamon, Oxford 1983.
3. Jakowlin, A.A. *Zeit. Physik. Chem.* 1899, 29, 613.
4. Harris, I. *Nature*, 1943, 151, 309.
5. Battino, R. ref.2, page 333.
6. Whitney, R.P.; Vivian, J.E. *Ind. Eng. Chem.* 1941, 33, 741.
7. Sherrill, M.S.; Izard, E.F. *J. Am. Chem. Soc.* 1928, 50, 1665; 1931, 53, 1667.
8. Fogg, P.G.T. ref. 2, Page 354.
9. Taylor, N.W.; Hildebrand, J.H. *J. Amer. Chem. Soc.* 1923, 45, 682.
10. Semb, J. *J. Am. Pharm. Assoc.* 1935, 24, 547.
11. Zetkin, V.I.; Kosorotov, V.I.; Zakharov, E.V.; Martynov, Yu. M.; Dzhagatspanyan, R.V. *Zh. Fiz. Khim.* 1970, 44, 830; Complete article deposited at VINITI, 1970, 1480-70.
12. Tsiklis, D.S.; Svetlova, G.M. *Zh. Fiz. Khim.* 1958, 32, 1476.
13. Pozin, M.E.; Opykhtina, M.A. *Zh. Prik. Khim. (Leningrad)* 1947, 20, 523.
14. Cervinka, M. *Chem. Prum.* 1960, 10, 249.
15. Zetkin, V.I.; Kosorotov, V.I.; Stul, B.Ya.; Dzahagatspanyan, R.V. *Khim. Prom.* 1971, 47, 102.
16. Lohse, M.; Deckwer, W.-D. *J. Chem. Eng. Data* 1981, 26, 159.
17. Egunov, A.V.; Konobeev, B.I.; Ryabov, E.A.; Gubanova, T.I. *Zh. Prik. Khim. (Leningrad)* 1973, 46, 1855.
18. Zetkin, V.I.; Kolesnikov, I.M.; Zakharov, E.V.; Dzhagatspanyan, R.V. *Khim. Prom.* 1966, 42, 624.
19. DuPont de Nemours and Co. (Inc.) *Chem. Eng. News* 1955, 33, 2366
20. Vdovichenko V.T.; Kondratenko V.I. *Khim. Prom.* 1967, 43, 290.
21. Gerrard, W. ref. 2, page 395.
22. Curda, M.; Holas, J. *Chem. Prum.* 1964, 14, 547.
23. Kalz, G.; Naumann, A. *Plaste Kautsch.* 1971, 18, 500.
24. Kogan, L.M.; Kol'tsov, N.S.; Litvinov, N.D. *Zh. Fiz. Khim.* 1963, 37, 1014.
25. Gjaldbaek, J.C.; Hildebrand, J.H. *J. Amer. Chem. Soc.* 1950, 72, 609.

Chapter 9
SOLUBILITIES OF THE HYDROGEN HALIDES

9.1 GENERAL BEHAVIOUR

The hydrogen halides have the following physical properties:

	HF	HCl	HBr	HI
M.pt/K	190.09	159.05	186.28	222.36
B.pt (1.013 bar)/K	292.66	188.20	206.43	237.75
Critical temperature/K	461.15	324.68	363.2	423.95
Critical pressure/bar	64.85	82.56	85.6	83.0
Critical volume /dm^3 mol^{-1}	0.0690	0.0810	0.0998	0.1351
Density of gas at 273.15 K, 1.013 bar/g dm^{-3}	—	1.6392	3.61633	5.78882
Relative molecular mass	20.006	36.461	80.912	127.912

The following vapour pressures have been published by Stull:[65]

	T/K	P/bar
HF	207.4	0.0133
	245.0	0.1333
	292.9	1.013
HCl	188.4	1.013
	241.5	10.132
	309.4	60.795
HBr	206.7	1.013
	264.8	10.132
	343.8	60.795
HI	238.1	1.013
	305.2	10.132
	400.7	60.795

These data fit the equations:

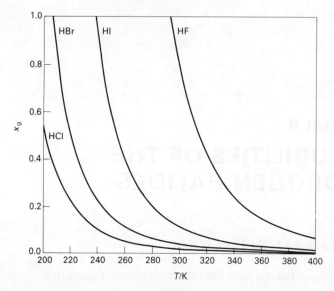

Figure 9.1. Reference lines for the hydrogen halides at a partial pressure of 1.013 bar.

Hydrogen fluoride

$$\ln(P/\text{bar}) = 9.6891 - 2674.88/[(T/\text{K}) - 16.46]$$

Hydrogen chloride

$$\ln(P/\text{bar}) = 10.4718 - 1967.00/[(T/\text{K}) - 0.33]$$

Hydrogen bromide

$$\ln(P/\text{bar}) = 9.5854 - 1755.74/[(T/\text{K}) - 23.28]$$

Hydrogen iodide

$$\ln(P/\text{bar}) = 9.0460 - 1771.49/[(T/\text{K}) - 41.99].$$

Reference lines based upon these equations are shown in Fig. 9.1.

9.2 SOLUBILITY OF HYDROGEN HALIDES IN WATER

9.2.1 Solubility of hydrogen chloride in water

Hydrogen chloride forms hydrates with one, two, three or six water molecules, which are stable at low temperatures.[1]

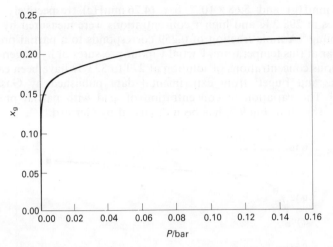

Figure 9.2. The variation with pressure of the mole fraction solubility of hydrogen chloride in water at 293.15 K.

The variation of vapour pressure of aqueous solutions of hydrogen chloride with temperature and pressure was investigated by Roscoe and Dittmar in 1859.[2] They found that the partial pressure of hydrogen chloride was negligible at 273.2 K at low concentrations to a mole fraction of acid of about 0.25. At higher concentrations the partial vapour pressure increased steeply with increase in concentration. Bates and Kirschman[3] measured the vapour pressure of solutions of mole fraction concentration 0.0551 to 0.152 at 298.2 K. The corresponding vapour pressures were 1.04×10^{-3} bar

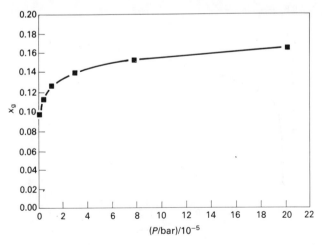

Figure 9.3. The variation with pressure of the mole fraction solubility of hydrogen bromide in water at 298.15 K.

(0.0078 mmHg) and 5.68×10^{-3} bar (4.26 mmHg), respectively. Vapour pressures at 298.2 K and higher concentrations were measured by Randall and Young.[4] A mole fraction of 0.259 corresponds to a partial pressure of 1.013 bar at this temperature. Partial vapour pressures of hydrogen chloride for various concentrations of solution at 273 to 323 K have been calculated by Fritz and Fuget[5] from experimental data published by Gaston and Gittler.[6] The variation in concentration of acid with partial pressure at 293.2 K, shown in Fig 9.2, has been discussed by Gerrard.[7]

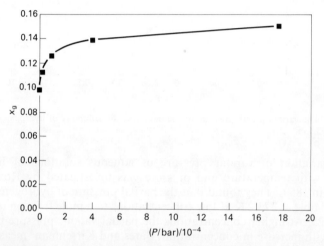

Figure 9.4. The variation with pressure of the mole fraction solubility of hydrogen iodide in water at 298.15 K.

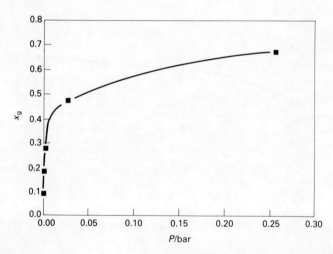

Figure 9.5. The variation with pressure of the mole fraction solubility of hydrogen fluoride in water at 303.15 K.

Vapour pressure data for aqueous solutions of hydrogen bromide and iodide at 298.2 K have also been published by Bates and Kirschman.[3] Variations in mole fraction solubilities with change of partial pressure of halide from these data are shown in Figs 9.3 and 9.4.

The vapour pressures of solutions of hydrogen fluoride between 273.25 and 343.2 K were measured by Munter et al.[8] The variation in mole fraction solubility with partial pressure of hydrogen fluoride is shown in Fig. 9.5.

9.3 SOLUBILITY OF HYDROGEN HALIDES IN NON-AQUEOUS SOLVENTS

Solubilities in non-aqueous solvents have been compiled and critically evaluated.[9] Many of the measurements on these systems have been carried out by Gerrard and his co-workers.[9]

9.3.1 Solubilities in non-aromatic hydrocarbons

Although there is some disagreement between solubilities of hydrogen chloride reported by different authors, the general pattern of behaviour is fairly clear. Mole fraction solubilities at a partial pressure of 1.013 bar lie close to the reference line. Solubilities in the shorter chain compounds tend to lie below the reference line and solubilities in higher alkanes above the reference line (Fig. 9.6). The solubility in the unsaturated compound 1-hexadecene has been measured and shows a higher mole fraction solubility

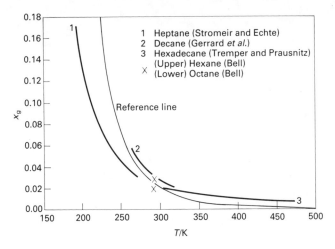

Figure 9.6. Mole fraction solubilities in alkanes of hydrogen chloride at a partial pressure of 1.013 bar.

than the corresponding saturated compound. Values at 293.2 K and 298.2 K are given in Table 9.1.

Table 9.1 Mole fraction solubilities of HCl in non-aromatic hydrocarbons at a partial pressure of HCl of 1.013 bar

Solvent	293.2 K	298.2 K	Ref.
Reference line value	0.024	0.021	
Pentane		0.0155	10
Hexane		0.0139	10
	0.0197		11
Cyclohexane	0.0154		11
	0.0179		12
Heptane		0.0156	10
		0.0215	13
Octane	0.0296		11
		0.0166	10
2,2,4-Trimethylpentane		0.0165	10
Decane	0.0321	0.0298	14
Dodecane	0.0314		11
Hexadecane		0.0229	15
	0.0270		11
1-Hexadecene		0.0357	16

The few data available on solubilities of hydrogen bromide indicate that the pattern of solubilities is similar to that of hydrogen chloride if the difference in vapour pressures of the two compounds in the liquid state is taken into account. The mole fraction solubility of hydrogen fluoride in octane[17] is 0.0038 at 298.2 K and a partial pressure of 1.013 bar. This is very small compared with the reference line value of 0.84.

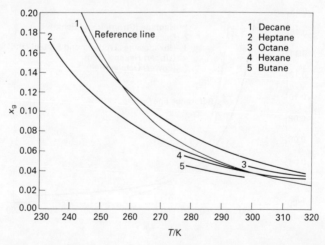

Figure 9.7. Mole fraction solubilities in alkanes of hydrogen bromide at a partial pressure of 1.013 bar.

Mole fraction solubilities of hydrogen bromide in hydrocarbons are shown in Table 9.2 and Fig. 9.7.

Table 9.2 Mole fraction solubilities of HBr in non-aromatic hydrocarbons at a partial pressure of HBr of 1.013 bar

Solvent	293.2 K	298.2 K	Ref.
Reference line value	0.047	0.041	
Butane		0.0332	18
Hexane	0.0422		18
		0.0416	19
Heptane	0.0449		20
Octane		0.0448	19
Decane	0.0557		20
		0.0519	19

9.3.2 Solubilities in aromatic hydrocarbons

Mole fraction solubilities of hydrogen chloride in benzene, methylbenzene and dimethylbenzenes at a partial pressure of 1.013 bar are higher than corresponding solubilities in non-aromatic hydrocarbons of the same carbon number. The general pattern of mole fraction solubilities in these aromatic solvents is shown in Fig. 9.8 and Table 9.3.

Various measurements have been made of the solubility of hydrogen chloride in benzene but there is poor agreement. Fogg[9] has given the following equation, based upon published data

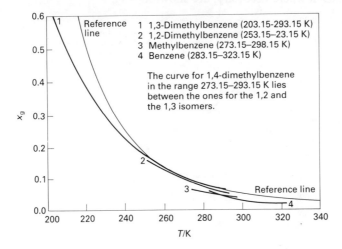

Figure 9.8. Mole fraction solubilities of hydrogen chloride in aromatic hydrocarbons at a partial pressure of HCl of 1.013 bar.

Table 9.3 Mole fraction solubilities of HCl in aromatic hydrocarbons at a partial pressure of HCl of 1.013 bar

Solvent	293.2 K	298.2 K	Ref.
Reference line value	0.024	0.021	
Benzene	0.0461		21
		0.0310	22
		0.047	23
	0.0425		11
Methylbenzene	0.0507		11
	0.0322		24
		0.0425	25
1,3-Dimethylbenzene	0.0668		20
1,4-Dimethylbenzene	0.0607		20

$$\ln x_g = -358.42 + 17853/(T/K) + 51.818\ln(T/K)$$

standard deviation in $x_g = 0.0031$
temperature range 283.2 to 323.2 K.

Mole fraction solubilities in methylbenzene are close to those in benzene. Mole fraction solubilities in the dimethylbenzenes are greater than those in benzene and increase in the order: 1,3-dimethylbenzene > 1,4-dimethylbenzene > 1,2-dimethylbenzene.

The solubility pattern for hydrogen bromide in aromatic hydrocarbons is similar to that of hydrogen chloride. Solubilities at a partial pressure of 1.013 bar lie above the reference line. Solubilities in benzene have been published by Brown and Wallace,[26] Ahmed et al.,[20] Kapustinskii and Mal'tsev,[27] and by O'Brien and Bobalek.[25] Values for a partial pressure of 1.013 bar, found by extrapolation where necessary, are shown in Fig. 9.9.

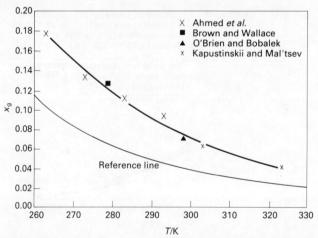

Figure 9.9. Mole fraction solubility in benzene of hydrogen bromide at a partial pressure of 1.013 bar.

Solubilities in methylbenzene,[20,25,26] 1,3-dimethylbenzene[20,26] and in 1,3,5-trimethylbenzene[26] have also been published and are plotted in Fig. 9.10. The available evidence indicates that the mole fraction solubilities at a partial pressure of 1.013 bar and temperatures close to 298.2 K increase in the order 1,3,5-trimethylbenzene > 1,3-dimethylbenzene > methylbenzene > benzene.

Solubilities of hydrogen fluoride in benzene at various temperatures and pressures have been published by Simons.[17] At partial pressures above about 0.005 bar (4 mmHg) mole fraction solubilities fall below the reference surface corresponding to variation in both pressure and temperature (Fig. 9.11).

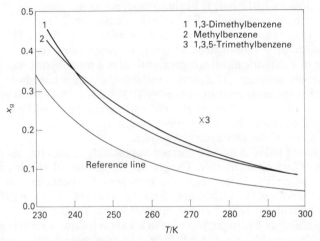

Figure 9.10. Mole fraction solubilities of hydrogen bromide in aromatic hydrocarbons at a partial pressure of HBr of 1.013 bar.

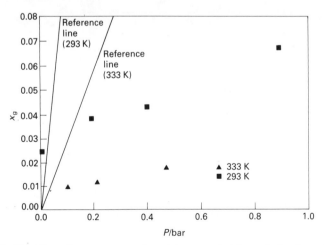

Figure 9.11. The effect of temperature and partial pressure on the mole fraction solubility of hydrogen fluoride in benzene.

9.3.3 Solubilities in alcohols

The mole fraction solubility of hydrogen chloride at a partial pressure of gas of 1.013 bar is high relative to hydrocarbons and relative to the reference line. Measurements by Gerrard et al.[28] from 273.2 to 288.2 K show that the solubility of hydrogen chloride in 1-butanol at pressures close to 1.013 bar is relatively insensitive to small changes in pressure (Fig. 9.12).

The variation with pressure of the solubilities of hydrogen chloride, bromide and iodide in all alkanols is likely to follow a similar pattern. It follows that the solubility in a volatile alkanol, measured under a total pressure close to 1.013 bar, is likely to be close to the solubility at a partial pressure of hydrogen halide of 1.013 bar (see Section 2.1).

There is little change of mole fraction solubility of hydrogen chloride with chain length of a monohydric alcohol. In most cases the mole fraction solubility in a dihydric alcohol is greater than in a monohydric alcohol of the same carbon number. The data available for 1,2-ethanediol, however, indicate that the mole fraction solubility at 298.2 K and a partial pressure of 1.013 bar is slightly less than in ethanol under the same conditions. Data for other diols indicate that the mole fraction solubility increases with the distance apart of the two hydroxyl groups.

Substituting halogen into an alkanol appreciably lowers the mole fraction solubility of hydrogen chloride. The greater the number of halogen atoms the lower the solubility. In contrast, the presence of a benzene ring has little effect on the solubility of hydrogen chloride. Available data for 298.2 K and a partial pressure of 1.013 bar are summarised in Table 9.4.

The solubility of hydrogen bromide in alkanols follows a similar pattern to that of hydrogen chloride. There is little change of mole fraction solubility at

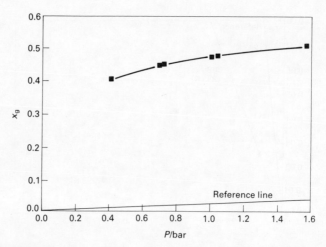

Figure 9.12. The variation with partial pressure of the mole fraction solubility of hydrogen chloride in 1-butanol at 288.15 K.

Table 9.4 Mole fraction solubilities in alcohols of hydrogen halides at a partial pressure of 1.013 bar. The temperature is 298.2 K unless otherwise stated

Solvent	x_g		Ref.
HCl in monohydric alkanols			
Methanol	0.420		28
	0.41		23
Ethanol	0.447		28
	0.45		23
1-Propanol	0.450		29
	0.449		28
2-Propanol	0.472		28
1-Butanol	0.452		28
	0.448		14
	0.451		32
2-Butanol	0.472		14
	0.477		28
2-Methyl-1-propanol	0.455		28
1-Pentanol	0.449		14
	0.454		32
	0.453		31
3-Pentanol	0.479		28
2-Methyl-1-butanol	0.471		28
1-Hexanol	0.454		31
	0.446		32
Cyclohexanol	0.472		31
4-Methyl-2-pentanol	0.477		14
	0.475		31
1-Heptanol	0.455		31
	0.458		32
4-Heptanol	0.478		28
2-Methylcyclohexanol	0.476		31
1-Octanol	0.454		32
	0.456		28
	0.450		14
	0.436		20
1-Nonanol	0.449		32
3,3,5-Trimethyl-1-hexanol	0.457		28
1-Decanol	0.458	0.362 (333.2 K)	34
	0.457	0.365 (333.2 K)	31
	0.449		32
1-Dodecanol	0.436	0.321 (333.2 K)	34
1-Tetradecanol	0.313 (333.2 K)		34
1-Hexadecanol	0.316 (333.2 K)		34
1-Octadecanol	0.309 (333.2 K)		34
HBr in monohydric alkanols			
1-Pentanol	0.489		34
1-Hexanol	0.475		34
1-Heptanol	0.497		34
1-Octanol	0.497		34
	0.514		20
1-Nonanol	0.508	0.325 (333.2 K)	34
1-Decanol	0.485	0.351 (333.2 K)	34

Table 9.4 (*continued*)

Solvent	x_g		Ref.
1-Dodecanol	0.471	0.351 (333.2 K)	34
1-Tetradecanol	0.301 (333.2 K)		34
1-Hexadecanol	0.316 (333.2 K)		34
HI in alkanols			
1-Octanol	0.527		20
HCl in dihydric alcohols			
2-Ethanediol	0.434		30
1,3-Propanediol	0.451		30
1,3-Butanediol	0.532		30
1,4-Butanediol	0.562		30
2,3-Butanediol	0.474		30
1,5-Pentanediol	0.587		30
HF in alcohols at 302.0 K			
1,2-Ethanediol	0.916		35
1,2,3-Propanetriol	0.946		35
HCl in substituted alkanols			
2-Chloroethanol	0.273		31
	0.278		14
2-Bromoethanol	0.269		31
2,2,2-Trichloroethanol	0.060		31
	0.054		33
2,2,2-Trifluoroethanol	0.044		31
1-Chloro-2-propanol	0.295		31
1-Bromo-2-propanol	0.309		31
1,3-Dibromo-2-propanol	0.078		31
2,3-Dibromo-1-propanol	0.162		31
Benzenemethanol (benzyl alcohol)	0.405		28
Phenylethanol	0.411		28
Phenyl-1-propanol	0.432		28
HBr in substituted alkanols			
2-Chloroethanol	0.377		33
2,2-Dichloroethanol	0.244		20
2,2,2-Trichloroethanol	0.123		20
	0.078		33

1.013 bar with increase in carbon number. Values are considerably in excess of the reference line for 1.013 bar. Mole fraction solubilities in 1-octanol from measurements by Ahmed et al.[20] and by Fernandes[34] are shown in Fig. 9.13 together with a regression line through the points.

The mole fraction solubility of hydrogen iodide in 1-octanol also follows a similar pattern (Fig. 9.14).

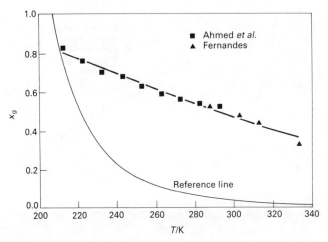

Figure 9.13. Mole fraction solubility in 1-octanol of hydrogen bromide at a partial pressure of 1.013 bar.

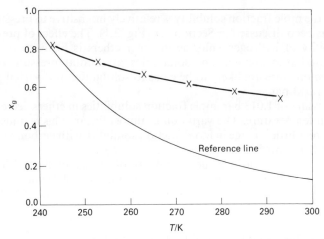

Figure 9.14. Mole fraction solubility in 1-octanol of hydrogen iodide at a partial pressure of 1.013 bar.

9.3.4 Solubilities in ethers

Gerrard and his co-workers[20,30,36] measured the solubility of hydrogen chloride in a large number of alkyl ethers, between 197 and 301 K at a total pressure equal to barometric pressure. The behaviour of ethers is similar to that of alkanols and solubilities are high relative to the reference line (Fig 9.15). The solubility in 1-methoxybutane was studied over a range of total pressure of 0.07 to 2.26 bar from 233.45 to 286.2 K. At pressures close to

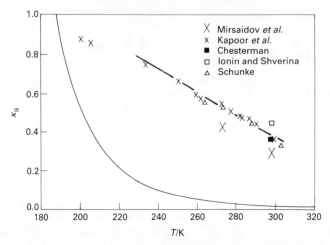

Figure 9.15. Mole fraction solubility of hydrogen chloride in 1,1'-oxybisethane at a total pressure of 1.013 bar.

1.013 bar the mole fraction solubility is relatively insensitive to pressure. This system has been discussed in Section 2.1 (Fig. 2.1). The effect of pressure on the solubility of hydrogen chloride in other ethers is likely to be similar. Experimental measurements of solubilities at a total pressure equal to barometric pressure are likely to be close to solubilities at a partial pressure of gas of 1.013 bar.

At a pressure of 1.013 bar, mole fraction solubilities in ethers decrease with increase in temperature. The variation is almost linear. This is unusual (Fig 9.15). There is little change in mole fraction solubility with carbon number of an alkyl ether (Table 9.5).

There have been various measurements of the solubility of hydrogen chloride in 1,1'-oxybisethane.[23,24,36,37,38] Mole fraction solubilities for a pressure of 1.013 bar from measurements by three of the groups[23,36,38] fit the equation[9]

$$\ln x_g = 288.44 - 11355/(T/\text{K}) - 44.115 \ln(T/\text{K}).$$

standard deviation in $x_g = 0.010$
temperature range 260 to 303 K.

Measurements are shown in Fig 9.15.

The presence of a benzene ring in an ether lowers the solubility of hydrogen chloride, especially if it is directly bonded to the oxygen in the ether link. Solubility in chloroethers is also lower than that in simple alkyl ethers (Table 9.5).

Solubilities of hydrogen bromide and hydrogen iodide in 1,1'-oxybisoc-

Table 9.5 Mole fraction solubilities of hydrogen halides in ethers at a total pressure of 1.013 bar

Solvent	T/K	x_g	Ref.
HCl in ethers			
Methoxyethane	251.7	0.644	36
1,1'-Oxybisethane	298.2	0.368	a
	283.2	0.478	a
1-Methoxypropane	297.4	0.378	36
	283.2	0.473	36
1-Methoxybutane	283.2	0.476	36
1,1'-Oxybispropane	283.2	0.473	36
1-Methoxyhexane	283.2	0.474	36
1,1'-Oxybishexane	283.2	0.472	36
1,1'-Oxybisoctane	283.2	0.468	36
Methoxybenzene	283.2	0.138	36
Ethoxybenzene	283.2	0.149	30
(Methoxymethyl)benzene	283.2	0.421	30
(Ethoxymethyl)benzene	283.2	0.419	30
1,1'-Oxybisbenzene	283.2	0.0824	30
1,1'-[Oxybis(methylene)]bisbenzene	283.2	0.352	30
Oxybis(chloromethane)	283.2	0.046	30
1-Chloro-1-(2-chloroethoxy)ethane	283.2	0.180	30
1,1'-Oxybis(2-chloroethane)	283.2	0.174	30
Tetrahydrofuran	283.2	0.581	30
1,4-Dioxane	283.2	0.517	30
HBr in ether			
1,1'-Oxybisoctane	283.2	0.668	20
HI in ether			
1,1'-Oxybisoctane	283.2	0.719	20
HF in ethers			
1,1'-Oxybis[3-methylbutane]	297.6	0.799 (0.989 bar)	35
1,1'-Oxybisbenzene	304.3	0.583 (0.989 bar)	35

[a] From equation above.
Experimental data have been extrapolated or interpolated where appropriate.

tane have been measured at various temperatures. The behaviour of hydrogen bromide is similar to that of hydrogen chloride when the differences in reference lines of the gases are taken into account (Fig 9.16).

The solubilities of hydrogen fluoride in 1,1'-oxybis(3-methylbutane) and in 1,1'-oxybisbenzene have been measured by Matuszak.[35]

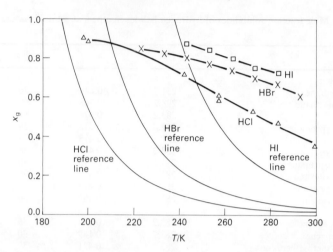

Figure 9.16. Mole fraction solubilities of hydrogen halides in 1,1'-oxybisoctane at a partial pressure of 1.013 bar.

9.3.5 Solubilities in carboxylic acids and their esters

Various groups have measured the solubility of hydrogen chloride in acetic acid at barometric pressure. Data published by four groups may be fitted to the equation:[9]

$$\ln x_g = 70.224 - 1056.9/(T/K) - 12.092\ln(T/K)$$

standard deviation in $x_g = 0.0095$
temperature range 253 to 353 K.

Solubility in acetic acid at low partial pressures of gas were measured by Rodebush and Ewart.[39] Gerrard and Macklen[28] measured solubilities in other carboxylic acids at various temperatures and a total pressure equal to barometric (Table 9.6).

Gerrard and Macklen[28] have measured the solubility of hydrogen chloride in numerous esters of alkanoic acids. Mole fraction solubilities are high, relative to the reference line. This behaviour is common to solvents containing oxygen. The measured values of mole fraction solubilities at a total pressure equal to barometric pressure are likely to be close to the mole fraction solubilities at a partial pressure of 1.013 bar. Mole fraction solubilities in straight chain alkyl acetates increase with chain length at this pressure. Solubilities in branched chain alkyl acetates are greater than in straight chain compounds. The presence of a benzene ring lowers solubility. Typical values are given in Table 9.6.

Gerrard and his co-workers[14,28,31,33] have shown that the presence of halogen in an ester molecule also lowers the solubility of hydrogen chloride (Table 9.6).

Table 9.6 Mole fraction solubilities of hydrogen halides at a partial pressure of 1.013 bar in carboxylic acids and their esters

Solvent	T/K	x_g	Ref.
HCl in esters			
Ethyl formate	298.2	0.170	28
Methyl acetate	298.2	0.302	28
Ethyl acetate	298.2	0.314	28
1-Methylethyl acetate	298.2	0.343	28
Propyl acetate	298.2	0.328	28
2-Methylpropyl acetate	298.2	0.338	28
1-Methylpropyl acetate	298.2	0.343	28
Butyl acetate	298.2	0.331	28
Octyl acetate	298.2	0.334	28
Phenyl acetate	298.2	0.190	28
Benzyl acetate	298.2	0.243	28
Ethyl propanoate	298.2	0.315	28
Ethyl butanoate	298.2	0.324	28
Ethyl chloroformate	298.2	0.0666	14
Propyl chloroformate	298.2	0.0794	33
Hexyl chloroformate	298.2	0.0943	33
Ethyl chloroacetate	298.2	0.157	31
Ethyl bromoacetate	298.2	0.172	28
Ethyl dichloroacetate	298.2	0.108	28
Ethyl trichloroacetate	298.2	0.0653	28
2-Chloroethyl acetate	298.2	0.209	28
2,2,2-Trichloroethyl acetate	298.2	0.153	28
Phenyl chloroacetate	298.2	0.150	33
HCl in carboxylic acids			
Formic acid	298.2	0.053	28
Acetic acid	298.2	0.109	a
	323.2	0.054	a
Propanoic acid	298.2	0.108	28
Butanoic acid	298.2	0.118	28
Hexanoic acid	293.2	0.160	20
Chloroacetic acid	323.2	0.0280	28
Trichloroacetic acid	323.2	0.0288	28
HBr in carboxylic acids			
Acetic acid	293.2	0.324	20
Hexanoic acid	293.2	0.398	20
HI in carboxylic acids			
Acetic acid	293.2	0.353	20[b]
Hexanoic acid	293.2	0.375	20

[a] From equation above.
[b] Extrapolated.

9.3.6 Solubilities of hydrogen chloride in halogenated organic solvents

Numerous measurements of the solubility of hydrogen chloride in tetrachloromethane have been reported.[11,40–46] There is appreciable scatter of the values of mole fraction solubility estimated for a partial pressure of 1.013 bar (Fig. 9.17). They all fall below the reference line except at temperatures above about 330 K. Fogg[9] has given the equation

$$\ln x_g = -206.48 + 9800.3/(T/\text{K}) + 29.732 \ln(T/\text{K})$$

standard deviation in $x_g = 0.0018$
temperature range 266 to 343 K.

Solubilities in trichloromethane have been measured by various workers.[11,40,41,43] Mole fraction solubilities tend to be greater than in tetrachloromethane but there is considerable disagreement between different authors (Fig. 9.18).

Solubility in dichloromethane at a total pressure equal to barometric pressure was measured by Vdovichenko and Kondratenko[43] from 263 to 298 K. The vapour pressure of the pure solvent is very high at 298 K (>0.56 bar). Reliable correction of the data to a partial pressure of 1.013 bar is not possible.

Solubilities of hydrogen chloride in 1,2-dichloroethane have been measured by various groups.[11,41,47–49] Mole fraction solubilities for a partial pressure of 1.013 bar fit the equation:

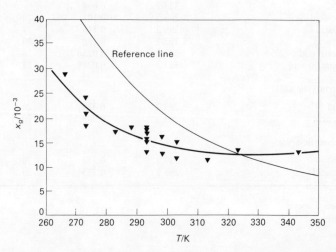

Figure 9.17. Mole fraction solubilities in tetrachloromethane of hydrogen chloride at a partial pressure of 1.013 bar.

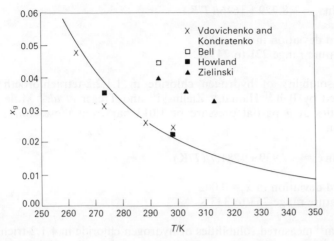

Figure 9.18. Mole fraction solubilities in trichloromethane of hydrogen chloride at a partial pressure of 1.013 bar.

$$\ln x_g = -9.355 + 1808/(T/K)$$

standard deviation in $x_g = 0.004$
temperature range 253 to 343 K (see Fig. 9.19).

The solubility of hydrogen chloride in 1,1-dichloroethane has been measured by Danov and Golubev[50] at various pressures in the range 204 to 243 K. Mole fraction solubilities at a partial pressure of 1.013 bar from this work may be fitted to an equation similar to the one given above, i.e.

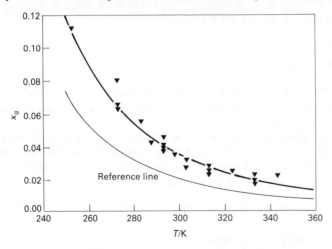

Figure 9.19. Mole fraction solubilities in 1,2-dichloroethane of hydrogen chloride at a partial pressure of 1.013 bar.

$$\ln x_g = -8.729 + 1646/(T/K)$$

standard deviation in $x_g = 0.005$
temperature range 221 to 243 K.

The solubility of hydrogen chloride in 1,1,2,2-tetrachloroethane was measured by Bell,[11] Hamai,[44] Zielinski[41] and Treger et al.[49] Mole fraction solubilities at a partial pressure of 1.013 bar from these sources fit the equation

$$\ln x_g = -7.939 + 1281.7/(T/K)$$

standard deviation in $x_g = 0.002$
temperature range 253 to 333 K.

Hamai[44] measured solubilities of hydrogen chloride in 1,1,2-trichloroethane and in pentachloroethane. Bell[11] also measured the solubility of HCl in pentachloroethane. Mole fraction solubilities for a partial pressure of 1.013 bar fit the equations:

1,1,2-Trichloroethane

$$\ln x_g = -9.906 + 1885.6/(T/K)$$

standard deviation in $x_g = 0.00003$
temperature range 285 to 293 K.

Pentachloroethane

$$\ln x_g = -8.566 + 1391.9/(T/K)$$

standard deviation in $x_g = 0.00005$
temperature range 285 to 293 K.

Mole fraction solubilities of hydrogen chloride for a partial pressure of 1.013 bar decrease as the number of chlorine atoms substituted in ethane increases (Fig. 9.20).

Data are given in the literature for the solubilities of hydrogen chloride in 1-chloroalkanes.[14,16,20,34] Mole fraction solubilities for a partial pressure of gas of 1.013 bar are almost independent of length of carbon chain and may be approximately represented by the equation:[9]

$$\ln x_g = -10.53 + 2213/(T/K)$$

standard deviation in $x_g = 0.011$
temperature range 243 to 433 K.

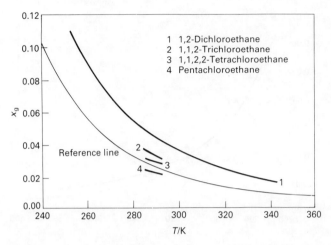

Figure 9.20. Mole fraction solubilities in chloro-alkanes of hydrogen chloride at a partial pressure of 1.013 bar.

Experimental values of the solubility in 1-chlorobutane for temperatures below 243 K are less than those given by this equation.

Solubilities of hydrogen chloride in chloroethenes[11,29,42,47–49] show a similar pattern to those in chloroethanes. Mole fraction solubilities for a partial pressure of 1.013 bar decrease, relative to the reference line, with increase in the number of chlorine atoms.

Solubilities of hydrogen chloride in bromoethanes follow a similar pattern with mole fraction solubilities decreasing, relative to the reference line, with increase in bromination. The mole fraction solubility in 1-bromobutane, 1-bromohexane and 1-bromooctane is similar to that in the 1-chloroalkanes. The single measurement of the mole fraction solubility in bromoethane by Bell[11] is higher than that in other 1-bromoalkanes.

Mole fraction solubilities in 1-iodohexane at 1.013 bar partial pressure are greater than those in 1-chloro and 1-bromo compounds of similar chain length.

The solubility in chlorobenzene was measured at barometric pressure by Bell,[11] Gerrard et al.[14] and by Zetkin et al.[21] O'Brien[51,52] also measured the solubility over a pressure range below 1.013 bar. The solubility at high pressures was measured by Strepikheev and Babkin.[53] Mole fraction solubilities at a partial pressure of 1.013 bar estimated from the published data are shown in Fig. 9.21 and fit the equation

$$\ln x_g = -82.804 + 5275.3/(T/K) + 10.802\ln(T/K)$$

standard deviation in $x_g = 0.0021$.
temperature range 273 to 373 K.

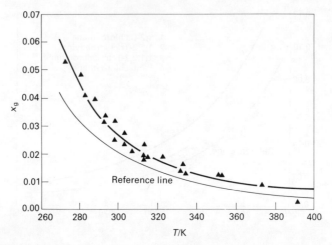

Figure 9.21. Mole fraction solubilities in chlorobenzene of hydrogen chloride at a partial pressure of 1.013 bar.

The solubility in bromobenzene was also measured by Bell,[11] Gerrard[14] and O'Brien[51]. Mole fraction solubilities at a partial pressure of 1.013 bar fit the equation:

$$\ln x_g = 231.01 - 8259.3/(T/K) - 36.315 \ln(T/K)$$

standard deviation in $x_g = 0.0012$
temperature range 273 to 316 K.

O'Brien and Byrne[51] also measured the solubility in fluorobenzene and in iodobenzene at pressures to a maximum of 0.35 bar. Extrapolated mole fraction solubilities at 298.2 K and 1.013 bar are in the order fluorobenzene > chlorobenzene > bromobenzene > iodobenzene.

Measurements[21,55] of the solubility in 1,2-dichlorobenzene, corrected to a partial pressure of 1.013 bar, fit the equation given by Fogg.[9]

$$\ln x_g = 18.049 - 106.766/(T/K) - 3.8027 \ln(T/K)$$

standard deviation in $x_g = 0.00090$
temperature range 288 to 438 K.

The solubilities in these and other solvents containing halogen are shown in Table 9.7.

Table 9.7 Mole fraction solubilities of hydrogen chloride and bromide at a partial pressure of 1.013 bar in halogenated solvents

Solvent	T/K	x_g	Ref.
Hydrogen Chloride			
Dichloromethane	273.2	0.022	43
Trichloromethane	293.2	0.044	11
	293.2	0.040	41
Tetrachloromethane	273.2	0.0223	a
	293.2	0.0158	a
	298.2	0.0149	a
1,2-Dichloroethane	298.2	0.0372	a
1,1,2-Trichloroethane	293.2	0.0310	44
1,1,2,2-Tetrachloroethane	298.2	0.026	a
Pentachloroethane	293.2	0.0225	44
	293.2	0.0214	11
1-Chlorobutane	273.2	0.0805	14
	293.2	0.056	14
1-Chlorohexane	293.2	0.054	14
	298.2	0.048	14
1-Chlorohexadecane	313.2	0.037	54
Tribromomethane	293.2	0.0306	11
Bromoethane	293.2	0.1019	11
1,2-Dibromoethane	293.2	0.0348	11
	293.2	0.0344	44
	298.2	0.0312	44
1,1,2,2-Tetrabromoethane	293.2	0.0236	11
1-Bromohexane	293.2	0.0537	14
	298.2	0.0483	14
Trichloroethene	293.2	0.0206	11
	293.2	0.0205	47
Tetrachloroethene	293.2	0.0163	11
	293.2	0.0160	42
Chlorobenzene	293.2	0.0319	a
	298.2	0.0283	a
Bromobenzene	293.2	0.0315	a
	298.2	0.0273	a
Iodobenzene	298.2	0.0232	51
Fluorobenzene	298.2	0.0308	51
1,2-Dichlorobenzene	293.2	0.0199	a
	298.2	0.0188	a
1,2,4-Trichlorobenzene	293.2	0.0251	21
	298.2	0.0217	21
Hydrogen Bromide			
Dichloromethane	273.2	0.069	20
	293.2	0.051	20
Trichloromethane	273.2	0.076	a
	293.2	0.050	a
	298.2	0.046	a
Tetrachloromethane	273.2	0.058	a
	293.2	0.037	a
	298.2	0.033	a
1-Chlorooctane	293.2	0.153	20
1-Bromooctane	293.2	0.091	20
1-Iodooctane	293.2	0.138	20

[a] From equation given in the text.

9.3.7 Solubilities of hydrogen bromide and iodide in organic solvents containing halogen

Solubilities of hydrogen bromide in chloromethanes fit into a clear pattern (Fig. 9.22). The mole fraction solubility in dichloromethane at a partial pressure of 1.013 bar, from measurements by Ahmed et al.,[20] is close to that in trichloromethane[20,40] but higher than that in tetrachloromethane.[20,40]

Mole fraction solubilities at a partial pressure of 1.013 bar fit the following equations:

Dichloromethane

$$\ln x_g = -9.054 + 1782/(T/K)$$

standard deviation in $x_g = 0.010$
temperature range 233 to 293 K.

Trichloromethane[9]

$$\ln x_g = -58.395 + 3772.6/(T/K) + 7.4877\ln(T/K)$$

standard deviation in $x_g = 0.0015$
temperature range 233 to 298 K.

Tetrachloromethane[9]

$$\ln x_g = 17.294 + 684.90/(T/K) - 4.0365\ln(T/K)$$

standard deviation in $x_g = 0.0061$
temperature range 233 K to 298 K.

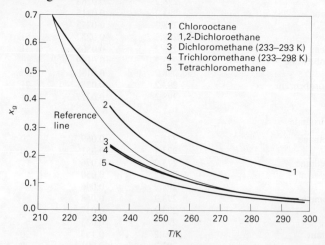

Figure 9.22. Mole fraction solubilities in chloroalkanes of hydrogen bromide at a partial pressure of 1.013 bar.

Solubilities of hydrogen bromide in several halogenated alkanes are shown in Table 9.7.

Gerrard and his co-workers[20,56] measured the solubility of hydrogen iodide in alkyl halides. Measurements were made over temperature ranges at barometric total pressure. At a particular temperature there is a tendency for the mole fraction solubility at a total pressure of 1.013 bar to increase with the length of the carbon chain and to increase from chloride to bromide to iodide (Table 9.8).

Solubilities of hydrogen bromide in halobenzenes from 233 K to 293 K were measured by Ahmed et al.[20] at barometric total pressure. At a fixed temperature, mole fraction solubilities increase from chlorobenzene to iodobenzene. This is contrary to the trend for solubilities of hydrogen chloride.

Table 9.8 Mole fraction solubilities of hydrogen iodide in haloalkanes at a total pressure of 1.013 bar and 293.2 K

Solvent	x_g	Ref.
1-Chloropropane	0.153	56
1-Chlorobutane	0.160	56
1-Chlorooctane	0.273	20
Bromoethane	0.141	56
1-Bromopropane	0.161	56
1-Bromobutane	0.187	56
1-Bromooctane	0.270	20
Iodomethane	0.213	56
Iodoethane	0.248	56
1-Iodopropane	0.192	56
1-Iodooctane	0.298	20

9.3.8 Solubilities of hydrogen halides in organic solvents containing nitrogen

The solubility of hydrogen chloride in nitrobenzene was measured by Zetkin et al.[57] and by Ahmed et al.[20] at barometric total pressure. O'Brien[22,52] made measurements to a maximum pressure of 0.698 bar. Mole fraction solubilities at a partial pressure of 1.013 bar fit the equation

$$\ln x_g = -39.613 + 3208.7/(T/K) + 4.5790\ln(T/K)$$

standard deviation in $x_g = 0.005$
temperature range 253 to 433 K.

Chlorination of the benzene nucleus reduces the mole fraction solubility. The presence of methyl groups increases the solubility (Table 9.9).

Ahmed et al.[20] measured the solubility of hydrogen bromide in nitrobenzene at 1.013 bar. Solubilities of hydrogen chloride and bromide in 1-methyl-2-nitrobenzene and in 1-methyl-3-nitrobenzene at pressures below 1.013 bar

Table 9.9 Mole fraction solubilities of hydrogen chloride and bromide at a partial pressure of 1.013 bar in nitrobenzene and substituted nitrobenzenes

Solvent	T/K	x_g	Ref.
Hydrogen chloride			
Nitrobenzene	298.2	0.063	a
	333.2	0.034	a
	373.2	0.020	a
1-Chloro-2-nitrobenzene	333.2	0.0250	57
	373.2	0.0150	57
1-Chloro-3-nitrobenzene	333.2	0.0228	57
	373.2	0.0140	57
1-Chloro-4-nitrobenzene	373.2	0.0175	57
1,2-Dichloro-4-nitrobenzene	333.2	0.0206	57
	373.2	0.0127	57
1,3-Dichloro-2-nitrobenzene	333.2	0.0222	57
	373.2	0.0112	57
1-Methyl-2-nitrobenzene	298.2	0.0719	58[b]
1-Methyl-3-nitrobenzene	298.2	0.0768	59[b]
Hydrogen bromide			
Nitrobenzene	298.2	0.0997	20[b]
1-Methyl-2-nitrobenzene	298.2	0.0941	25[b]
1-Methyl-3-nitrobenzene	298.2	0.0881	25[b]

[a] From equation given in the text.
[b] Estimated by extrapolation.

were measured by O'Brien et al.[25,58,59] The data indicate that the presence of a methyl group attached to the benzene ring decreases the mole fraction solubility of hydrogen bromide but increases that of hydrogen chloride.

9.3.9 Solubilities of hydrogen chloride in solvents containing sulfur, phosphorus, boron or silicon

Gerrard and his co-workers[60-63] have measured the solubilities of hydrogen chloride in many sulfur compounds at a total pressure equal to barometric pressure.

The mole fraction solubility in sulfuric acid at 273.2 K is less than 0.0196. It is appreciably higher in esters of sulfonic acid where one of the hydroxyl groups of sulfuric acid is replaced by an alkyl or aryl group and the other hydroxyl group is esterified. It is even higher in sulfones where both hydroxyl groups are replaced by alkyl groups. Solubility is also high in thiols and thioethers.

The replacement by chlorine of a hydroxyl group in sulfuric acid also enhances the solubility of hydrogen chloride but the effect is less than that of an alkyl or aryl group. Substituting a chlorine atom into the benzene ring of benzenesulphonic acid, butyl ester, lowers the solubility of hydrogen chloride.

Selected solubilities in sulfur compounds are given in Table 9.10.

Table 9.10 Mole fraction solubilities of hydrogen halides in solvents containing sulfur, phosphorus or boron

Solvent	T/K	x_g	Ref.
HCl in sulfur compounds			
Sulfuric acid	273.2	<0.0196	61
Methanesulfonic acid, butyl ester	273.2	0.338	61
	298.2	0.203	61
Ethanesulfonic acid, butyl ester	273.2	0.387	61
	298.2	0.240	61
Benzenesulfonic acid, butyl ester	273.2	0.295	61
	298.2	0.176	61
4-Methylbenzenesulfonic acid, butyl ester	273.2	0.308	61
	298.2	0.199	61
1,1'-Sulfonylbispropane	273.2	0.502	61
	298.2	0.383	61
1,1'-Sulfonylbisbutane	298.2	0.385	61
Sulfuryl chloride	273.2	0.0310	61
Methanesulfonyl chloride	298.2	0.0421	61
Ethanesulfonyl chloride	273.2	0.089	61
	298.2	0.053	61
Benzenesulfonyl chloride	298.2	0.0403	61
Chlorosulfuric acid, butyl ester	273.2	0.105	61
	298.2	0.0539	61
4-Chlorobenzenesulfonic acid, butyl ester	273.2	0.174	61
	298.2	0.0766	61
1-Butanethiol	273.2	0.111	60
Benzenethiol	273.2	0.085	60
1,1'-Thiobisbutane	273.2	0.390	60
1,1'-Thiobisbenzene	273.2	0.126	60
Thiophene	273.2	0.0329	60
Tetrahydrothiophene	273.2	0.402	60
HCl in phosphorus compounds			
Phosphoric acid, tributyl ester	298.2	0.684	14[a]
	298.2	0.624	67
Phosphorous acid, triphenyl ester	298.2	0.349	14
Phosphorous trichloride	273.2	0.0263	20
HCl in boron compounds			
Boric acid, triethyl ester	298.2	0.160	33
tributyl ester	298.2	0.172	14
tripentyl ester	298.2	0.170	33
Trichloroborane	263.2	0.017	20
HBr in sulfur compounds			
1-Butanethiol	273.2	0.259	60
Benzenethiol	273.2	0.153	60
1,1'-Thiobisbutane	273.2	0.717	60
1,1'-Thiobisbenzene	273.2	0.187	60
Thiophene	273.2	0.123	60
HBr in boron compounds			
Boric acid, tripentyl ester	298.2	0.279	33

Table 9.10 (continued)

Solvent	T/K	x_g	Ref.
HI in sulfur compounds			
1-Butanethiol	273.2	0.401	60
Benzenethiol	273.2	0.281	60
1,1'-Thiobisbutane	273.2	0.743	60
1,1'-Thiobisbenzene	273.2	0.333	60
HF in sulfur compounds			
Fluorosulfuric acid	299.8	0.874	64
	333.2	0.382	64

[a] Estimated by extrapolation to the stated temperature.

Solubilities at barometric total pressure in solvents containing silicon have been reported for a number of temperatures by Gerrard et al.[14,20,33] Solubility in the tetra(4-methyl-2-pentyl) ester of silicic acid was measured at several pressures in the temperature range 253 to 278 K. Solubility at pressures close to 1.013 bar is relatively insensitive to small changes in pressure (Fig. 9.23). Solubilities in this solvent at a total pressure of 1.013 bar are close to those in the tetraethyl and tetrapropyl esters and to those in the tetramethyl ester at temperatures below about 240 K. Solubilities in each of the four solvents show a marked decrease with increase in temperature above about 283 K (Fig. 9.24).

Gerrard and his co-workers have also studied the solubility in a variety of solvents containing boron[14,20,33,62] or phosphorus.[14,20,29,33,62] Mole fraction solubilities are high in esters of boric acid and phosphoric acid.

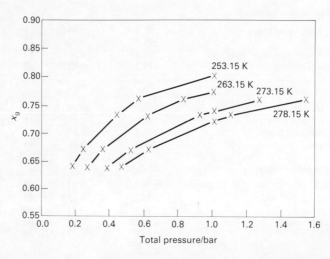

Figure 9.23. Variation with temperature and total pressure of the solubility of hydrogen chloride in tetra(4-methyl-2-pentyl) silicate.

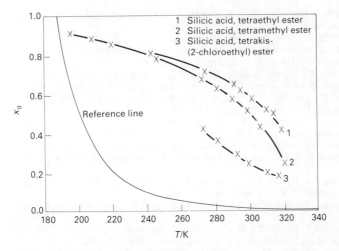

Figure 9.24. The solubilities of hydrogen chloride in esters of silicic acid at a total pressure of 1.013 bar.

9.3.10. Solubilities of hydrogen bromide, iodide and fluoride in solvents containing sulfur or boron

Frazer and Gerrard[60] measured the solubility of hydrogen bromide and iodide in several thiols and organic sulfides. The pattern of solubility is similar to that of hydrogen chloride in these solvents when the differences in reference lines are taken into account.

The solubility of hydrogen fluoride at a pressure of 1.013 bar in fluorosulfuric acid at 299.8 K to 333.2 K has been reported by Hartman.[64]

Gerrard et al.[33] have measured the solubility of hydrogen bromide in tripentyl borate at barometric pressure in the range 273 K to 307 K. Mole fraction solubilities are greater than those of hydrogen chloride because of the higher volatility of the latter. When the difference in the reference line is taken into account the behaviour of the two gases is similar.

REFERENCES

1. Yoon, Y.K.; Carpenter, G.B. *Acta Cryst.* 1959, 12, 17; Lundgren, J.-O.; Olovsson, I. *Acta Cryst.* 1967, 23, 966; Vuillard, C.G. *Compt. Rend.* 1955, 241, 1308.
2. Roscoe, H.E.; Dittmar, W. *Lieb. Ann.* 1859, 112, 336.
3. Bates, S.J.; Kirshman, H.D. *J. Amer. Chem. Soc.* 1919, 41, 1991.
4. Randall, M.; Young, L.E. *J. Amer. Chem. Soc.* 1928, 50, 989.
5. Fritz, J.J.; Fuget, C.R. *Chem. Eng. Data Ser.* 1956, 1, 10.
6. Gaston, J.; Gittler, F.L. *J. Amer. Chem. Soc.* 1955, 77, 3173.
7. Gerrard, W. *Solubility of Gases and Liquids*, 1976, Plenum Press, New York, p.157.
8. Munter, P.A.; Aepli, O.T.; Kossatz, R.A. *Ind. Eng. Chem.* 1949, 41, 1504.
9. *Solubility Data Series, Volume 42, Hydrogen Halides in Non-aqueous Solvents*, ed. P.G.T. Fogg and W. Gerrard, Pergamon Press, Oxford, 1989.

10. Ryabov, V.G.; Solomonov, A.B.; Ketov, A.N.; Bugaichuk, A.M. *Zh. Fiz. Khim.* 1979, 53, 2915.
11. Bell, R.P. *J. Chem. Soc.* 1931, 1371.
12. Tsiklis, D.S.; Svetlova, G.M. *Zh. Fiz, Khim.* 1958, 32, 1476.
13. Stromeir, W.; Echte, A. *Z. Elektrochem.* 1957, 61, 549.
14. Gerrard, W.; Mincer, A.M.A.; Wyvill, P.L. *J. Appl. Chem.* 1959, 9, 89.
15. Tremper, K.L.; Prausnitz, J.M. *J. Chem. Eng. Data* 1976, 21, 295.
16. Scher, M.; Gill, W.N.; Jelinek, R.V. *Ind. Eng. Chem. Fundam.* 1963, 2, 107.
17. Simons, J. H. *J. Am. Chem. Soc.* 1931, 53, 83.
18. Fontana, C. M.; Herold, R.J. *J. Am. Chem. Soc.* 1948, 70, 2881.
19. Boedeker, E.R.; Lynch, C.C. *J. Am. Chem. Soc.* 1950, 72, 3234.
20. Ahmed, W.; Gerrard, W.; Maladkar, V.K. *J.Appl. Chem.* 1970, 20, 109.
21. Zetkin, V.I.; Kosorotov, V.I.; Stul, B.Ya.; Dzhagatspanyan, R.V. *Khim. Prom.* 1971, 47, 102.
22. O'Brien, S.J.; Kenny, C.L.; Zeurcher, R.A. *J. Am. Chem. Soc.* 1939, 61, 2504.
23. Chesterman, D. R. *J. Chem. Soc.* 1935, 906.
24. Mirsaidov, U.; Dzhuraev, Kh.Sh.; Semenenko, K.N. *Dokl. Akad. Nauk. Tadzh. SSR* 1975, 18, 30.
25. O'Brien, S.J.; Bobalek, E.G. *J. Am. Chem. Soc.* 1940, 62, 3227.
26. Brown, H.C.; Wallace, W.J. *J. Am. Chem. Soc.* 1953, 75, 6268.
27. Kapustinskii, A.F.; Mal'tsev, B.A. *Zh. Fiz. Khim.* 1940, 14, 105.
28. Gerrard, W.; Macklen, E.D. *J. Appl. Chem.* 1956, 6, 241.
29. Cook, T.M. *Thesis,* 1966, University of London.
30. Gerrard, W.; Macklen, E.D. *J. Appl. Chem.* 1960, 10, 57.
31. Gerrard, W.; Macklen, E.D. *J. Appl. Chem.* 1959, 9, 85.
32. Ionin, M.V.; Kurina, N.V.; Sudoplatova, A.E. *Tr. po. Khim. i Khim. Tekhnol.* 1963, (1), 47.
33. Gerrard, W.; Mincer, A.M.A.; Wyvill, P.L. *J. Appl. Chem.* 1960, 10, 115.
34. Fernandes, J.B.; Sharma, M.M. *Indian Chem. Eng.* 1965, 7, 38.
35. Matuszak, M.P. *U.S. Patent* 2 520 947 September 5, 1950 (*Chem. Abs.* 1950, 44, 11044g)
36. Kapoor, K.P.; Luckcock, R.G.; Sandbach, J. A. *J. Appl. Chem. Biotech.* 1971, 21, 97.
37. Ionin, M.V.; Shverina, V.G. *Zh. Obshch. Khim.* 1965, 35, 209.; *J. Gen. Chem. USSR,* 1965, 35, 211.
38. Schunke, J. *Z. Phys. Chem.* 1894, 14, 331.
39. Rodebush, W.H.; Ewart, R.H. *J. Am. Chem. Soc.* 1932, 54, 419.
40. Howland, J.J.; Miller, D.R.; Willard, J.E. *J. Am. Chem. Soc.* 1941, 63, 2807.
41. Zielinski, A.Z. *Przem. Chem.* 1958, 37, 338.
42. Curda, M.; Holas, J. *Chem. Prumysl.* 1964, 14, 547.
43. Vdovichenko, V. T.; Kondratenko, V.I. *Khim. Prom.* 1967, 43, 290.
44. Hamai, S. *Bull. Chem. Soc. Japan* 1935, 10, 5.
45. Khodeeva, S.M.; Rozovskii, M.B. *Zh. Fiz. Khim.* 1975, 49, 1396; *Russ. J. Phys. Chem.* 1975, 49, 824.
46. Chesterman, D.R. *J. Chem. Soc.* 1935, 906.
47. Abdullaev, A.I.; Aliev, A.M.; Mamedov, M.B. *Uch. Zap. Azerb. Gos. Univ., Ser. Khim. Nauk* 1968, No. 3, 80.
48. Hannaert, H.; Haccuria, M.; Mathieu, M.P. *Ind. Chim. Belge* 1967, 32, 156.
49. Treger, Yu. A.; Flid, R.M.; Pimenov, I.F.; Avet'yan, M.G. *Zh. Fiz. Khim.* 1967, 41, 2967.
50. Danov, S.M.; Golubev, Yu.D. *Khim. Prom. (Moscow)* 1968, 44 (2), 116.
51. O'Brien, S.J.; Byrne, J.B. *J. Am. Chem. Soc.* 1940, 62, 2063.
52. O'Brien, S.J. *J. Am. Chem. Soc.* 1941, 63, 2709.
53. Strepikheev, Yu.A.; Babkin, B.M. *Khim. Prom. (Moscow)* 1953, 39, (1), 38.
54. Fernandes, J.B.; Sharma, M.M. *Indian Chem. Eng.* 1965, 38.
55. Lavrova, E.M.; Tudorovskaya, G.L. *Zh. Prikl. Khim (Leningrad)* 1977, 50, 2105.; *J. Appl. Chem. (USSR)* 1977, 50, 2005.
56. Maladkar, V.R. *Thesis,* 1970, University of London.
57. Zetkin, V.I.; Kolesnikov, I.M.; Zakharov, E.V. Dzhagatspanyan, B.V. *Khim. Prom. (Moscow)* 1966, 42, (8), 624.
58. O'Brien, S.J.; Kenny, C.L. *J. Am. Chem. Soc.* 1940, 62, 1189.

59. O'Brien, S.J.; King, C.V. *J. Am. Chem. Soc.* 1949, 71, 3632.
60. Frazer, M.J.; Gerrard, W. *Nature*, 1964, 204, 1299.
61. Charalambous, J.; Frazer, M.J.; Gerrard, W. *J. Chem. Soc.* 1964, 1520.
62. Ahmed, W. *Thesis*, 1970, University of London.
63. Borissov, R.S.; Ionin, M.V. *Tr. Gor'k. Politekh. Inst.* 1973, 29, 11.
64. Hartman, B.F. *U.S. Patent* 2 434 040 January 6, 1948.
65. Stull, D.R. *Ind. Eng. Chem.* 1947, 39, 517.

Chapter 10

SOLUBILITY OF HYDROGEN SULFIDE

10.1 GENERAL BEHAVIOUR

Hydrogen sulfide has the following physical properties:

Melting point	= 187.7 K
Boiling point (1.013 bar)	= 212.88 K
Critical temperature	= 373.6 K
Critical pressure	= 90.10 bar
Critical volume	= 0.09771 dm^3 mol^{-1}
Density of gas at 273.15 K, 1.013 bar	= 1.5392 g dm^{-3}
Relative molecular mass	= 34.076

Details of physical properties of hydrogen sulfide, relevant to the Girdler-Sulfide process for heavy water production, have been published by Neuberg et al.[1]

The vapour pressure of the pure liquid, over the whole of the liquid range, is expressed by the Cox equation:[2]

$$\log_{10}(P/\text{bar}) = A(1 - T_b/T) + 0.00572$$

where P = the saturated vapour pressure of liquid H_2S
T = temperature
T_b = boiling point (1.013 bar), i.e. 212.88 K
A = a temperature dependent constant

A may be expressed as a function of the reduced temperature T_r:

$$A = 5.6958 - 2.5610 T_r + 1.3958 T_r^2$$

The reference line for a partial pressure of 1.013 bar, based upon the equation for vapour pressure, is shown in Fig. 10.1.

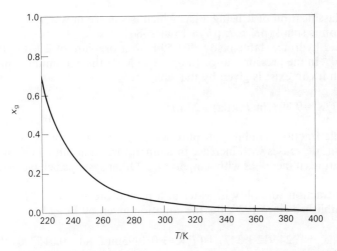

Figure 10.1. Reference line for hydrogen sulfide at a partial pressure of 1.013 bar.

The vapour pressure of liquid hydrogen sulfide may also be expressed as an Antoine type equation satisfactory to about 318 K,[3] i.e.

$$\log_{10}(P/\text{bar}) = 4.2541 - 820.58/[(T/\text{K}) - 19.77]$$

Hydrogen sulfide is a reactant or a byproduct in many industrial processes and occurs widely in nature as a product of decay. As an unwanted component of liquid and gaseous fuels it can cause corrosion and pollution.

Behaviour towards solvents is similar to that of sulfur dioxide. Solubility is high, relative to the reference line, in electron donor solvents. It may be low, relative to the reference line, in strongly hydrogen bonded solvents such as water. Solubility measurements from the literature have been compiled and evaluated in a volume of the Solubility Data Series.[4]

10.2 SOLUBILITY OF HYDROGEN SULFIDE IN AQUEOUS SOLVENTS

Hydrogen sulfide exists in aqueous solution in equilibrium with its ions. The presence of other electrolytes in solution may increase the number of equilibria involving ions, and accurate estimation of the concentration of undissociated hydrogen sulfide in solution may not be possible. When reporting the solubility of hydrogen sulfide in aqueous solution it is usual to consider that the number of moles of the gas in solution is equal to the number of moles of undissociated H_2S plus the number of moles of HS^- and of S^{2-} which originate from the dissolved gas.

Solubility in water is very low, relative to the reference line, despite the usual classification of it being a gas which is soluble in water. The extent of ionisation is small ($pK_1 \simeq 7$; $pK_2 \simeq 15$ at 298 K).[5]

A solid hydrate, stable below 301.59 K at a pressure of 2 MPa, has been reported.[6] In the pressure range 1.80 to 2.35 MPa the maximum temperature at which it can exist is given by the equation[7]

$$T/K = 9.3987\ln(P/kPa) + 230.15.$$

The mole fraction solubility in pure water, at a partial pressure of H_2S of 1.013 bar, decreases with increase in temperature to about 450 K. At higher temperatures it increases with temperature. Other systems behave in a similar manner.

Mole fraction solubility in water at a partial pressure of 1.013 bar, from work published since 1930, fits the equation given by Fogg.[8]

$$\ln x_g = -24.912 + 3477.1/(T/K) + 0.3993\ln(T/K) + 0.015700 T/K$$

standard deviation in $x_g = 0.000065$
temperature range 273 to 603 K (see Fig. 10.2).

The addition of a salt to water usually reduces the solubility of hydrogen sulfide. Addition of an acid usually increases the solubility. Solubilities at low concentrations of electrolyte can often be fitted to empirical relationships based upon a Sechenov equation. A general form of the equation is

$$s = s'\exp(-kc)$$

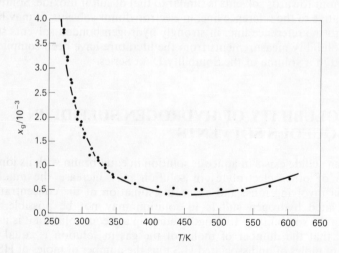

Figure 10.2. Mole fraction solubility in water of hydrogen sulfide at a partial pressure of 1.013 bar.

where s is the solubility of the gas in the solution of the electrolyte, s' is the solubility in pure water in the same units and at the same temperature, c is the concentration of the electrolyte and k is an empirical constant. The equation can be written

$$\ln(s/s') = -kc.$$

The effects of different electrolytes may be expressed as

$$\ln s = \ln s' - (k_1 c_1 + k_2 c_2 + \ldots)$$

where k_1, k_2, etc. are constants for electrolytes 1,2, etc. at concentrations $c_1, c_2 \ldots$

Sometimes additional terms, containing concentrations of electrolytes raised to higher powers, are introduced to improve agreement with experimental measurements. Alternative methods of defining Sechenov parameters have been discussed by Clever.[9]

Dede and Becker[10] measured the solubility of hydrogen sulfide in solutions of sodium perchlorate at 293.2 K and a total pressure equal to barometric. They found that log(solubility) varied linearly with concentration of salt from 0 to 4 mol dm^{-3}. Similar measurements with calcium chloride and sodium sulfate gave a non-linear variation over the range of salt concentration from 0 to 2 mol dm^{-3}. In these cases additional terms, involving concentrations raised to higher powers, are needed for an empirical relationship between gas solubility and concentration of salt.

Gamsjäger, Rainer and Schindler[11] measured solubilities in solutions of sodium perchlorate at 298.2 K. The measurements were in line with those of Dede and Becker at the slightly lower temperature. When perchloric acid is also present, solubilities fit the equation:

$$\log_{10}(s/\text{mol dm}^{-3}) = -0.9918 - 0.0590[\text{Na}^+] + 0.0081[\text{H}^+] - 0.0001[\text{H}^+]^4$$

for ionic strengths to 3 mol dm^{-3}.

There are several measurements of solubilities of hydrogen sulfide in solutions containing hydrochloric acid. Measurements by Gamsjäger and Schindler[12] at 298 K agree with earlier measurements by Kendall[13] to within about 2%. Solubilities in hydrochloric acid, in the presence of sodium chloride and in the presence of other salts, have been investigated by several groups. Although the usual effect of salts is to decrease solubility, the presence of zinc chloride, at least to concentrations of 4.7 mol kg^{-1} of water, increases solubility.

The solubility in solutions of hydrogen iodide at concentrations to 9.21 mol dm^{-3} was measured by Pollitzer[14] and approximately fits the equation

$$\log_{10}(s/\text{mol dm}^{-3}) = -9979 + 0.04494 c/\text{mol dm}^{-3}$$

where c is concentration of hydrogen iodide.

Equilibria between hydrogen sulfide and potassium carbonate solution and sodium carbonate solution were investigated by Litvinenko.[15] He also investigated equilibria with potassium carbonate solution in the presence of gaseous carbon dioxide. These were also investigated by Dryden.[16] Berl and Rittener[17] investigated equilibria involving sodium carbonate and bicarbonate in the presence of carbon dioxide together with hydrogen sulfide. Several equilibria co-exist in such systems as these. Ratios of equilibrium concentrations depend on temperature and on ionic strength. There is no obvious inconsistency between the various sets of measurements but precise comparison is not possible because of different experimental conditions.

Van Krevelen, Hoftijzer and Huntjens[18] were the first to make a detailed study of the equilibria involving hydrogen sulfide, ammonia and water. Other work has since been published and data extend over the temperature range 293 K to 373 K and a wide range of concentrations and pressures of hydrogen sulfide and ammonia. In these systems various chemical species exist in equilibrium in the liquid phase. Ammonium, sulfide and hydrosulfide ions are present together with ammonia and hydrogen sulfide. Studies have also been made on these systems in the presence of ammonium sulfate and ammonium chloride.[18] Measurements on such systems as these can be partially correlated with theoretical models.[18-20]

The situation becomes even more complex when the phase equilibria of the $NH_3 + H_2S + CO_2 + H_2O$ system is investigated. Not only are ammonium, hydrosulfide, sulfide, carbonate and bicarbonate ions present but also carbamate ions and its parent acid, carbamic acid (NH_2COOH). This system was investigated by Badger and Silver,[21] Dryden,[16] van Krevelen et al.[18] and Wilson et al.[22] over the range 298 to 533 K. The different measurements appear to be consistent but again direct comparison is not possible because of different conditions. Some experimental data can, however, be correlated with a simple model for the system which has been put forward by van Krevelen.[18] Wilson et al. have also published a correlation model.[22]

There have been many studies of the behaviour of hydrogen sulfide and its mixtures with carbon dioxide towards aqueous solutions of ethanolamine, diethanolamine and other alkanolamines. This work has been stimulated by the need to find suitable absorbents to remove hydrogen sulfide from effluent gases and fuel gases.

Weakly basic alkanolamines take part in acid-base equilibria in the aqueous phase. The equilibria respond readily to changes of pressure of carbon dioxide and hydrogen sulfide and to changes of temperature.

The ratio of dissolved hydrogen sulfide to alkanolamine increases with partial pressure of hydrogen sulfide but decreases with increase in partial pressure of carbon dioxide. The ratio also decreases with increase in concentration of alkanolamine. Dissolved hydrogen sulfide and carbon dioxide are expelled and the absorbent regenerated when a solution is heated.

Bottoms[23] published work on $H_2S + H_2O +$ diethanolamine in 1931. Muhlbauer and Monaghan[24] published studies on the system $H_2S + CO_2 + H_2O$ + monoethanolamine in 1957. There have been many studies of these and other related systems since these papers were published. Much of this later work has been carried out by Mather and his co-workers. Detailed compilations and evaluation of work on these systems have been published.[4]

There have also been attempts to predict the solubility of hydrogen sulfide, carbon dioxide, and mixtures of the two gases, in alkanolamine solutions from the physical properties of the components. Several sets of measurements have been partially correlated with a model developed by Kent and Eisenberg.[25] Improvements to this model have been described by Deshmukh and Mather[26] and by Dingman et al.[27] Reliable prediction of solubilities from models cannot yet be achieved.

10.3 SOLUBILITY OF HYDROGEN SULFIDE IN NON-AQUEOUS SOLVENTS

10.3.1 Solubilities in non-aromatic hydrocarbons

Many of the measurements of solubility have been made at a partial pressure close to barometric pressure and at temperatures close to 298.2 K. However, a sufficient number of measurements have been made at other temperatures for a general picture of the behaviour of the gas to emerge. Fogg[28] has reviewed the available data.

Numerous phase equilibria studies have been carried out on mixtures of hydrogen sulfide with alkanes. Mixtures with methane have been investigated by Reamer et al.[29] and Kohn and Kurata.[30] Two liquid phases can co-exist at low temperatures. Mixtures with ethane[31] and with propane[32-34] show azeotropic behaviour.

Mixtures with 2-methylpropane,[35] butane[36] and decane[42] have also been studied at pressures above barometric. The variation with pressure of the solubility in decane is shown in Fig. 10.3. However, most of the measurements of solubilities in higher alkanes have been made at pressures close to 1.013 bar.

Makranczy et al.[37] measured solubilities in straight-chain alkanes from pentane to hexadecane. King and Al-Najjar[38] measured solubilities in hexane, octane, decane, dodecane, tetradecane and hexadecane. Measurements by the two groups agree to within a few per cent at a partial pressure of hydrogen sulfide of 1.013 bar at 298.2 K and 313.2 K. Mole fraction solubility increases with carbon number. This pattern is not substantiated by other workers[39-43] (Table 10.1).

Mole fraction solubilities in cyclohexane were measured by Tsiklis and Svetlova[44] from 283 to 313 K and pressures from 0.133 to 1.067 bar. Phase equilibria of hydrogen sulfide with methyl and other alkyl cyclohexanes have

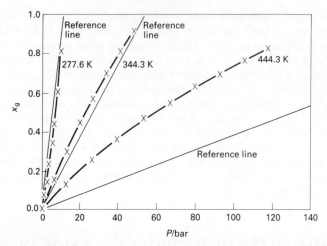

Figure 10.3. The dependence on temperature and partial pressure of the mole fraction solubility of hydrogen sulfide in decane.

also been studied. The mole fraction solubility for a partial pressure of hydrogen sulfide of 1.013 bar tends to increase with length of side chain (Table 10.1).

10.3.2 Solubilities in aromatic hydrocarbons

The measurement by Bell[45] of the solubility in benzene at 293.2 K and a total pressure equal to barometric is consistent with solubility measurements by Gerrard.[41] The solubility reported by Patyi et al.[46] is inconsistent with these other values. Solubility data are given in Table 10.1. There is greater consistency in data for methylbenzene. Measurements by Gerrard[41] and by Bell[45] at barometric pressure are consistent with measurements by Ng et al.[40] at higher pressures.

Solubility in 1,2-dimethylbenzene has been measured by Gerrard[41] at barometric pressure. Huang and Robinson[47] measured the solubility in 1,3-dimethylbenzene at higher total pressures. Approximate values of solubilities at a partial pressure of 1.013 bar can be obtained by extrapolation of these measurements (Table 10.1).

Solubilities in 1,3,5-trimethylbenzene over a pressure and temperature range have been reported by Eakin and DeVaney[48] and by Huang and Robinson.[47] Measurements are consistent, and also agree with the value of mole fraction solubility at 298.2 K and 1.013 bar reported by Patyi et al.[46]

Patyi et al.[46] also measured mole fraction solubilities in ethylbenzene, propylbenzene, and (1-methylethyl)benzene (isopropylbenzene) at 298.2 K and 1.013 bar. They are close to that for 1,3,5-trimethylbenzene under the same conditions.

Table 10.1 Mole fraction solubilities of hydrogen sulfide at a partial pressure of 1.013 bar in hydrocarbons

Solvent	T/K	x_g	Ref.
Pentane	298.2	0.0421	37
Hexane	293.2	0.0341	45
	298.2	0.0429	37
	298.2	0.0412	38
	298.2	0.0372	39
Heptane	298.2	0.0439	37
	298.2	0.035	40[a]
Octane	293.2	0.0440	45
	298.2	0.0451	37
	298.2	0.0437	38
Nonane	298.2	0.0465	37
Decane	298.2	0.0481	37
	298.2	0.0465	38
	298.2	0.039	41[a]
	298.2	0.040	42[b]
Undecane	298.2	0.0496	37
Dodecane	293.2	0.0513	45
	298.2	0.0511	37
	298.2	0.0495	38
Tridecane	298.2	0.0528	37
Tetradecane	298.2	0.0541	37
	298.2	0.0530	38
Pentadecane	298.2	0.0558	37
Hexadecane	298.2	0.0573	37
	298.2	0.0573	38
	300	0.0397	43
Cyclohexane	293.2	0.0338	45
	300	0.038	44
Methylcyclohexane	300	0.034	71[a]
Ethylcyclohexane	300	0.036	47[a]
Propylcyclohexane	300	0.043	47[a]
(1-Methylethyl)cyclohexane	300	0.041	48[a]
Benzene	293.2	0.057	41
	293.2	0.0563	45
	298.2	0.0358	46
Methylbenzene	293.2	0.066	41
	293.2	0.0672	45
1,2-Dimethylbenzene	293.2	0.070	41
1,3,5-Trimethylbenzene	298.2	0.0487	46
Ethylbenzene	298.2	0.042	46
Propylbenzene	298.2	0.052	46
(1-Methylethyl)benzene	298.2	0.053	46

[a] Extrapolated
[b] Interpolated

10.3.3 Solubilities in compounds containing oxygen

Solubilities in methanol,[49–51] ethanol,[41] butanol[51] and octanol[41] have been measured. Mole fraction solubilities for a partial pressure of 1.013 bar increase from ethanol to octanol but, in all cases, solubilities fall below the

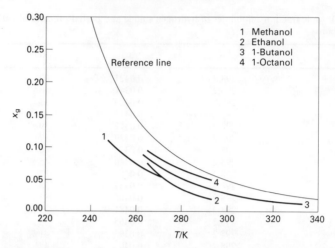

Figure 10.4. Mole fraction solubilities in alkanols of hydrogen sulfide at a partial pressure of 1.013 bar.

reference line. The mole fraction solubility in methanol is less than that in ethanol at temperatures below 273.2 K but greater at higher temperatures (Fig. 10.4.).

The solubility in 1,2-ethanediol (ethylene glycol) measured by Gerrard[41] agrees with that measured by Short et al.[51] The mole fraction solubility in this solvent is less than in ethanol. The single measurement of solubility in 2,2'-[1,2-ethanediylbis(oxy)]bisethanol (tri(ethylene glycol)) reported by Byeseda et al.[52] is consistent with Blake's[53] measurements at 273.2 to 373.2 K and pressures from 1.013 to 11.355 bar. In this case the mole fraction solubility at 1.013 bar lies above the reference line and higher than values for monohydric alcohols under the same conditions. Solubilities in poly(ethylene glycols) have also been reported.[54]

Mole fraction solubility in acetic acid[51] at 298 K and 333 K and a partial pressure of 1.013 bar falls below the reference line. Hexanoic acid[41] at 265 to 293 K behaves in a similar way but the extrapolated value at 298 K is higher than the corresponding value for acetic acid. As in the case of alcohols, increase in chain length increases mole fraction solubility. Mole fraction solubilities in acetic anhydride, 2-propanone, and in oxybisoctane, all lie above the reference line (Fig. 10.5 and Table 10.2).

A number of workers measured solubility in 4-methyl-1,3-dioxolan-2-one (propylene carbonate). Sweeney[55] and also Lenoir et al.[56] measured limiting values of Henry's law constants by gas chromatography. Henry's law constants measured by a volumetric method were reported by Rivas and Prausnitz.[57] Shakhova et al.[58] measured solubilities at partial pressures from 0.27 to 1.013 bar. Issacs et al.[59] measured solubilities at pressures to 49.6 bar. Mole fraction solubilities at a partial pressure of 1.013 bar from measurements by the last three groups fit the equation given by Fogg.[60]

SOLUBILITY OF HYDROGEN SULFIDE

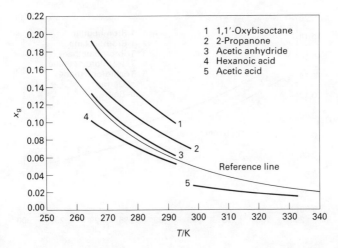

Figure 10.5. Mole fraction solubilities of hydrogen sulfide in solvents containing oxygen (partial pressure of hydrogen sulfide = 1.013 bar).

Table 10.2 Mole fraction solubilities of hydrogen sulfide at a partial pressure of 1.013 bar in solvents containing carbon, hydrogen and oxygen

Solvent	T/K	x_g	Ref.
Methanol	298.2	0.0276	51
Ethanol	293.2	0.019	41
1-Butanol	298.2	0.0315	51
1-Octanol	293.2	0.051	41
1,2-Ethanediol	293.2	0.013	41
2,2'-[1,2-ethanediylbis(oxy)]bisethanol	297.1	0.062	52
Acetic acid	298.2	0.0287	51
Hexanoic acid	293.2	0.054	41
2-Propanone	298.2	0.0698	51
Oxybisoctane	293.2	0.099	41
4-Methyl-1,3-dioxolan-2-one	298.2	0.0380	58
	298.2	0.0396	57

$$\ln x_g = -8.955 + 1704/(T/K)$$

standard deviation in values of $x_g = 0.003$
temperature range 263 to 343 K.

10.3.4 Solubilities in halogenated hydrocarbons

Solubility in 1-bromooctane at 1.007 bar from 265 to 293 K was measured by Gerrard.[41] The mole fraction solubility estimated for a partial pressure of 1.013 bar lies above the reference line (Fig. 10.6).

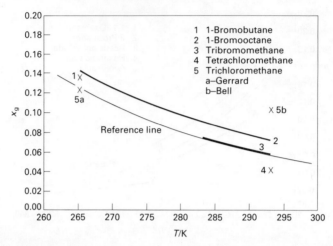

Figure 10.6. Mole fraction solubilities of hydrogen sulfide in chloroalkanes (partial pressure of hydrogen sulfide = 1.013 bar).

Mole fraction solubility in 1-bromobutane at 265.2 K and a partial pressure of 1.013 bar, from Gerrard's measurement, is 0.135. The corresponding value for 1-bromooctane is 0.144. Mole fraction solubility in bromoethane at 293.2 K and a partial pressure of 1.013 bar, from Bell's[45] measurement, is 0.126 compared with a value of 0.074 for 1-bromooctane at this temperature.

The few data available in the literature suggest that the mole fraction solubility is lower if more than one bromine atom is present in an organic solvent molecule. The values of the mole fraction solubility in tribromomethane and 1,2-dibromoethane at a partial pressure of 1.013 bar at 283.2 and 293.2 K from measurements by Gerrard agree with measurements by Bell. Mole fraction solubilities in the two solvents differ by only about 2% but are about 20% lower than values for 1-bromooctane under the same conditions. Bell's value for the mole fraction solubility in 1,1,2,2-tetrabromoethane at 293.2 K and 1.013 bar is about 60% lower than the corresponding value for 1-bromooctane.

Mole fraction solubilities in alkanes substituted by chlorine are greater than those in the corresponding bromo-compounds under the same conditions. Bell[45] and Gerrard[41] measured solubilities in trichloromethane. The mole fraction solubility for a partial pressure of 1.013 bar at 293.13 K is 0.103 compared with 0.059 for the solubility in tribromomethane. The smaller chlorine atoms offer less steric hindrance to hydrogen bond formation with hydrogen sulfide than do the larger bromine atoms. The mole fraction solubility in 1,1,2,2-tetrachloroethane at 293.2 K and 1.013 bar is also greater than in the corresponding bromo-compound.

Mole fraction solubilities tend to decrease with decrease in the number of hydrogen atoms in the molecule of solvent (Table 10.3). This is consistent

Table 10.3 Mole fraction solubilities of hydrogen sulfide at a partial pressure of 1.013 bar in halocarbons

Solvent	T/K	x_g	Ref.
Trichloromethane	293.2	0.103	45
Tetrachloromethane	293.2	0.0419	45
1,2-Dichloroethane	293.2	0.0719	45
1,1,2,2-Tetrachloroethane	293.2	0.0702	45
Pentachloroethane	293.2	0.0514	45
Tetrachloroethene	293.2	0.0372	45
Trichloroethene	293.2	0.0482	45
Tribromomethane	293.2	0.0581	45
	293.2	0.059	41
Bromoethane	293.2	0.126	45
1,2-Dibromoethane	293.2	0.0608	45
	293.2	0.058	41
1,1,2,2-Tetrabromoethane	293.2	0.0446	45
1-Bromooctane	293.2	0.074	41
Chlorobenzene	293.2	0.055	41
	298.2	0.0505	51
Bromobenzene	293.2	0.056	41
Iodobenzene	293.2	0.057	41

with the hypothesis that solubility in halo-alkanes is influenced by hydrogen bond formation.

There is some discrepancy between solubilities in chlorobenzene reported by different authors. Measurements have been made in the temperature range 263 K to 333 K at barometric pressure by Bell,[45] Gerrard,[41] Patyi et al.[46] and Short et al.[51] The most recent measurements were made by Short using modern flow techniques to ensure equilibrium between phases. Gerrard's values lie close to those of Short except at 265.2 and 267.2 K. Values given by Patyi and by Bell are at least 30% lower than Short's values.

There is also discrepancy between values for bromobenzene. Gerrard's[41] measurements are again higher than those of Bell[45] and Patyi et al.[46] when allowance is made for differences in temperature. Mole fraction solubility at 1.013 bar from Gerrard's measurements is close to that for chlorobenzene under the same conditions as given by Short et al.[51]

Gerrard also measured solubilities in iodobenzene. Mole fraction solubility is again close to values for chlorobenzene given by Short. The apparent pattern of solubilities in chlorobenzene, bromobenzene and iodobenzene, based upon the work of Gerrard and Short et al., is shown in Fig. 10.7.

10.3.5 Solubility of hydrogen sulfide in solvents containing nitrogen

Mole fraction solubilities in nitrogen bases are high relative to the reference line. Solubility in benzenamine has been measured by Gerrard[41] at a total pressure of 1.006 bar and temperatures from 265 to 293 K. Bancroft and

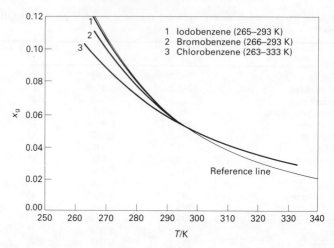

Figure 10.7. Mole fraction solubilities of hydrogen sulfide in halobenzenes (partial pressure of hydrogen sulfide = 1.013 bar).

Beldon[61] measured the solubility at pressures from 0.14 bar to 1.55 bar at 295 K. The value is about 7% less than an extrapolated value from Gerrard's measurements. Patyi et al.[46] measured solubility at 1.013 bar and 298.2 K. The value is about 24% lower than that from Gerrard. Lenoir et al.[56] published a limiting value of Henry's law constant measured by chromatography at 298.2 K. The corresponding value of the mole fraction solubility at 1.013 bar is about 24% greater than Gerrard's value. Gerrard's measurements are self-consistent over a temperature range and fit the equation

$$\ln x_g = -115.33 + 6496.9/(T/K) + 15.906\ln(T/K)$$

standard deviation in $x_g = 0.001$
temperature range 265 to 293 K.

Gerrard[41] measured solubility in N,N-dimethylbenzenamine at 1.013 bar at temperatures from 278 to 293 K. The measurement by Patyi et al.[46] at 298.2 K is again inconsistent with measurements by Gerrard. Patyi also measured solubilities in methylbenzenamine and in N-ethylbenzenamine.

Gerrard[41] measured solubilities in pyridine and in quinoline at 265 K to 293 K.

Several groups[56,57,62–64] have measured solubilities of hydrogen sulfide in 1-methyl-2-pyrrolidinone. Work by Yarym-Agaev et al.[62] extended from 273 to 399 K and 1.013 to 20.27 bar (Fig. 10.8). Values of the solubility at 298.2 K and a partial pressure of 1.013 bar from the different sources are in poor agreement.

Gerrard[41] measured the solubility in nitrobenzene at barometric pressure

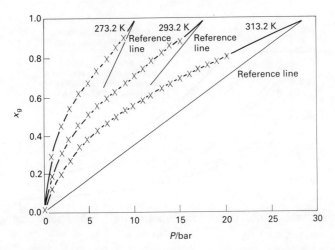

Figure 10.8. Variation with pressure and temperature of the mole fraction solubility of hydrogen sulfide in 1-methyl-2-pyrrolidinone.

Table 10.4 Mole fraction solubilities of hydrogen sulfide at a partial pressure of 1.013 bar in solvents containing nitrogen or sulfur

Solvent	T/K	x_g	Ref.
Acetonitrile	298.2	0.0476	39
Benzonitrile	293.2	0.081	41
Benzenamine	293.2	0.061	41
	295.2	0.053	61
	298.2	0.043	46
	298.2	0.054	41[a]
N-Methylbenzenamine	298.2	0.045	46
N,N-Dimethylbenzenamine	293.2	0.083	41
	298.2	0.056	46
Pyridine	293.2	0.093	41
Quinoline	293.2	0.089	41
N,N-Dimethylformamide	293.2	0.131	41
	297.1	0.097	52
	298.2	0.116	39
	298.2	0.109	66
1-Methyl-2-pyrrolidinone	298.2	0.133	57
	298.2	0.156	62
	298.2	0.131	63
Nitrobenzene	293.2	0.053	41
Tetrahydrothiophene 1,1-dioxide	297.1	0.069	52
	298.2	0.053	57[a]

[a] Extrapolated.

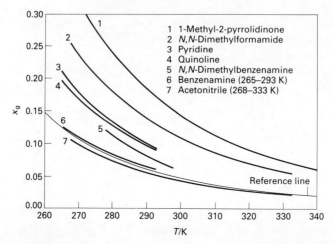

Figure 10.9. Mole fraction solubilities, at a partial pressure of 1.013 bar, of hydrogen sulfide in organic compounds of nitrogen.

from 265 to 293 K. The values are likely to be more reliable than those from chromatographic measurements by Lenoir et al.[56] and Devyatykh et al.[65]

Solubility in acetonitrile at various temperatures was measured by Hayduk and Pahlevanzadeh[39] and in benzonitrile by Gerrard.[41]

At least four groups[39,41,52,66] have published values of the solubility in N,N-dimethylformamide. Mole fraction solubilities at a partial pressure of 1.013 bar from measurements by Gerrard[41] and Hayduk and Pahlevanzadeh[39] fit the equation

$$\ln x_g = 8.8031 + 1341.2/(T/K) - 2.7110 \ln(T/K)$$

standard deviation in $x_g = 0.003$.
temperature range 265 to 333 K.

The overall pattern of solubilities in organic nitrogen compounds is shown in Fig. 10.9 and Table 10.4.

10.3.6 Solubility of hydrogen sulfide in solvents containing phosphorus or sulfur

Five groups[54–56,67,68] have measured solubility in tributyl phosphate. Limiting values of Henry's law constant were published or may be estimated from the data which were given. Values from different sources are given in Table 10.5.

Lenoir et al.[56] also reported limiting values of Henry's law constants for other alkyl phosphates. These were measured by gas chromatography so are subject to some uncertainty.

Table 10.5 Limiting values at zero pressure of Henry's law constant for hydrogen sulfide in tributyl phosphate

T/K	Henry's law constant/bar	Ref.
325.2	9.19	56
298.2	4.14	55
323.2	8.27	55
298.2	4.03	67
293.2	3.54	54
293.2	3.38	68

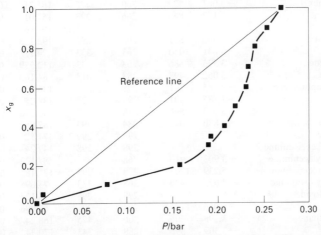

Figure 10.10. Variation with pressure of the mole fraction solubility of hydrogen sulfide in carbon disulfide at 190 K.

Rivas and Prausnitz[57] measured the solubility in tetrahydro-1,1-thiophene at an unspecified partial pressure over the range 303 to 373 K and reported values of Henry's law constant (Table 10.4).

The total pressures of mixtures of hydrogen sulfide and carbon disulfide were measured at 153 to 213 K by Gattow and Krebs.[69] The vapour pressure of carbon disulfide is low under the conditions of the experiment and the total pressure is close to the partial pressure of hydrogen sulfide (Fig. 10.10).

Hydrogen sulfide dissolves in molten sulfur and exists in equilibrium with unstable polyhydrogen sulfides. The system has been investigated by Fanelli[70] at a total pressure equal to barometric. Dissolution of gas increases with temperature until it reaches a maximum at about 644 K.

10.4 TABLES SUMMARISING THE SOLUBILITY OF HYDROGEN SULFIDE

Data available from the literature have been used to prepare the following tables which summarise the solubility behaviour of hydrogen sulfide.

Table 10.6 Mole fraction solubilities of hydrogen sulfide in various solvents at a partial pressure of hydrogen sulfide of 1.013 bar in alphabetical order

	A	B	Temp. range Low	Temp. range High	M_r of solvent	x_g at 298 K[a]
Acetic acid	−4.205	794	298	333	60.042	0.029
Acetone	−3.879	811.8	263	298	58.075	0.070
Acetonitrile	−4.507	946	268	333	41.053	0.048
Aniline	−4.758	1026	265	298	93.13	0.048
Benzenamine	−4.758	1026	265	298	93.13	0.048
Benzene	−3.935	792.7	278	293	78.114	0.053
1,1′-Bicyclohexyl	−3.081	429	300	475	166.31	0.023
Bromobenzene	−4.127	845	266	293	157.01	0.051
Bromoethane	b	b	293.15	293.15	108.966	0.126
1-Bromooctane	−3.905	811	265	293	193.128	0.065
1-Butanol	−5.041	1050	263	333	74.118	0.030
Chlorobenzene	−3.593	686	263	333	112.559	0.051
Cyclohexane	−3.052	492	283	313	84.162	0.040
Decahydronaphthalene	−2.785	391	298	323	138.25	0.034
Decalin	−2.785	391	298	323	138.25	0.034
Decane	−3.340	598	288	343	142.286	0.046
1,2-Dibromoethane	−4.880	1070	284	293	187.862	0.051
Diethyl ether	b	b	299.2	299.2	74.12	0.100
N,N-Diethylbenzenamine	b	b	298	298	149.24	0.060
N,N-Dimethylacetamide	−3.903	910	268	333	87.122	0.146
N,N-Dimethylbenzenamine	−5.239	1203	278	298	121.18	0.063
N,N-Dimethylformamide	−4.027	920	268	333	73.095	0.116
Dimethylsulfoxide	b	b	298	298	78.124	0.092
1,4-Dioxane	−5.066	1185	283	293	88.096	0.081
Diphenylmethane	−3.198	506	300	475	168.24	0.032
Dodecane	−3.309	598	288	343	170.34	0.050
Ethyl acetate	−4.077	898	268	333	88.106	0.087
Ethanol	−7.174	1603	265	293	46.069	0.016
1,2-Ethanediol	−4.898	899	263	333	60.058	0.013
N-Ethylbenzenamine	b	b	298	298	121.18	0.055
Ethylbenzene	b	b	298.15	298.15	106.17	0.042
Hexadecane	−3.080	532	293	475	226.448	0.051
Hexane	−3.398	600	288	303	86.178	0.041
Isopropylbenzene	b	b	298.15	298.15	120.5	0.053
Mesitylene	−3.495	648	298	478	120.19	0.048
Methyl salicylate	−4.304	905	265	293	152.134	0.054
N-Methylbenzenamine	b	b	298	298	107.16	0.045
Methylbenzene	−3.811	768	265	478	92.14	0.058
1,1′-Methylenebisbenzene	−3.198	506	300	475	168.24	0.032
(1-Methylethyl)benzene	b	b	298.15	298.15	120.95	0.053
1-Methylnaphthalene	−3.193	506	300	475	142.2	0.032
1-Methyl-2-pyrrolidinone	−3.986	943	260	399	99.13	0.150
Octane	−3.428	617	288	303	114.232	0.044
1,1′-Oxybisethane	b	b	299.2	299.2	74.12	0.100
1,1′-Oxybisoctane	−3.755	805	265	293	242.442	0.088
Oxybispropanol	−5.045	1113	298	343	134.16	0.049
Pentane	−3.313	576	278	344	72.146	0.042
2-Propanone	−3.879	811.8	263	298	58.075	0.070
Propylbenzene	b	b	298.15	298.15	120.19	0.052
Sulfinylbismethane	b	b	298	298	78.124	0.092
1,1,2,2-Tetrabromoethane	b	b	293.15	293.15	345.654	0.045

Table 10.6 (*continued*)

	A	B	Temp. range Low	Temp. range High	M_r of solvent	x_g at 298 K[a]
1,1,2,2-Tetrachloroethane	b	b	293.15	293.15	167.85	0.070
Tetradecane	−3.305	605	288	343	198.394	0.053
Tetrahydrofuran	−3.553	763	263	298	72.102	0.101
Tetrahydrothiophene, 1,1-dioxide	−3.899	783	303	373	120.156	0.053
Tribromomethane	−4.451	943	283	293	252.731	0.052
Trichloromethane	−1.881	262	265	293	119.378	0.099
Tri(ethylene glycol)	−3.689	743	273	372	150.154	0.064
1,3,5-Trimethylbenzene	−3.495	648	298	478	120.19	0.048
Tripropyl phosphate	−4.281	1092	298	343	224.217	0.241
Water	b	b	298.15	298.15	18.015	0.00183

[a] Mole fraction solubilities, x_g, estimated for 298.15 K except where indicated.
[b] Solubilities measured at one temperature only, except in the case of water. The value for this temperature is given.
Experimental measurements have been fitted to equations of the form:

$$\log_{10} x_g = A + B/(T/K)$$

Values of A and B are given.

Table 10.7 Mole fraction solubilities of hydrogen sulfide in various solvents at a partial pressure of hydrogen sulfide of 1.013 bar in increasing order

	A	B	Temp. range Low	Temp. range High	M_r of solvent	x_g at 298 K[a]
Water	b	b	298.15	298.15	18.015	0.002
1,2-Ethanediol	−4.898	899	263	333	60.058	0.013
Ethanol	−7.174	1603	265	293	46.069	0.016
1,1′-Bicyclohexyl	−3.081	429	300	475	166.31	0.023
Acetic acid	−4.205	794	298	333	60.042	0.029
Butanol	−5.041	1050	263	333	74.118	0.030
Acetonitrile	b	b	298	298	41.053	0.031
1,1′-Methylenebisbenzene (diphenylmethane)	−3.198	506	300	475	168.24	0.032
1-Methylnaphthalene	−3.193	506	300	475	142.2	0.032
Decahydronaphthalene (decalin)	−2.785	391.1	298	323	138.25	0.034
Cyclohexane	−3.052	492	283	313	84.162	0.040
Hexane	−3.398	600	288	303	86.178	0.041
Pentane	−3.313	576	278	344	72.146	0.042
Ethylbenzene	b	b	298.15	298.15	106.17	0.042
Octane	−3.428	617	288	303	114.232	0.044
1,1,2,2-Tetrabromoethane	b	b	293.15	293.15	345.654	0.045
N-Methylbenzenamine	b	b	298	298	107.16	0.045
Decane	−3.340	598	288	343	142.286	0.046
1,3,5-Trimethylbenzene (mesitylene)	−3.495	648	298	478	120.19	0.048

Table 10.7 (*continued*)

	A	B	Temp. range Low	Temp. range High	M_r of solvent	x_g at 298 K[a]
Benzenamine (aniline)	−4.758	1026	265.15	298.15	93.13	0.048
Oxybispropanol	−5.045	1113	298	343	134.16	0.049
Dodecane	−3.309	598	288	343	170.34	0.050
Hexadecane	−3.080	532	293	475	226.448	0.051
Chlorobenzene	−3.593	686	263	333	112.559	0.051
Bromobenzene	−4.127	845	266	293	157.01	0.051
1,2-Dibromoethane	−4.880	1070	284	293	187.862	0.051
Tribromomethane	−4.451	943	283	293	252.731	0.052
Propylbenzene	[b]	[b]	298.15	298.15	120.19	0.052
Benzene	−3.935	793	278	293	78.114	0.053
Tetradecane	−3.305	605	288	343	198.394	0.053
(1-Methylethyl)-benzene (isopropylbenzene)	[b]	[b]	298.15	298.15	120.95	0.053
Tetrahydrothiophene, 1,1-dioxide	−3.899	783	303	373	120.156	0.053
Methyl salicylate	−4.304	905	265	293	152.134	0.054
N-Ethylbenzenamine	[b]	[b]	298	298	121.18	0.055
Methylbenzene	−3.811	768	265	478	92.14	0.058
N,N-Diethylbenzenamine	[b]	[b]	298	298	149.24	0.060
N,N-Dimethylbenzenamine	−5.239	1203	278	298	121.18	0.063
Tri(ethylene glycol)	−3.689	743	273	372	150.154	0.064
1-Bromooctane	−3.905	811	265	293	193.128	0.065
2-Propanone (acetone)	−3.879	812	263	298	58.075	0.070
1,1,2,2-Tetrachloroethane	[b]	[b]	293.15	293.15	167.85	0.070
1,4-Dioxane	−5.066	1185	283	293	88.096	0.081
1,1′-Oxybisoctane	−3.755	805	265	293	242.442	0.088
Sulfinylbismethane (dimethylsulfoxide)	[b]	[b]	298	298	78.124	0.092
Trichloromethane	−1.8809	261.98	265	293	119.378	0.099
1,1′-Oxybisethane (diethyl ether)	[b]	[b]	299.2	299.2	74.12	0.100
Tetrahydrofuran	−3.553	763	263	298	72.102	0.101
N,N-Dimethylformamide (DMF)	−4.162	960	265	298	73.095	0.114
Bromoethane	[b]	[b]	293	293	108.966	0.126
1-Methyl-2-pyrrolidinone	−3.986	943	260	399	99.13	0.150
Tripropyl phosphate	−4.281	1092	298	343	224.217	0.241

[a] Mole fraction solubilities, x_g, estimated for 298.15 K except where indicated.
[b] Solubilities measured at one temperature only, except in the case of water. The value for this temperature is given.

Experimental measurements have been fitted to equations of the form:

$$\log_{10} x_g = A + B/(T/K)$$

Values of A and B are given.

SOLUBILITY OF HYDROGEN SULFIDE

Table 10.8 Weight ratio solubilities of hydrogen sulfide in various solvents at a partial pressure of hydrogen sulfide of 1.013 bar in increasing order

	M_r of solvent	Temp. range of measurements/K Low	High	Wt. of H_2S at 298.15 K/100 g of solvent[a]
Water[b]	18.015	298.15	298.15	0.097
1,1,2,2-Tetrabromoethane[b]	345.654	293.15	293.15	0.46
1,1'-Bicyclohexyl	166.310	300	475	0.48
1,1'-Methylenebisbenzene (diphenylmethane)	168.240	300	475	0.66
Tribromomethane	252.731	283.15	293.15	0.73
1,2-Ethanediol	60.058	263	333	0.75
1-Methylnaphthalene	142.200	300	475	0.79
Hexadecane	226.448	293	475	0.80
Decahydronaphthalene (decalin)	138.250	298	323	0.86
Tetradecane	198.394	288	343	0.96
1,2-Dibromoethane	187.862	284.15	293.15	0.98
Dodecane	170.340	288	343	1.05
Decane	142.286	288	343	1.16
Bromobenzene	157.010	266.15	293.15	1.17
Ethanol	46.069	265	293	1.20
1-Bromooctane	193.128	265.15	293.15	1.23
Methyl salicylate	152.134	265.15	293.15	1.28
Oxybispropanol	134.160	298	343	1.30
1,1'-Oxybisoctane	242.442	265.15	293.15	1.36
Octane	114.232	288	303	1.37
Ethylbenzene[b]	106.170	298.15	298.15	1.41
1,3,5-Trimethylbenzene	120.190	298	478	1.42
Butanol	74.118	263	333	1.43
N,N-Diethylbenzenamine[b]	149.240	298	298	1.46
N-Methylbenzenamine[b]	107.160	298	298	1.50
1,1,2,2-Tetrachloroethane[b]	167.850	293.15	293.15	1.53
Propylbenzene[b]	120.190	298.15	298.15	1.56
(1-Methylethyl)benzene[b]	120.950	298.15	298.15	1.58
Tetrahydrothiophene, 1,1-dioxide	120.156	303	373	1.60
Chlorobenzene	112.559	263.15	333.15	1.63
N-Ethylbenzenamine[b]	121.180	298	298	1.64
Cyclohexane	84.162	283	313	1.67
Acetic acid	60.042	298	333	1.68
Hexane	86.178	288	303	1.70
Benzenamine (aniline)	93.130	265.15	298.15	1.85
N,N-Dimethylbenzenamine	121.180	278	298	1.87
Pentane	72.146	278	344	2.05
Methylbenzene	92.140	265	478	2.29
Benzene	78.114	278.15	293.15	2.44
Acetonitrile[b]	41.053	298	298	2.66
Trichloromethane	119.378	265.15	293.15	3.15
1,4-Dioxane	88.096	283.15	293.15	3.42
2-Propanone (acetone)	58.075	263	298	4.40
Sulfinylbismethane[b]	78.124	298	298	4.42
Bromoethane[b]	108.966	293.15	293.15	4.51
Tripropyl phosphate	224.217	298	343	4.82
1,1'-Oxybisethane[b]	74.120	299.20	299.20	5.11

Table 10.8 (*continued*)

	M_r of solvent	Temp. range of measurements/K Low	High	Wt. of H$_2$S at 298.15 K/100 g of solvent[a]
Tetrahydrofuran	72.102	263	298	5.33
N,N-Dimethylformamide (DMF)	73.095	265.15	298.15	6.01
1-Methyl-2-pyrrolidinone	99.130	260	373	6.08[a]

[a] Solubilities expressed as the weight of hydrogen sulfide dissolved in 100 g of solvent have been estimated for 298.15 K by fitting experimental data to equations of the form:

$$\log_{10} x_g = A + B/(T/K).$$

[b] Solubilities measured at one temperature only, except in the case of water. The value for this temperature is given.

Table 10.6 Mole fraction solubilities for a partial pressure of 1.013 bar at or near to 298 K with solvents arranged in alphabetical order.

Table 10.7 Mole fraction solubilities for a partial pressure of 1.013 bar at or near to 298 K arranged in order.

Table 10.8 Weight ratio solubilities for a partial pressure of 1.013 bar at or near to 298 K arranged in order.

In many cases experimental values correspond to a given total pressure of hydrogen sulfide and solvent. Approximations discussed earlier have been made to convert to solubilities corresponding to a partial pressure of gas of 1.013 bar. Details of original measurements on the solubility of hydrogen sulfide are given in Solubility Data Series Volume 32.[4]

Where possible, values of the mole fraction solubility for a partial pressure of 1.013 bar, x_g, have been fitted to equations of the type:

$$\log_{10} x_g = A + B/(T/K)$$

Values for 298 K have then been calculated from such equations. Estimated values of A and B are given in the tables but the corresponding equations may not be reliable outside the temperature range of the original experimental measurements.

Estimations of precision and reliability have not been included as the intention is to present a general overall behaviour of the gas. In general, the more volatile the solvent the less reliable are the corrections for the solvent vapour pressure.

REFERENCES

1. Neuburg, H.J.; Atherley, J.F.; Walker, L.G. *Girdler-Sulfide Process Physical Properties, AECL-5702*, Atomic Energy of Canada Ltd., 1977.

2. Cox, E.R. *Ind. Eng. Chem.* 1936, 28, 613.
3. Ref. 1, Page 31.
4. *Solubility Data Series Vol. 32, Hydrogen Sulfide, Deuterium Sulfide and Hydrogen Selenide*, ed. P.G.T. Fogg and C.L. Young, Pergamon, Oxford, 1988
5. Mellor, J.W., *A Comprehensive Treatise on Inorganic and Theoretical Chemistry*, Longmans, London, 1922–37.
6. Selleck, F.T.; Carmichael, L.T.; Sage, B.H. *Ind. Eng. Chem.* 1952, 44, 2219.
7. Burgess, M.P.; Germann, R.P. *Am. Inst. Chem. Engnrs. J.* 1969, 15, 273.
8. Fogg, P.G.T. Ref. 4, Page 1.
9. Clever, H.L. *J. Chem. Eng. Data* 1983, 23, 401.
10. Dede, L.; Becker, Th. *Zeit. Anorg. Allgem. Chem.* 1926, 152, 185.
11. Gamsjäger, H.; Rainer, W.; Schindler, P. *Monatsh. Chem.* 1967, 98, 1782.
12. Gamsjäger, H.; Schindler, P. *Helv. Chim. Acta* 1969, 52, 1395.
13. Kendall, J.; Andrews, J.C. *J. Amer. Chem. Soc.* 1921, 43, 1545.
14. Pollitzer, F. *Zeit. Anorg. Chem.* 1909, 64, 121.
15. Litvinenko, M.S. *Zh. Prikl. Khim.* 1952, 25, 516; *J. Appl. Chem. (USSR)* 1952, 25, 579.
16. Dryden, I.G.C. *J. Soc. Chem. Ind.* 1947, 66, 59.
17. Berl, A.; Rittener, A. *Zeit. Angew. Chem.* 1907, 20, 1637.
18. van Krevelen, D.W.; Hoftijzer, P.J.; Huntjens, F.J. *Recl. Trav. Chim. Pays-Bas* 1949, 68, 191.
19. Edwards, T.J.; Newman, J.; Prausnitz, J.M. *Amer. Inst. Chem. Eng. J.* 1975, 21(2), 248.
20. Beutier, D.; Renon, H. *Ind. Eng. Chem. Proc. Des. Dev.* 1978, 17, 220.
21. Badger, E.H.M.; Silver, L. *J. Soc. Chem. Ind.* 1938, 57, 110.
22. Wilson, G.M.; Gillespie, P.C.; Owens, J.L. *Proc. 64th Ann. Conv. Gas Processors Association*, 1985, 282.
23. Bottoms, R.R. *Ind. Eng. Chem.* 1931, 23, 501.
24. Muhlbauer, H.G.; Monaghan, P.R. *Oil and Gas J.* 1957, 55(17), 139.
25. Kent, R.L.; Eisenberg, B. *Hydrocarbon Processing* 1976, 55(2), 87.
26. Deshmukh, R.D.; Mather, A.E. *Chem. Eng. Sci.* 1981, 36, 355.
27. Dingman, J.C.; Jackson, J.L.; Moore, T.F. *Proc. 62nd Ann. Conv. Gas Processors Association*, 1983, 256.
28. Fogg, P.G.T. Ref.4, page 166.
29. Reamer, H.H.; Sage, B.H.; Lacey, W.N. *Ind. Eng. Chem.* 1951, 43, 976.
30. Kohn, J.P.; Kurata, F. *Am. Inst. Chem. Engnrs. J.* 1958, 4, 211.
31. Kay, W.B.; Brice, D.B. *Ind. Eng. Chem.* 1953, 45, 615.
32. Kay, W.B.; Rambosek, G.M. *Ind. Eng. Chem.* 1953, 45, 221.
33. Brewer, F.; Rodewald, N.; Kurata, F. *Am. Inst. Chem. Engnrs. J.* 1961, 7, 13.
34. Gilliland, E.R.; Scheeline, H.W. *Ind. Eng. Chem.* 1940, 32, 48.
35. Besserer, G.J.; Robinson, D.B. *J. Chem. Eng. Japan* 1975, 8, 11.
36. Robinson, D.B.; Hughes, R.E.; Sandercock, J.A.W. *Can. J. Chem. Eng.* 1964, 42 (4), 143.
37. Makranczy, J.; Megyery-Balog, K.; Rusz, L.; Patyi, L. *Hung. J. Ind. Chem.* 1976, 4, 269.
38. King, M.B.; Al-Najjar, H. *Chem. Eng. Sci.* 1977, 32, 1241.
39. Hayduk, W.; Pahlevanzadeh, H. *Can. J. Chem. Eng.* 1987, 65, 299.
40. Ng, H.-J.; Kalra, H.; Robinson, D.B.; Kubota, H. *J. Chem. Eng. Data* 1980, 25, 51.
41. Gerrard, W. *J. Appl. Chem. Biotechnol.* 1972, 22, 623.
42. Reamer, H.H.; Selleck, F.T.; Sage, B.H.; Lacey, W.N. *Ind. Eng. Chem.* 1953, 45, 1810.
43. Tremper, K.K.; Prausnitz, J.M. *J. Chem. Engng. Data* 1976, 21, 295.
44. Tsiklis, D.S.; Svetlova, G.M. *Zh. Fiz. Khim.* 1958, 32, 1476.
45. Bell, R.P. *J. Chem. Soc.* 1931, 1371.
46. Patyi, L.; Furmer, I.E.; Makranczy, J.; Sadilenko, A.S.; Stepanova, Z.G.; Berengarten, M.G. *Zh. Prikl. Khim.* 1978, 51, 1296; *J. Appl. Chem. (USSR)* 1978, 51, 1240.
47. Huang, S. S.-S.; Robinson, D.B. *J. Chem. Eng. Data* 1985, 30, 154.
48. Eakin, B.E.; DeVaney, W.E. *Am. Inst. Chem. Engnrs. Symp. Ser.* 1974, No. 140, 70, 80.
49. Bexdel, L.S.; Teodorovich, V.P. *Gazovaya Prom.* 1958, No.8, 38.
50. Yorizane, M.; Sadamoto, S.; Masuoka, H.; Eto, Y. *Kogyo Kagaku Zasshi* 1969, 72, 2174.
51. Short, I.; Sahgal, A.; Hayduk, W. *J. Chem. Eng. Data* 1983, 28, 63.
52. Byeseda, J.J.; Deetz, J.A.; Manning, W.P. *Proc. Lawrance Reid Gas Cond. Conf.* 1985.
53. Blake, R.J. *Oil and Gas J.* 1967, 65(2), 105.

54. Härtel, G.H. *J. Chem. Eng. Data* 1985, 30, 57.
55. Sweeney, C.W. *Chromatographia* 1984, 18, 663.
56. Lenoir, J.-Y.; Renault, P.; Renon, H. *J. Chem. Eng. Data* 1971, 16, 340.
57. Rivas, O.R.; Prausnitz, J.M. *Am. Inst. Chem. Engnrs. J.* 1979, 25, 975.
58. Shakhova, S.F.; Bondareva, T.I. *Khim. Prom.* 1966, (10), 753.
59. Isaacs, E.E.; Otto, F.D.; Mather, A.E. *Can. J. Chem. Eng.* 1977, 55, 751.
60. Fogg, P.G.T. Ref. 4, Page 176.
61. Bancroft, W.D.; Belden, B.C. *J. Phys. Chem.* 1930, 34, 2123.
62. Yarym-Agaev, N.L.; Matvienko, V.G.; Povalyaeva, N.V. *Zh. Prikl. Khim.* 1980, 53, 2456; *J. Appl. Chem. USSR* 1980, 53, 1810.
63. Murrieta-Guevara, F.; Rodriguez, A.T. *J. Chem. Eng. Data* 1984, 29, 456.
64. Rivas, O.R.; Prausnitz, J.M. *Ind. Eng. Chem. Fundam.* 1979, 18, 289.
65. Devyatykh, G.G.; Exheleva, A.E.; Zorin, A.D.; Zueva, M.V. *Zh. Neorgan. Khim.* 1963, 8(6), 1307. *Russ. J. Inorg. Chem.* 1963, 8, 678.
66. DuPont de Nemours and Co. (Inc) *Chem. Eng. News*, 1955, 33, 2366.
67. Vei, D.; Furmer, I.E.; Sadilenko, A.S.; Efimova, N.M.; Stepanova, Z.G.; Gracheva, N.V. *Gazov. Prom.-st.* 1975, 7(7), 47.
68. Sergienko, I.D.; Kosyakov, N.E.; Yushko, V.L.; Khokhlov, S.F.; Pushkin, A.G. *Vop. Khim. Tekhnol.* 1973, 29, 57.
69. Gattow, G.; Krebs, B. *Zeit. Anorg. Allgem. Chem.* 1963, 325, 15.
70. Fanelli, R. *Ind. Eng. Chem.* 1949, 41, 2131.

Chapter 11
SOLUBILITY OF CARBON DIOXIDE

11.1 GENERAL BEHAVIOUR

Carbon dioxide has the following physical properties:

Sublimation point at 1.013 bar = 194.4 K
Triple point 5.176 bar; 216.6 K
Critical temperature = 304.3 K
Critical pressure = 73.9 bar
Critical volume = 0.0940 dm^3 mol^{-1}
Relative molecular mass = 44.009
Density of gas at 273.15 K, 1.977 bar = 1.977 g dm^{-3}

The vapour pressure of liquid carbon dioxide is 57.23 bar at 294.26 K. The following equation is given in the International Critical Tables for the vapour pressure of liquid carbon dioxide between 216.6 K and 304.3 K:

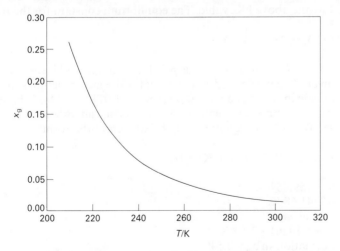

Figure 11.1. Reference line for carbon dioxide at a partial pressure of 1.013 bar.

$$\log_{10}(P/\text{atm}) = 1.8630096 - (AK/T)[3.0067 - 9.03453 \times 10^{-3}A$$
$$+ 2.37353 \times 10^{-4} \times A^2 - 3.7788 \times 10^{-6} \times A^3 + 3.27304 \times 10^{-8} \times A^4$$
$$- 1.11383 \times 10^{-10} \times A^5]$$

where $A = (T_c - T)/\text{K}$; T_c = critical temperature.

The reference line for solubilities at a partial pressure of 1.013 bar, based upon the vapour pressure equation, is shown in Fig. 11.1.

The following simple Antoine type equation approximately fits the vapour pressures in the range 273.2 K to 303.2 K:

$$\log_{10}(P/\text{atm}) = 5.696 - 1519/[(T/\text{K}) + 91.9]$$

11.2 SOLUBILITY OF CARBON DIOXIDE IN AQUEOUS SOLVENTS

11.2.1 Solubility of carbon dioxide in water

When carbon dioxide dissolves in water it exists in ionic equilibrium with hydrogen ions and bicarbonate ions

$$CO_2 + H_2O \rightleftharpoons H^+ + HCO_3^-$$

$pK_a = 6.352$ at 298.15 K[1]

Only about 0.35% of CO_2 exists in the form of carbonic acid at 298.2 K and a partial pressure of CO_2 of 1.013 bar. Carbonic acid is a stronger acid than is indicated by the above pK_a value. The equilibrium constant for the reaction:

$$H_2CO_3 \rightleftharpoons H^+ + HCO_3^-$$

is 1.32×10^{-4} at 298.2 K and the true pK_a for carbonic acid is therefore 3.88.[2]

Wilhelm et al.[3] have derived an equation for the mole fraction solubility of carbon dioxide in water at a partial pressure of 1.013 bar. This is based upon 17 data points from measurements by Morrison and Billett,[4] Murray and Riley,[5] and Weiss.[6] The equation may be written in the form:

$$\ln x_g = A + B/T + C\ln(T/\text{K}) + DT$$

$A = -159.854$
$B = 8741.68$ K
$C = 21.6694$
$D = -1.10261 \times 10^{-3}$ K^{-1}
standard deviation in $x_g = 0.54\%$
temperature range 273 to 353 K.

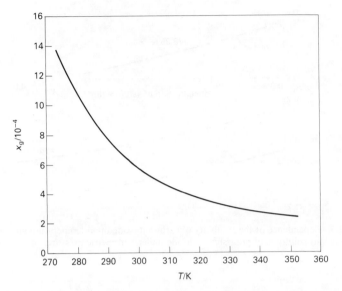

Figure 11.2. The mole fraction solubility in water of carbon dioxide at a partial pressure of 1.013 bar.

The equation is plotted in Fig. 11.2.

Above about 450 K the mole fraction solubility for a partial pressure of 1.013 bar increases with rise in temperature. Several other gases, such as hydrogen sulfide, oxygen, nitrogen, hydrogen and helium, behave in a similar way. The mole fraction solubilities in water at a partial pressure of 1.013 bar have minimum values at temperatures which are characteristic of the gas.[7,8]

11.2.2 Solubility of carbon dioxide in water containing other solutes

There have been many studies of the variation of solubility of carbon dioxide in water with concentration of added strong electrolyte.

Markham and Kobe[9] measured the solubility of carbon dioxide in aqueous solutions of various salts at pressures close to barometric pressure. Temperatures were from 273.4 to 313.2 K and concentrations of salt from 0.1 mol dm^{-3} to 8 mol dm^{-3}. Solubilities were reported as Bunsen coefficients, α, which, in this case, were taken to be the volume of gas, corrected to a partial pressure of 1.013 bar and a temperature of 273.2 K, dissolved by one volume of solution. They were also given as the corrected volume of gas dissolved by a solution containing one gram of water.

Values of $\ln \alpha$ are plotted against the molar concentration of the salt in Figs 11.3 and 11.4. The decrease in solubility differed from salt to salt but it was demonstrated that the effects were approximately an additive function of ion concentration. The solubilities in solutions of sodium nitrate are close to

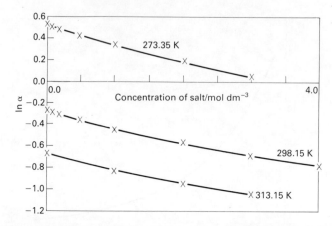

Figure 11.3. Dependence of the solubility of carbon dioxide upon temperature and concentration of aqueous solutions of potassium chloride under barometric pressure.

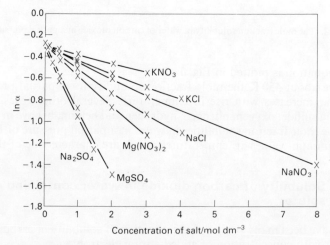

Figure 11.4. The solubility of carbon dioxide at barometric pressure in solutions of salts at 298.15 K.

those estimated by subtracting the effect of potassium chloride from the sum of the effects of sodium chloride and potassium nitrate. For a concentration of 0.1 mol dm^{-3} the predicted value of the Bunsen coefficient is 0.7448 and the experimental value 0.7415. In the case of a concentration of 1 mol dm^{-3} the predicted and experimental values are 0.6364 and 0.6352, respectively. For a concentration of 3 mol dm^{-3} the two values are 0.4705 and 0.4659, respectively.

The slope of a plot of ln α against the concentration of salt may be equated with a Sechenov coefficient, k, as defined by the equation

$$\ln\alpha = \ln\alpha_0 + km$$

where α_0 is the Bunsen coefficient for dissolution in pure water and m is the concentration of salt in units of mol dm^{-3}. As well as varying from salt to salt the slopes, and hence the Sechenov coefficients, vary with concentration.

Equilibria in the system $CO_2 + H_2O + NH_3$ have been investigated by numerous workers. Important tables of measurements were published by Badger and Silver,[10] Dryden[11] and van Krevelen et al.[12] There have been attempts to use theoretical models of the behaviour of electrolyte solutions to explain observed behaviour of the system and to predict the behaviour under conditions in which experimental data are not available.[12,13]

The following equilibria exist in aqueous solutions of ammonia and carbon dioxide:

$$NH_3 + CO_2 + H_2O \rightleftharpoons NH_4^+ + HCO_3^-$$
$$NH_3 + HCO_3^- \rightleftharpoons NH_2COO^- + H_2O$$
$$NH_3 + HCO_3^- \rightleftharpoons NH_4^+ + CO_3^{2-}$$

The proportions of different species in solution depend on temperature, total pressure of gas above the solution and relative proportions of ammonia and carbon dioxide in the gas. Simple mass action expressions, in which quotients of concentrations are assumed to be constant, do not lead to meaningful results for the concentrations which are of practical interest. Predicting and interpreting the behaviour of the system depends, to a large extent, upon the proper understanding of the variation of the activity coefficients of the species in solution. Interactions in the gas phase between carbon dioxide and ammonia, and at high temperatures between these gases and steam are also significant.[14]

There is also interest in the systems involving the additional presence of hydrogen sulfide and/or sulfur dioxide. When a mixture of CO_2, NH_3, H_2S and SO_2 dissolves in water there are a total of fifteen different species in solution. Twenty independent equations would need to be solved to predict the concentrations of these species from the composition of the gas phase.[15]

Over many years there has been interest in the equilibria in the systems consisting of alkali carbonate + CO_2 + H_2S. A study of the potassium carbonate system and brief details of earlier work have been published by Dryden.[11]

There have been many studies of equilibria between aqueous solutions of alkanolamines and carbon dioxide, in the presence and absence of hydrogen sulfide. Solutions of alkanolamines find extensive use in the removal of these gases from natural gas and from gas from refineries. The absorption of carbon dioxide at a particular temperature and partial pressure depends upon the concentration and nature of the alkanolamine and upon the partial pressure of hydrogen sulfide. Various ionized and unionized species exist in solution. Deshmukh and Mather[16] and others have published methods of predicting equilibrium concentrations of these species under certain partial pressures of carbon dioxide and hydrogen sulfide.

Kiss et al.[17] measured the solubility of carbon dioxide at a partial pressure of 1.013 bar in water containing various concentrations of ethanol, 1,2,3-propanetriol, 2-propanone or urea. Temperatures were 273.55, 285.65 and 298.2 K. 2-Propanone solutions were studied to a maximum concentration of 0.26 mol dm^{-3}, other solutions to a lower maximum concentration. At low concentrations of organic liquid, except in the case of 2-propanone at 285.65 K and 298.2 K, solubility of gas decreases with increase in concentration of organic component. At higher concentrations of ethanol, urea or 2-propanone at 273.55 the solubility passes through a minimum and then increases. Solubility in the 2-propanone solutions at 285.65 K and 298.2 K increases with concentration of 2-propanone at all concentrations. Solubility in 1,2,3-propanetriol must also pass through a minimum at a concentration higher than that at which measurements were made. Further studies of the solubility of carbon dioxide in mixtures of 2-propanone and water have been carried out by Kir'yakov et al.[18]

11.3 SOLUBILITY OF CARBON DIOXIDE IN ORGANIC SOLVENTS

11.3.1 Solubility in hydrocarbons

Solubilities in non-cyclic alkanes have been measured by various workers. Makranczy et al.[19] reported solubilities at 298.2 and 313.2 K, corrected to a partial pressure of 1.013 bar, in straight chain alkanes from pentane to hexadecane. Their measurements show a decrease in mole fraction solubility with increase in chain length of the hydrocarbon. This is not in agreement with measurements reported by numerous other workers, and the weight of evidence indicates that mole fraction solubilities at 298.2 K and 1.013 bar increase with chain length and, in the range hexane to hexadecane, may be represented by the empirical equation:

$$x_g = 0.01156 + 9.28 \times 10^{-6} C^2$$

where C is the carbon number.

This equation is based upon measurements by King and Al-Najjar;[20] Gjaldbaek;[21] Wilcock et al.;[22] Hayduk et al.;[23] Lenoir et al.;[24] Lin and Parcher;[25] Chai and Paulaitis;[26] and Tremper and Prausnitz[27] (Fig. 11.5).

Solubilities in octacosane and hexatriacontane, from measurements at higher temperatures by Lin and Parcher are consistent with the trend towards increase in mole fraction solubility with increase in carbon number. The solubility in eicosane and in 2,6,10,15,19,23-hexamethyltetracosane from measurements by Chai and Paulaitis are also consistent with this trend.

Measurements by Hiraoka and Hildebrand[28] indicate that the mole

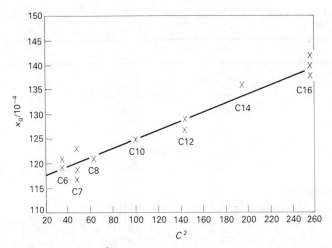

Figure 11.5. Mole fraction solubilities of carbon dioxide in straight chain alkanes plotted against the square of the carbon number of the solvent. The line corresponds to the equation given in the text. (Partial pressure of carbon dioxide = 1.013 bar; temperature = 298.15 K.)

fraction solubility in 2,2,4-trimethylpentane at 298.2 K and a partial pressure of 1.013 bar is greater than that in the linear isomer. Mole fraction solubility in the non-linear 2,6,10,15,19,23-hexamethyltetracosane or squalane is, however, consistent with extrapolation of solubilities of the lower, linear, alkanes when allowance is made for temperature.

Tremper and Prausnitz[27] measured solubilities in hexadecane over the range 300 to 475 K. Mole fraction solubilities at a partial pressure of 1.013 bar decrease with increase in temperature (Fig. 11.6).

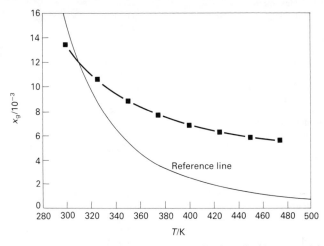

Figure 11.6. The mole fraction solubility in hexadecane of carbon dioxide at a partial pressure of 1.013 bar.

Table 11.1 Mole fraction solubilities of carbon dioxide in hydrocarbons at a partial pressure of 1.013 bar

Solvent	T/K	x_g	Ref.
Pentane	298.2	0.01258	19
Hexane	298.2	0.01207	19
	298.2	0.01215	35
	298.2	0.0119	20
Heptane	298.2	0.0123	21[a]
	298.2	0.01177	19
	298.2	0.0119	20
Octane	298.2	0.0121	20
	298.2	0.01212	29[a]
	298.2	0.01153	19
2,2,4-Trimethylpentane	298.2	0.01387	28[a]
Nonane	298.2	0.01126	19
Decane	298.2	0.01106	19
	298.2	0.01247	29[a]
	300	0.0119	36
	298.2	0.0125	20
Undecane	298.2	0.01053	19
Dodecane	298.2	0.01089	19
	298.2	0.0129	20
	298.2	0.01270	23
Tridecane	298.2	0.01080	19
Tetradecane	298.2	0.01073	19
	298.2	0.0136	20
Pentadecane	298.2	0.01067	19
Hexadecane	298.2	0.01064	19
	298.2	0.01401	26
	298.2	0.0142	20
	298.2	0.0138	27[b]
	298.2	0.01380	25
Eicosane	314.3	0.01323	26
2,6,10,15,19,23-Hexamethyl-tetracosane (squalane)	298.6	0.02205	26
Octacosane	353.2	0.0122	25
Hexatriacontane	353.2	0.0146	25
Cyclohexane	298.2	0.00759	21,34,35, 43–45[c]
Methylcyclohexane	298.2	0.00934	38[a]
(Z)-1,2-Dimethylcyclohexane	298.2	0.00938	46[a]
(E)-1,2-Dimethylcyclohexane	298.2	0.01020	46[a]
Cyclooctane	298.2	0.00690	29[a]
1,1′-Bicyclohexyl	300	0.00781	27
1-Methyl-4-(methylethenyl)-cyclohexane (d-limonene)	298.2	0.0118	31
Benzene	298.2	0.00912	21,31-36[c]
Methylbenzene	298.2	0.0107	34
		0.0105	39
		0.0103	38
1,2-Dimethylbenzene	298.2	0.0100	41
1,3-Dimethylbenzene	298.2	0.0107	41
1,4-Dimethylbenzene	298.2	0.0109	41
(1-Methylethyl)benzene	298.2	0.0101	31
1,1′-Methylenebisbenzene	300	0.00813	27
	298.2	0.00831	27[a]
	300	0.0080	36

Table 11.1 (*continued*)

Solvent	T/K	x_g	Ref.
1-Methylnaphthalene	300	0.00675	26[a]
	300	0.00659	36
1,2,3,4-Tetrahydronaphthalene	300	0.00750	34[a]
	300	0.00694	36

[a]Interpolated value.
[b]Extrapolated value.
[c]Average value.

Carbon dioxide has an appreciably lower solubility in cyclic alkanes than in non-cyclic alkanes at a partial pressure of 1.013 bar and 298.2 K (Table 11.1). Six independent measurements of the solubility in cyclohexane under these conditions correspond to mole fraction solubilities ranging from 0.00736 to 0.00772 with an average value of 0.00759. This may be compared with the corresponding average value for hexane of 0.0120. Measurements by Wilcock et al.[29] at this temperature and pressure indicate that the mole fraction solubility in cyclooctane is 0.00690 compared with 0.0121 in octane.

High pressure data for cyclohexane have also been published. Anderson et al.[30] measured the solubility of carbon dioxide at temperatures from 348.2 to 423.2 K and pressures from 19.8 to 104.3 bar.

The solubilities of carbon dioxide in methylcyclohexane, (E)- and (Z)-1,2-dimethylcyclohexane, and 1,1'-bicyclohexyl have also been measured. Mole fraction solubilities at 298.2 K and a partial pressure of 1.013 bar are given in Table 11.1. Of interest is the significant difference in solubilities in the two isomers of 1,2-dimethylcyclohexane.

The mole fraction solubility in 1-methyl-4-(1-methylethenyl)cyclohexene from Just's[31] work is 0.0118 at 298.2 K and 1.013 bar. This is higher than the corresponding values for saturated cyclic compounds discussed above but is lower than the value for decane which has the same carbon number.

Solubility in benzene at pressures close to barometric and temperatures close to 298 K has been measured by several groups.[21,31-36] Mole fraction solubilities corrected to a partial pressure of 1.013 bar fit the equation:

$$\ln x_g = -73.824 + 3804.8/(T/K) + 9.8929\ln(T/K)$$

standard deviation in $x_g = 0.00019$.
temperature range 283 to 313 K.

Phase equilibria between carbon dioxide and benzene at temperatures from 303.2 to 333.2 K and pressures from 8.5 to 53.4 bar have been studied by Wan and Dodge.[37]

Solubility in methylbenzene at a partial pressure of 1.013 bar and temperatures close to 298.2 K was measured by Field et al.,[38] Gjaldbaek and Anderson,[39] and Krauss and Gestrich.[34] Values agree within about 4 per cent.

Solubilities given by Field et al. and those given by Williams[40] for lower temperatures fit the equation

$$\ln x_g = -13.391 + 1512.9/(T/K) + 0.6580 \ln(T/K)$$

standard deviation in $x_g = 0.0013$
temperature range 203.2 to 313.2 K (see Fig. 11.7).

The mole fraction solubility in 1,3-dimethylbenzene from 283.0 to 313.1 K at a partial pressure of 1.013 bar has been published by Byrne et al.[41] Values are within about 2% of those calculated from data published by Just[31] and Horvath et al.[36] Other measurements by Byrne et al. show that the corresponding mole fraction solubility in this temperature range is greater in the 1,4 isomer and less in the 1,2 isomer.

Mole fraction solubility in (1-methylethyl)benzene from data published by Just[31] is close to that in methylbenzene at a partial pressure of gas of 1.013 bar and temperatures close to 298.2 K. The mole fraction solubility in 1,1'-methylenebisbenzene is less than in benzene or methylbenzene under these conditions.[27,36]

Mole fraction solubilities in naphthalene derivatives are lower than in benzene derivatives under the conditions at which measurements have been made. The estimated mole fraction solubility in 1-methylnaphthalene at a partial pressure of 1.013 bar at 300 K from data published by Horvath et al.[36] is close to the value estimated from data published by Chai and Paulaitis.[26] The agreement between corresponding values for 1,2,3,4-tetrahydronaphthalene from data published by Horvath et al. and Krauss and Gestrich[34] is not as good.

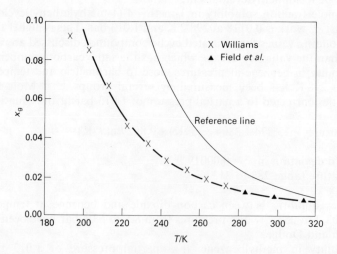

Figure 11.7. The mole fraction solubility in methylbenzene of carbon dioxide at a partial pressure of 1.013 bar.

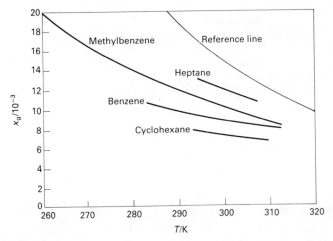

Figure 11.8. The mole fraction solubility in hydrocarbons of carbon dioxide at a partial pressure of 1.013 bar.

Solubilities in some of the hydrocarbons mentioned above are shown in Fig. 11.8.

11.3.2 Solubility in organic solvents containing carbon, hydrogen and oxygen

Measurements by Makranczy et al.[42] indicate that the mole fraction solubility in straight chain alkanols from methanol to dodecanol at 298.2 K and a partial pressure of carbon dioxide of 1.013 bar increases linearly with carbon number. This finding is not supported by the measurements of Just[31] on the solubility in 1-pentanol or those of Wilcock et al.[22] on the solubilities in 1-octanol and 1-decanol, which differ very markedly from those of Makranczy. Makranczy et al. indicate that solubility at 298.2 K and 1.013 bar in the higher straight chain alkanols is greater than in the corresponding alkanes. Measurements by other workers indicate that the reverse is true (Fig. 11.9).

Solubilities in methanol and ethanol were measured by Just[31] from 288 to 298 K and by Won et al.[43] at 298.2 K. Just also measured the solubility in 1-propanol at 298.2 K. In these cases there is better agreement with Makranczy et al. (Table 11.2).

Solubility in 2-methyl-1-propanol at a partial pressure of 1.013 bar over the range 273.98 to 328.00 K measured by Battino et al.[44] is close to that measured by Just[31] at 288 to 298 K (Fig. 11.10 and Table 11.2).

Chromatographic measurement of solubility in benzenemethanol by Lenoir et al.[45] at 298.2 K indicates a limiting value of Henry's law constant of 128 atm. Extrapolating to 1.013 bar gives a mole fraction solubility of 0.00781. This is close to the value for alkanols of similar carbon number.

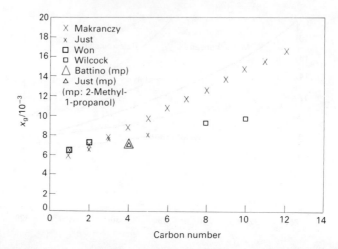

Figure 11.9. Mole fraction solubilities of carbon dioxide in alkanols plotted against the carbon number of the alkanol. All data points, with the exception of two points for 2-methyl-1-propanol, correspond to published solubilities in straight chain primary alkanols. (Partial pressure of carbon dioxide = 1.013 bar; temperature = 298.15 K.)

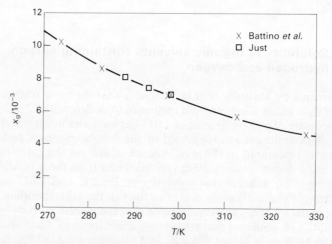

Figure 11.10. The variation with temperature of the mole fraction solubility of carbon dioxide in 2-methyl-1-propanol at a partial pressure of gas of 1.013 bar.

An early measurement by Cauquil[46] of the solubility of carbon dioxide in cyclohexanol corresponds to a mole fraction solubility at a partial pressure of 1.013 bar of 0.00286 at 298.2 K. This is low compared with straight chain alkanols of the same carbon number.

The solubility in 1,2-ethanediol is very low compared with that in monohydric alcohols. The estimated mole fraction solubility at 298.2 K and

Table 11.2 Mole fraction solubilities of carbon dioxide in solvents containing carbon, hydrogen and oxygen at a partial pressure of 1.013 bar

Solvent	T/K	x_g	Ref.
Methanol	298.2	0.00587	42
	298.2	0.00635	31
	298.2	0.00643	43
Ethanol	298.2	0.00689	42
	298.2	0.00645	31
	298.2	0.00728	43
1-Propanol	298.2	0.00762	31
	298.2	0.00782	42
1-Butanol	298.2	0.00883	42
2-Methyl-1-propanol	298.2	0.00697	44
	298.2	0.00698	31
1-Pentanol	298.2	0.00807	31
	298.2	0.00974	42
1-Hexanol	298.2	0.0108	42
1-Heptanol	298.2	0.01175	42
1-Octanol	298.2	0.01277	42
	298.2	0.00930	22
1-Nonanol	298.2	0.01378	42
1-Decanol	298.2	0.01473	42
	298.2	0.00973	22
1-Undecanol	298.2	0.01571	42
1-Dodecanol	298.2	0.01664	42
Benzenemethanol	298.2	0.00781	45
Cyclohexanol	298.2	0.00286	46
1,2-Ethanediol	298.2	0.00382	45
	297.1	0.00306	47
1,2,3-Propanetriol	298.2	0.00009	31
2-Propanone	293.2	0.0207	31
	293.2	0.0202	48
	293.2	0.0211	33
	298.2	0.0187	31
	298.2	0.0211	39
Benzaldehyde	298.2	0.0117	31
1,1′-Oxybisethane	288.2	0.0233	51
Tetrahydrofuran	298.2	0.027	49[b]
	273.2	0.040	49
2-Methoxyethanol	298.2	0.0100	49[a]
Oxybispropanol	298.2	0.00826	45
Methyl acetate	298.2	0.0208	31
	298.2	0.0226	39
Ethyl acetate	298.2	0.0230	49[b]
Propyl acetate	298.2	0.0245	39
2-Methylpropyl acetate	298.2	0.0250	31
Pentyl acetate	298.2	0.0246	31
Pentyl formate	298.2	0.0212	31
Acetic acid	298.2	0.0109	31
Propanoic acid	298.2	0.0123	31
Butanoic acid	298.2	0.0130	31
Acetic anhydride	298.2	0.0199	31
4-Methyl-1,3-dioxolan-2-one	298.2	0.0121	52
Phenol	323.2	0.0047	45
3-Methylphenol	300.5	0.0059	36

[a] Extrapolated.
[b] Interpolated.

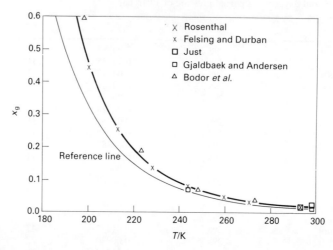

Figure 11.11. The mole fraction solubility in 2-propanone of carbon dioxide at a partial pressure of 1.013 bar.

1.013 bar from gas chromatographic measurements by Lenoir et al.[45] is 0.00382. A direct measurement of the volume of gas absorbed by a given volume of solvent carried out by Byeseda et al.[47] corresponds to a mole fraction solubility of 0.00306 at 1.013 bar and 297.1 K. Measurements by Just[31] indicate an even lower mole fraction solubility of 0.0009 in 1,2,3-propanetriol under these conditions. Such a low value as this must be treated with some suspicion until further measurements are made.

The mole fraction solubility in 2-propanone at a partial pressure of 1.013 bar for 293.2 and 298.2 K may be calculated from data published by Just,[31] Rosenthal[33] and Gjaldbaek and Andersen[39] (Table 11.2). Felsing and Durban.[48] measured solubility at a total pressure of 1.013 bar over the range 200 to 293 K. Bodor et al.[49] made measurements at 198 to 318.2 K and pressures from 0.077 bar to 0.972 bar.

Values of mole fraction solubility at a partial pressure of 1.013 bar from the available data are shown in Fig. 11.11 and fit the equation

$$\ln x_g = -10.313 + 1905.1/(T/K) + 0.0017\ln(T/K)$$

standard deviation in $x_g = 0.0055$
temperature range 200 to 298 K.

Rosenthal measured the effect on the solubility of carbon dioxide of adding sodium iodide to 2-propanone (Fig. 11.12).

Bodor[50] published solubilities in acetaldehyde for the range 198 to 248.2 K and pressure range 0.332 to 1.013 bar. Mole fraction solubility under these

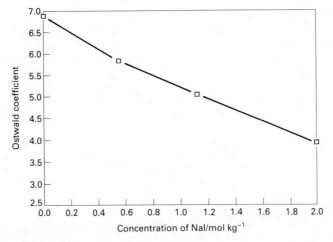

Figure 11.12. The effect of addition of sodium iodide on the solubility of carbon dioxide in 2-propanone at 293.15 K.

conditions is lower than in 2-propanone. At 248.2 K and a partial pressure of 1.013 bar the mole fraction solubility in acetaldehyde is 0.047 compared with 0.072 in 2-propanone. Solubility in benzaldehyde was measured by Just[31] (see Table 11.2).

Measurements by Christoff[51] show that mole fraction solubility in 1,1'-oxybisethane at a partial pressure of 1.013 bar from 273 to 288 K is close to that in 2-propanone under the same conditions.

Solubility in tetrahydrofuran in the range 198 to 318.2 K and pressure range 0.105 to 0.980 bar was measured by Bodor et al.[49] The mole fraction solubility is very high under these conditions in comparison with mole fraction solubilities in other solvents (Fig. 11.13). Extrapolated data indicate that the mole fraction solubility at 298.2 K and a partial pressure of 1.013 bar is approximately 0.027, compared with a value of 0.0211 for dissolution in 2-propanone.

The presence of hydroxy groups lowers the mole fraction solubility in hydroxy ethers. Bodor et al.[49] measured solubility in 2-methoxyethanol and Lenoir et al.[45] solubility in oxybispropanol (Table 11.2).

Solubilities in ethyl acetate over the range 198 to 318.2 K and from 0.096 to 1.065 bar were also measured by Bodor et al.[49] Solubility at 298.2 K and 1.013 bar is higher than in 2-propanone. Measurements of solubilities in other esters of carboxylic acids at barometric pressure and temperature of, or close to, 298.2 K have been published by Just[31] and by Gjaldbaek and Andersen.[39] Gjaldbaek and Andersen's value for the solubility in methyl acetate is likely to be more reliable than the earlier value given by Just which is about 8% lower.

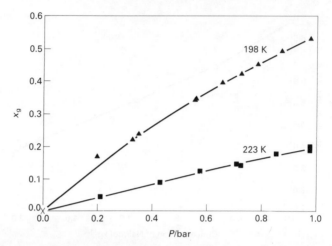

Figure 11.13. The variation with pressure of the mole fraction solubility of carbon dioxide in tetrahydrofuran at 198 K and 223 K.

The available data indicate that at a partial pressure of 1.013 bar and 298.2 K mole fraction solubilities in methyl, ethyl, propyl, and pentyl acetate increase with carbon number of the alkyl group.

Measurements by Just[31] show that mole fraction solubilities in acetic acid, propanoic acid and butanoic acid at 1.013 bar and 298.2 K also increase with the carbon number of the acid. They show that the mole fraction solubility in acetic anhydride is greater than in acetic acid.

Mole fraction solubilities in 4-methyl-1,3-dioxolan-2-one at a partial pressure of 1.013 bar can be calculated from Henry's law constants published by Rivas and Prausnitz[52] and fit the equation

$$\ln x_g = -56.037 + 3733.7/(T/K) + 6.8635\ln(T/K)$$

standard deviation in $x_g = 0.00021$
temperature range 263 to 373 K.

Chromatographic measurements by Sweeney[53] at 298.2 and 323.2 K are consistent with this equation. Similar measurements by Lenoir et al.[45] at 298 to 343 K are about 20% higher than values from the equation.

Lenoir et al.[45] also measured the solubility of carbon dioxide in phenol by chromatography. The estimated mole fraction solubility at 1.013 bar and 323.2 K is 0.00467. This can be compared with the value of 0.00592 at 300.5 K for dissolution in 3-methylphenol from measurements by Horvath et al.[36]

Solubilities in various solvents containing carbon, hydrogen and oxygen are shown in Fig. 11.14.

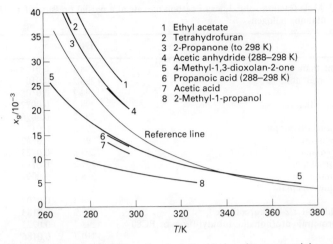

Figure 11.14. Mole fraction solubilities of carbon dioxide in solvents containing oxygen at a partial pressure of gas of 1.013 bar.

11.3.3 Solubility of carbon dioxide in solvents containing halogens

The solubility of carbon dioxide in tetrachloromethane at 298.2 K and a pressure close to 1.013 bar by Gjaldbaek[21] agrees within about 2% with a value determined by Horiuti.[54] The earlier measurement by Just[31] is about 14% lower than Horiuti's value and less likely to be reliable. Mole fraction solubility at 1.013 bar and 298.2 K in the polar trichloromethane is higher than in tetrachloromethane. Gjaldbaek and Andersen's[39] value is about 12% higher than the earlier one given by Just.[31]

There is much better agreement between data for 1,2-dibromoethane published by Gjaldbaek and Andersen[39] and by Just.[31] Mole fraction solubility at 298.2 K and a partial pressure of 1.013 bar from the two sources are, respectively, 0.0772 and 0.0761. Other mole fraction solubilities in alkyl halides from measurements by Just are given in Table 11.3.

Solubilities in fluorinated alkanes are higher than in chlorinated alkanes. There is close agreement between the measurement of the solubility in hexadecafluoroheptane (perfluoroheptane) by Gjaldbaek[21] and by Kobatake and Hildebrand[55] (Table 11.3). A selection of data for other fluorinated alkanes is also given in the table. Variation in solubility with temperature in a fluorinated alkane, Caroxin-F, is shown in Fig. 11.15.

Solubility in chlorobenzene at 298.2 K and 1.013 bar measured by Gjaldbaek and Andersen[39], is about 4% higher than the earlier measurement by Just.[31] Just also measured solubilities in bromobenzene and iodobenzene. Mole fraction solubilities at a partial pressure of 1.013 bar in the temperature range 288 K to 298 K are in the order chloro- > bromo- > iodobenžene. Mole

Table 11.3 Mole fraction solubilities of carbon dioxide at a partial pressure of 1.013 bar in solvents containing halogens

Solvent	T/K	x_g	Ref.
Trichloromethane	298.2	0.0128	39
Tetrachloromethane	298.2	0.0107	21
	298.2	0.0105	54
1,2-Dibromoethane	298.2	0.00772	39
	298.2	0.00761	31
1-Chloro-2-methylpropane	298.2	0.0133	31
1-Chloropentane	298.2	0.0142	31
1-Bromopentane	298.2	0.0123	31
1,2-Dichloroethane	298.2	0.0113	31
1,2-Dibromopropane	298.2	0.00976	31
Hexadecafluoroheptane	298.2	0.02088	55
	298.3	0.0209	21
1,1,2-Trichloro-1,2,2-trifluoroethane	298.2	0.0183	28[a]
Heptafluorotetrahydro(nonafluorobutyl)furan or FC80	298.2	0.0223	56
	310.2	0.0186	56
	310.2	0.0188	57
1,1,2,2,3,3,4,4-Octafluoro-1,4-bis(tetrafluoro-1-(trifluoromethyl)-ethoxy)butane or Caroxin-D	298.2	0.0248	56
	310.2	0.0227	56
1,1,1,2,2,3,3,4,4,5,5,6,6-Tridecafluoro-6-(1,2,2,2-tetrafluoro-1-(trifluoromethyl)ethoxy) hexane (Caroxin-F)	298.2	0.0250	56
	310.2	0.0232	56
Chlorobenzene	298.2	0.00982	39
	298.2	0.00938	31
Bromobenzene	298.2	0.00788	31
Iodobenzene	298.2	0.00592	31
Hexafluorobenzene	297.98	0.0232	58
(Chloromethyl)benzene	298.2	0.00907	31
(Trichloromethyl)benzene	298.2	0.00950	31

[a]Interpolated.

fraction solubility in hexafluorobenzene[58] is high compared with available values for other halobenzenes and haloalkanes (Table 11.3 and Fig. 11.16).

Solubilities in (chloromethyl)benzene and in (trichloromethyl)benzene were also measured by Just (Table 11.3).

11.3.4 Solubility of carbon dioxide in solvents containing nitrogen

The solubility of carbon dioxide in nitrogen bases is less than may be expected. The mole fraction solubility in benzenamine at 298.2 K and a partial pressure of 1.013 bar is 0.00491 from data published by Gjaldbaek and Andersen.[39] This value is very close to that from data reported by Just[31] and about half the value of 0.00912 for benzene under the same conditions. Mole fraction solubility in the methyl benzenamines is also appreciably less than in methylbenzene. Mole fraction solubility in pyridine is, however,

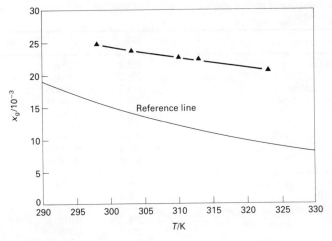

Figure 11.15. The mole fraction solubility in Caroxin-F of carbon dioxide at a partial pressure of 1.013 bar.

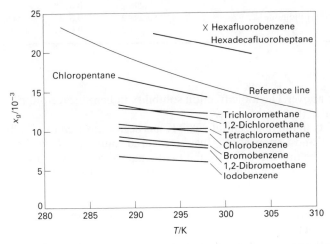

Figure 11.16. Mole fraction solubilities of carbon dioxide in solvents containing halogens at a partial pressure of gas of 1.013 bar.

greater than in benzene at this temperature and pressure. The solubility in quinoline is close to that in benzene. The compound formed when all the hydrogen atoms in tributylamine are replaced by fluorine i.e. 1,1,2,2,3,3,4,4,4-nonafluoro-*N*,*N*-bis(nonafluorobutyl)-1-butamine is a better solvent than the simple amines (Table 11.4).

Mole fraction solubility at 298.2 K and 1.013 bar in propanenitrile is high compared with solubility in alkanes and in most of the nitrogen compounds

Table 11.4 Solubility of carbon dioxide at a partial pressure of 1.013 bar in solvents containing nitrogen.

Solvent	T/K	x_g	Ref.
Benzeneamine	298.2	0.00493	31
	298.2	0.00491	39
2-Methylbenzenamine	298.2	0.00605	31
3-Methylbenzenamine	298.2	0.00634	31
Pyridine	298.2	0.0119	31
	298.2	0.0121	39
Quinoline	300.0	0.00939	36
	298.2	0.00906	26[a]
	300.0	0.00882	26[b]
1,1,2,2,3,3,4,4,4-Nonafluoro-N,N-bis (nonafluorobutyl)-1-butanamine	298.2	0.0200	55
Propanenitrile	298.2	0.0169	39
Benzeneacetonitrile	298.2	0.0105	39
Nitrobenzene	298.2	0.01015	39
	298.2	0.0103	31
N,N-Dimethylformamide	293.2	0.0164	33
	297.1	0.0154	47
	293.2	0.0157	59
	297.1	0.0142	59[b]
	298.2	0.0138	59[b]
1-Methyl-2-pyrrolidinone	298.2	0.0143	60[b]
	298.2	0.0151	52

[a]Interpolated.
[b]Extrapolated.

discussed above. The mole fraction solubility in benzeneacetonitrile is about 10% greater than in benzene.

Solubility in nitrobenzene at 298.2 K has been measured by Gjaldbaek and Andersen[39] and Lenoir et al.[45] Just[31] measured solubilities over the range 288 to 298 K. The mole fraction solubility at a partial pressure of 1.013 bar from measurements by Gjaldbaek and Andersen is 0.01015, which is in good agreement with the value of 0.0103 from measurements by Just. A value estimated from the limiting value of Henry's law constant obtained by Lenoir et al. from chromatographic measurements is about 30% greater and likely to be incorrect.

Solubility in N,N-dimethylformamide over the range 278 K to 313 K and 0.27 to 1.20 bar has been measured by Haidegger et al.[59] Rosenthal[33] reported the solubility at 293.2 K and partial pressure of 1.013 bar. Byeseda et al.[47] reported the solubility at 297.1 K and a partial pressure of 1.016 bar. The value of the mole fraction solubility at 298.2 K and partial pressure of 1.013 bar from Rosenthal's work is about 4% greater than the value from Haidegger et al. The mole fraction solubility at 297.1 and 1.013 bar from Byeseda's work is about 8% greater than that from Haidegger et al. Rosenthal also investigated the effect on solubility of adding sodium iodide.

Solubility in 1-methyl-2-pyrrolidinone at 298.2 and 323.2 K over the range 0.23 to 2.04 bar has been measured by Murrieta-Guevara and Rodriguez.[60]

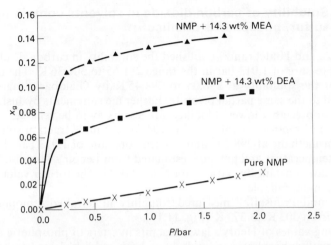

Figure 11.17. Mole fraction solubility of carbon dioxide at 298.15 K in 1-methyl-2-pyrrolidinone (NMP) and in mixtures of NMP with ethanolamine (MEA) and diethanolamine (DEA).

Rivas and Prausnitz[52] reported Henry's law constants over the range 303 to 373 K. The two sets of data agree to within about 5% (Table 11.4). The system was also studied by Lenoir et al.[45] and by Sweeney,[53] who published limiting values of Henry's law constants from chromatographic measurements. Murrieta-Guevara and Rodriguez also investigated the effect of adding ethanolamine and diethanolamine, which enhance the solubility of carbon dioxide (Fig. 11.17). Rivas and Prausnitz made similar measurements on mixtures of diglycolamine and 1-methyl-2-pyrrolidinone. Solubilities in compounds containing nitrogen are shown in Fig. 11.18.

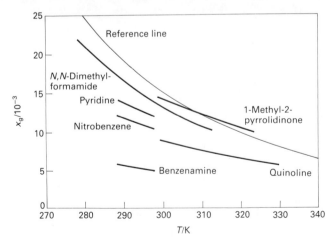

Figure 11.18. Mole fraction solubilities of carbon dioxide in solvents containing nitrogen at a partial pressure of gas of 1.013 bar.

11.3.5 Solubility of carbon dioxide in solvents containing sulfur, phosphorus or silicon

Kobatake and Hildebrand[55] published the solubility in carbon disulfide at a partial pressure of 1.013 bar in the range 281.65 to 306.36 K. The measurements of the solubility at 298.05 to 298.45 K by Gjaldbaek[21] agree when corrected to the same partial pressure. Earlier measurements by Just,[31] which indicate appreciably lower solubility, are less likely to be reliable.

Dymond[61] reported a value of 0.00908 for the mole fraction solubility in sulfinylbismethane at 298.2 K and a partial pressure of 1.013 bar. The value for this temperature and pressure, estimated from Lenoir's[45] limiting value of Henry's law constant by extrapolation, is 0.00943. The former value is likely to be the more reliable.

Rivas and Prausnitz[52] measured solubility in tetrahydrothiophene 1,1-dioxide from 303 K to 373 K (Fig. 11.19).

Limiting values of Henry's law constants in esters of phosphoric acid and in hexamethylphosphoric triamide have been published by Lenoir et al.[45] Sweeney[53] published limiting values of Henry's law constants for dissolution in tributyl phosphate. Approximate values of mole fraction solubilities at a partial pressure of 1.013 bar may be estimated from these constants (Table 11.5).

The solubility in octamethylcyclotetrasiloxane at 292 to 313 K and a partial pressure of 1.013 bar has been published by Wilcock et al.[62] The mole fraction solubility is high compared with the reference line (Fig. 11.19).

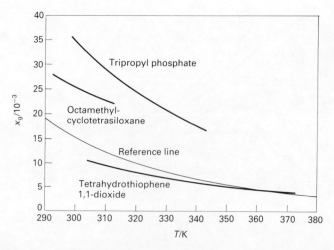

Figure 11.19. Mole fraction solubilities of carbon dioxide in solvents containing silicon, phosphorus or sulfur at a partial pressure of gas of 1.013 bar.

Table 11.5 Solubility of carbon dioxide in solvents containing sulfur, phosphorus or silicon at a partial pressure of 1.013 bar

Solvent	T/K	x_g	Ref.
Carbon disulfide	298.2	0.00328	55
	298.1	0.00330	21
Sulfinylbismethane	298.2	0.00908	61
	298.2	0.00943	24[a]
Tetrahydrothiophene 1,1-dioxide	303.2	0.00109	52
Hexamethylphosphoric triamide	298.2	0.0329	24
Trimethyl phosphate	325.2	0.0115	24
Triethyl phosphate	325.7	0.0184	24
Tripropyl phosphate	298.2	0.0356	24
	323.2	0.0236	24
Tributyl phosphate	298.2	0.0345	53[a]
	323.2	0.0217	53
	325.2	0.0233	24
Tris(2-methylpropyl) phosphate	325.2	0.0260	24
Octamethylcyclotetrasiloxane	298.2	0.0261	62

[a]Solubilities from Refs 24 and 53 were estimated from limiting values of Henry's law constant by assuming that the mole fraction varies linearly with pressure to 1.013 bar.

REFERENCES

1. Harned, H.S.; Davies, R. *J. Am. Chem. Soc.* 1943, 65, 2030.
2. Berg, D.; Patterson, A. *J. Am. Chem. Soc.* 1953, 75, 5197.
3. Wilhelm, E.; Battino, R.; Wilcock, R.J. *Chem. Rev.* 1977, 77, 219.
4. Morrison, T.J.; Billett, F. *J. Chem. Soc.* 1952, 3819.
5. Murray, C.N.; Riley, J.P. *Deep-Sea Res.* 1971, 18, 533.
6. Weiss, R.F. *Mar. Chem.* 1975, 2, 203.
7. Kozintseva, T.N. *Geokhimicheskie Issledovaniia v Oblasti Povyshennykh.* Edited by N.I. Khitarov, Akad.nauk SSSR. Institut geokhimii i analiticheskoi khimii. 1965, 121.
8. Ellis, A.J.; Golding, R.M. *Amer. J. Sci.* 1963, 261, 47.
9. Markham, A.E.; Kobe, K.A. *J. Am. Chem. Soc.* 1941, 63, 449.
10. Badger, E.H.M.; Silver, L. *J. Soc. Chem. Ind.* 1938, 57, 110.
11. Dryden, I.G.C. *J. Soc. Chem. Ind.* 1947, 66, 59.
12. van Krevelen, D.W.; Hoftijzer, P.J.; Huntjens, F.J. *Recl. Trav. Chim. Pays-Bas* 1949, 68, 191.
13. Wilson, G.M.; Gillespie, P.C.; Owens, J.L. *Proc. 64th Ann. Conv. Gas Processors Association*, 1985, 282.
14. Roberts, B.E.; Tremaine, P.R. *Can. J. Chem. Eng.* 1985, 63, 294.
15. Maurer, G. *Thermodynamics of Aqueous Systems with Industrial Applications*, ACS Symposium Series 133, 1980, 140.
16. Deshmukh, R.D.; Mather, A.E. *Chem. Eng. Sci.* 1981, 36, 355.
17. Kiss, A. v.; Lajtai, I.; Thury, G. *Z. Anorg. Allgem. Chem.* 1937, 233, 346.
18. Kir'yakov, V.N.; Usyukin, I.P.; Shleinikov, V.M. *Neftepererabotka i Neftekhim.* 1966, (9), 40.
19. Makranczy, J.; Megyery-Balog, K.; Rusz, L.; Patyi, L. *Hung. J. Ind. Chem.* 1976, 4, 269.
20. King, M.E.; Al-Najjar, H. *Chem. Eng. Sci.* 1977, 32, 1214.
21. Gjaldbaek, J.H. *Acta Chem. Scand.* 1953, 7, 537.
22. Wilcock, R.J.; Battino, R.; Danforth, W.F.; Wilhelm, E. *J. Chem. Thermodyn.* 1978, 10, 817.

23. Hayduk, W.; Walter, E.B.; Simpson, P. *J. Chem. Eng. Data* 1972, 17, 59.
24. Lenoir, J.-Y.; Renault, P.; Renon, H. *J. Chem. Eng. Data* 1971, 16, 340.
25. Lin, P.J.; Parcher, J.F. *J. Chromatog. Sci.* 1982, 20, 33.
26. Chai, C.-P.; Paulaitis, M.E. *J. Chem. Eng. Data* 1981, 26, 277.
27. Tremper, K.K.; Prausnitz, J.M. *J. Chem. Eng. Data* 1976, 21, 295.
28. Hiraoka, H.; Hildebrand, J.H. *J. Phys. Chem.* 1964, 68, 213.
29. Wilcock, R.J.; Battino, R.; Wilhelm, E. *J. Chem. Thermodyn.* 1978, 9, 111.
30. Anderson, J.M.; Barrick, M.W.; Robinson, R.L. *J. Chem. Eng. Data* 1986, 31, 172
31. Just, G. *Z. Phys. Chem.* 1901, 37, 342.
32. Byrne, J.E.; Battino, R.; Danforth, W.F. *J. Chem. Thermodyn.* 1974, 6, 245.
33. Rosenthal, W. *Thes. Fac. Sci. Univ Strasbourg (France)* 1954
34. Krauss, W.; Gestrich, W. *Chem.-Tech. (Heidelberg)*, 1977, 6, 513.
35. Patyi, L.; Furmer, I.E.; Makranczy, J.; Sadilenko, A.S.; Stepanova, Z.G.; Berengarten, M.G. *Zh. Prikl. Khim.* 1978, 51, 1296.
36. Horvath, M.J.; Sebastian, H.M.; Chao, K.-C. *Ind. Eng. Chem. Fundam.* 1981, 20, 394.
37. Wan, S.-W.; Dodge, B.F. *Ind. Eng. Chem.* 1940, 32, 95.
38. Field, L.R.; Wilhelm, E.; Battino, R. *J. Chem. Thermodyn.* 1974, 6, 237.
39. Gjaldbaek, J.C.; Andersen, E.K. *Acta Chem. Scand.* 1954, 8, 1398.
40. Williams, D.L. *U.S. Atomic Energy Commission Report LA-1484*, 1952.
41. Byrne, J.E.; Battino, R.; Wilhelm, E. *J. Chem. Thermodyn.* 1974, 7, 515.
42. Makranczy, J.; Rusz, L.; Balog-Megyery, K. *Hung. J. Ind. Chem.* 1979, 7, 41.
43. Won, Y.S.; Chung, D.K.; Mills, A.F. *J. Chem. Eng. Data* 1981, 26, 140.
44. Battino, R.; Evans, F.D.; Danforth, W.F.; Wilhelm, E. *J. Chem. Thermodyn.* 1971, 3, 743.
45. Lenoir, J.-Y.; Renault, P.; Renon, H. *J. Chem. Eng. Data* 1971, 16, 340.
46. Cauquil, G. *J. Chim. Phys.* 1927, 24, 53.
47. Byeseda, J.J.; Deetz, J.A.; Manning, W.P. *Proc. Lawrance Reid Gas Cond. Conf.* 1985.
48. Felsing, W.A.; Durban, S.A. *J. Am. Chem. Soc.* 1926, 48, 2885.
49. Bodor, E.; Bor, Gy.; Maleczkine, M.; Mesko, G.; Mohai, B.; Siposs, G. *Veszpremi. Vegyip. Egyet. Kozlemen.* 1957, 1, 63; 77; 89; 99.
50. Bodor, E.; Mohai, B.; Pfeifer, Gy. *Veszpremi. Vegyip. Egyet. Kozlemen.* 1959, 3, 205.
51. Christoff, A. *Z. Phys. Chem.* 1912, 79, 456.
52. Rivas, O.R.; Prausnitz, J.M. *Ind. Eng. Chem. Fundam.* 1979, 18, 289; *Am. Inst. Chem. Engnrs. J.* 1979, 25, 975.
53. Sweeney, C.W. *Chromatographia*, 1984, 18, 663.
54. Horiuti, J. *Sci. Pap. Inst. Phys. Chem. Res. (Jpn)* 1931/32, 17, 125.
55. Kobatake, Y.; Hildebrand, J.H. *J. Phys. Chem.* 1961, 65, 331.
56. Tham, M.K.; Walker, R.D. Jr.; Modell, J.H. *J. Chem. Eng. Data* 1973, 18, 385.
57. Navari, R.M.; Rosenblum, W.I.; Kontos, H.A.; Patterson, J.L. *Res. Exp. Med.* 1977, 170, 169.
58. Evans, D.F.; Battino, R. *J. Chem. Thermodyn.* 1971, 3, 753.
59. Haidegger, F.; Szebenyi, I.; Szekely, A. *Magy. Kem. Foly.* 1958, 64, 365.
60. Murrieta-Guevara, F.; Rodriguez, A.T. *J. Chem. Eng. Data* 1984, 29, 456.
61. Dymond, J.H. *J. Phys. Chem.* 1967, 71, 1829.
62. Wilcock, R.J.; McHale, J.L.; Battino, R.; Wilhelm, E. *Fluid Phase Equilib.* 1978, 2, 225.

Chapter 12

SOLUBILITY OF THE OXIDES OF NITROGEN

12.1 GENERAL PROPERTIES OF OXIDES OF NITROGEN, STABLE AND GASEOUS AT 298 K

Gas	M.pt /K	B.pt/ K	Crit. temp. /K	Crit. press. /bar	Crit.vol. /dm^3 mol^{-1}	M_r	Density of gas at 273.15 K; 1.013 bar/g dm^{-3}
Nitrous oxide N_2O	107.8	183.7	309.6	72.4	0.0967	44.013	1.977
Nitric oxide NO	109.6	121.5	180.1	65	0.058	30.006	1.3402
Nitrogen dioxide NO_2	261.9	294.2	431.2	101	0.082	46.005	—
Dinitrogen tetroxide N_2O_4	—	—	—	—	—	92.011	—

Nitrogen dioxide and dinitrogen tetroxide exist in equilibrium in the gas and the liquid phase. In the solid the substance exists entirely as the tetroxide. The degree of dissociation to the dioxide in the gas phase varies from about 20% at 300 K to about 90% at 373 K.[1] The degree of dissociation in solution in various solvents was measured by Cundell.[2] The solubilities of nitrous oxide and of nitric oxide have been compiled and evaluated for the Solubility Data Series.[3]

12.2 SOLUBILITY OF NITROUS OXIDE

Stull[4] has reported the following values of the vapour pressure of liquid nitrous oxide:

T/K	P/bar
254.4	20.26
281.2	40.53
300.6	60.79

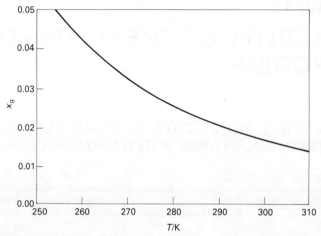

Figure 12.1. The reference line for nitrous oxide at a partial pressure of 1.013 bar.

Other data for lower temperatures were also published. The data given fit the equation

$$\ln(P/\text{bar}) = 8.7337 - 1113.69/[(T/\text{K}) - 59.81].$$

The reference line for a partial pressure of 1.013 bar, based upon vapour pressure data published by Stull, is shown in Fig. 12.1.

12.2.1 Solubility of nitrous oxide in water and aqueous solutions

An equation for variation of the mole fraction solubility with temperature in water at a partial pressure of 1.013 bar has been published by Battino[5] and is based upon twenty three data points from eleven papers. It may be written in the form

$$\ln x_g = -158.621 + 8882.8/(T/\text{K}) + 21.253 \ln(T/\text{K})$$

standard deviation in $x_g = 1.2\%$
temperature range 273 to 313 K.

Battino could not take into account recent data published by Weiss and Price.[6] The following equation is based upon data given in this paper and a single data point published by Sada et al.[7]

$$\ln x_g = -177.065 + 9703.3/(T/K) + 24.0105\ln(T/K)$$

standard deviation in $x_g = 7.342 \times 10^{-7}$ (about 0.2%)

temperature range 273 K to 313 K (Fig. 12.2).

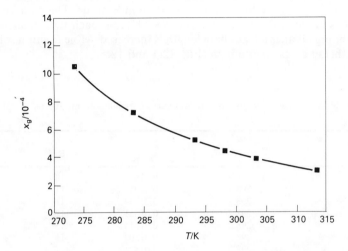

Figure 12.2. The mole fraction solubility in water of nitrous oxide at a partial pressure of 1.013 bar.

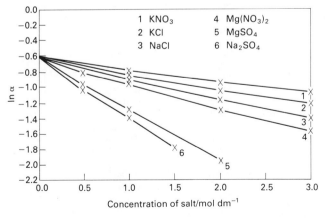

Figure 12.3. The effect of electrolytes on the solubility of nitrous oxide in water at 298.15 K (see Ref. 27).

There have been numerous studies of the solubility of nitrous oxide in aqueous solutions of electrolytes. The effect of various electrolytes is shown in Fig. 12.3. Measurements made before 1981 have been evaluated by Young[8] who has calculated Sechenov salt effect parameters.

12.2.2 Solubility of nitrous oxide in organic solvents

Data published before 1980 have been evaluated by Young.[9] Selected mole fraction solubilities for a partial pressure of 1.013 bar are shown in Table 12.1. At a partial pressure of 1.013 bar the mole fraction solubility in alkanes is close to the reference line and greater than in benzene. The mole fraction solubilities in alcohols are low but, at 298.2 K, approach the values for the coresponding alkanes as the chain length is increased. Mole fraction solubilities in ethers and esters are high (Fig. 12.4 and 12.5).

Table 12.1 Mole fraction solubilities of nitrous oxide in organic solvents at a partial pressure of 1.013 bar

Solvent	T/K	x_g	Ref.
Pentane	294.3	0.0194	10
	298.2	0.01917	11
Hexane	298.2	0.01844	11
	298.2	0.01851	12
Heptane	298.2	0.01814	13[a]
	298.2	0.01809	11
Octane	298.2	0.01805	11
2,2,4-Trimethylpentane	298.2	0.02121	13[a]
Nonane	298.2	0.01798	11
Decane	298.2	0.01795	11
Undecane	298.2	0.01768	11
Dodecane	298.2	0.01734	11
Tridecane	298.2	0.01759	11
Tetradecane	298.2	0.01725	11
Pentadecane	298.2	0.01739	11
Hexadecane	298.2	0.01763	11
Cyclohexane	298.2	0.00853	12
Benzene	298.2	0.01349	14
	298.2	0.01352	13[a]
Methanol	298.2	0.00535	15[a]
	298.2	0.00525	16
Ethanol	298.2	0.00686	16
	298.2	0.00738	11
	298.2	0.00674	15
1-Propanol	298.2	0.00840	11
1-Butanol	298.2	0.00949	11
3-Methyl-1-butanol	298.2	0.0104	15[a]
1-Pentanol	298.2	0.01048	11
1-Hexanol	298.2	0.01157	11
1-Heptanol	298.2	0.01265	11

Table 12.1 (*continued*)

Solvent	T/K	x_g	Ref.
1-Octanol	298.2	0.01376	11
1-Nonanol	298.2	0.01487	11
1-Decanol	298.2	0.01591	11
1-Undecanol	298.2	0.0170	11
1-Dodecanol	298.2	0.0180	11
2-Propanone	298.2	0.0181	14
1,1'-Oxybisethane	298.2	0.0321	10
Methyl acetate	298.2	0.02045	14
Pentyl acetate	294.3	0.0305	10
	294.3	0.0302	15[a]
	298.2	0.0288	15[a]
Acetic acid	294.3	0.0113	10
	294.3	0.0113	15[a]
	298.2	0.0105	15[a]
Benzaldehyde	298.2	0.0123	15[a]
Benzenamine	298.2	0.00528	15[a]
Pyridine	298.2	0.0112	15[a]
Tetrachloromethane	298.2	0.01698	14
	298.2	0.01686	13
Trichloromethane	294.3	0.0182	10
	294.3	0.0182	15[a]
	298.2	0.0168	15[a]
1,2-Dibromoethane	298.2	0.00938	15[a]
Chlorobenzene	298.2	0.01326	14

[a]Interpolated.

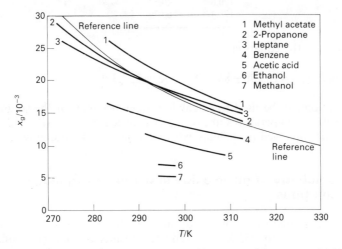

Figure 12.4. Mole fraction solubilities of nitrous oxide, at a partial pressure of 1.013 bar, in hydrocarbons and solvents containing carbon, hydrogen and oxygen.

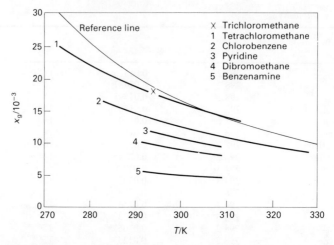

Figure 12.5. Mole fraction solubilities of nitrous oxide, at a partial pressure of 1.013 bar, in solvents containing nitrogen or halogens.

12.3 SOLUBILITY OF NITRIC OXIDE

Stull[4] has published the following vapour pressures of nitric oxide:

T/K	P/bar
156.4	20.26
170.0	40.53
178.4	60.79

These data fit the equation

$$\ln(P/\text{bar}) = 23.770 - 8174.4/[(T/K) + 237.39]$$

This equation can be used to calculate a reference line for the solubility of nitric oxide at temperatures above the critical temperature (Fig. 12.7). The uncertainties involved in such an extrapolation must be borne in mind.

12.3.1 Solubility of nitric oxide in water and aqueous solutions

Winkler[17] measured the solubility of nitric oxide in water over the temperature range 273 K to 353 K and reported results as Bunsen coefficients. Measurements at single temperatures have been reported by Usher[18] and

Armor.[19] Both these measurements agree with Winkler's work within about 1.3%.

Mole fraction solubilities at a partial pressure of 1.013 bar from data in the three papers mentioned above fit the equation

$$\ln x_g = -69.656 + 5577.2/(T/K) + 5.7049\ln(T/K) + 2.7435 \times 10^{-2} T/K$$

standard deviation in $x_g = 2.9 \times 10^{-7}$

temperature range 273 to 353 K.

This equation indicates that the mole fraction solubility is likely to pass through a minimum value at about 358.7 K. The solubility curve corresponding to the equation is shown in Fig. 12.6.

Nitric oxide is appreciably soluble in aqueous solutions of ferrous salts. The following equilibrium is established:

$$Fe^{2+} + NO \rightleftharpoons FeNO^{2+}.$$

Studies on these systems have been evaluated by Gerrard.[21]

12.3.2 Solubilities of nitric oxide in non-aqueous solvents

Shaw and Vosper[22] have measured solubilities at barometric pressure in seven organic solvents. Solubility in cyclohexane has been measured by Tsiklis and Svetlova.[23] There are several earlier papers dealing with the

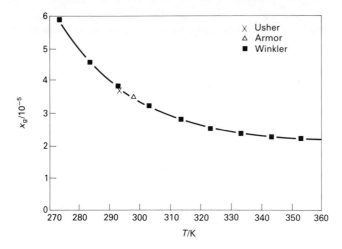

Figure 12.6. The mole fraction solubility in water of nitric oxide at a partial pressure of 1.013 bar.

Table 12.2 Mole fraction solubilities of nitric oxide at 293.2 K and a partial pressure of 1.013 bar

Solvent	x_g	Ref.
Cyclohexane	0.00196	23[a]
Hexane	0.00238	22[b]
Methylbenzene	0.00118	22[b]
1,3-Dimethylbenzene	0.00135	22[b]
1,1'-Oxybisethane	0.00253	22[b]
Ethyl acetate	0.00171	22[b]
Tetrachloromethane	0.00138	22[b]
Acetonitrile	0.000777	22[b]

[a] Extrapolated to 1.013 bar.
[b] Extrapolated to 293.2 K.

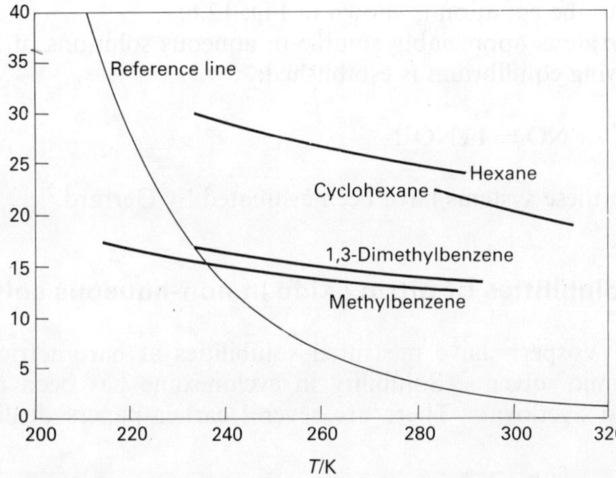

Figure 12.7. Mole fraction solubilities of nitric oxide, at a partial pressure of 1.013 bar, in hydrocarbons.

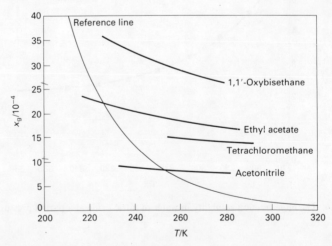

Figure 12.8. Mole fraction solubilities of nitric oxide, at a partial pressure of 1.013 bar, in solvents containing nitrogen, oxygen or chlorine.

solubility of this gas but these are of uncertain reliability. Mole fraction solubilities from data published by Shaw and Vosper and by Tsiklis and Svetlova are given in Table 12.2 and shown graphically in Figs 12.7 and 12.8. All solubility measurements were made above the critical temperature. Most values of mole fraction solubilities lie well above the reference line based on extrapolation of the vapour pressure equation.

12.4 SOLUBILITY OF NITROGEN DIOXIDE/ DINITROGEN TETROXIDE

There are few data in the literature relating to the solubility of nitrogen dioxide. The gas reacts with water to give nitrous acid and nitric acid.

$$N_2O_4 + H_2O \rightleftharpoons HNO_3 + HNO_2$$

The degree of dissociation of the oxide in a range of dilute solutions in organic solvents was measured by Cundall.[24] Solutions in hydrocarbons are said to be dangerously explosive.[25] Phase equilibria involving the compound have been investigated by Chang et al.[26] and by Whittaker et al.[28]

REFERENCES

1. Sidgwick, N.V. *Chemical Elements and their Compounds*, Oxford University Press, Oxford, 1950.
2. Cundell, J.T. *J. Chem. Soc.* 1891, 59, 1076; 1895, 67, 808.
3. *Solubility Data Series Vol. 8, Oxides of Nitrogen*, ed. C.L. Young, Pergamon, Oxford, 1981.
4. Stull, *Ind. Eng. Chem.* 1947, 39, 517.
5. Battino, R. ref. 3, page 1.
6. Weiss, R.F.; Price, B.A. *Marine Chemistry*, 1980, 8, 347.
7. Sada, E.; Kumazawa, H.; Butt, M.A. *J. Chem. Eng.Data* 1977, 22, 277.
8. Young, C.L. ref. 3, page 27.
9. Young, C.L. ref. 3, page 160.
10. Hsu, H.; Campbell, D. *Aerosol Age*, 1964, December, 34.
11. Makranczy, J.; Megyery-Balog,K.; Rusz,L.; Patyi,L. *Hung. J. Ind. Chem.* 1976, 4(2), 269.
12. Patyi, L.; Furmer, I.E.; Makranczy, J.; Sadilenko, A.S. Stepanova, Z.G.; Berengarten,M.G. *Zh. Prikl. Khim.* 1978, 51, 1296.
13. Yen, L.C.; McKetta, J.J. *J. Chem. Eng. Data* 1962, 7, 288.
14. Horiuti, J. *Sci. Pap. Inst. Phys. Chem. Res. (Japan)* 1931/32, 17, 125.
15. Kunerth, W. *Phys. Rev.* 1922, 19, 512.
16. Sada, E.; Kito, S.; Ito, Y. *Eng. Chem. Fundam.* 1975, 14, 232.
17. Winkler, L.W. *Ber.* 1901, 34, 1408.
18. Usher, F.L. *Z. Physik. Chem.* 1908, 62, 622.
19. Armor, J.N. *J. Chem. Eng. Data* 1974, 19, 82.
20. Battino, R. ref. 3, page 260.
21. Gerrard, W. ref. 3, page 265.
22. Shaw, A. W.; Vosper, A. J. *J. Chem. Soc. Faraday Trans. I* 1977, 73, 1239.
23. Tsiklis, D.S.; Svetlova, G.M. *Zh. Fiz. Khim.* 1958, 32, 1476.
24. Cundall, J.T. *J. Chem. Soc.* 1891, 59, 1076; 1895, 67, 808.
25. Partington, J.R. *General and Inorganic Chemistry*, Macmillan, London, 1946, 583.
26. Chang, E.T.; Gokcen, N.A.; Poston, T.M. *J. Chem. Eng. Data* 1971, 16, 404.
27. Markham, A.E.; Kobe, K.A. *J. Amer. Chem. Soc.* 1941, 63, 449.
28. Whittaker, A.G.; Sprague, R.W.; Skolnik, S.; Smith, G.B.L. *J. Phys. Chem.* 1966, 70, 2394.

Chapter 13

SOLUBILITIES OF CARBON MONOXIDE, NITROGEN, OXYGEN AND HYDROGEN

13.1 GENERAL BEHAVIOUR

All solubility measurements on these gases have been made at temperatures well above the critical. Depending on the system, solubilities may decrease, pass through a minimum or increase as the temperature is raised. As a general rule the lower the boiling point of a liquefied gas under a pressure of 1.013 bar the lower the solubility of the gas in a particular solvent at a temperature above the critical temperature. Reference lines need to be based upon hypothetical values of the vapour pressure of the liquefied gases, but they can assist in the correlation of solubilities of different systems.

13.2 SOLUBILITY OF CARBON MONOXIDE

Carbon monoxide has the following physical properties:

Melting point	= 66.2 K
Boiling point (1.013 bar)	= 81.7 K
Critical temperature	= 133 K
Critical pressure	= 35 bar
Critical volume	= 0.0931 dm^3 mol^{-1}
Density at 273.15 K; 1.013 bar	= 1.25003 g dm^{-3}
Relative molecular mass	= 28.010

Vapour pressures of liquid carbon monoxide from measurements by Michels et al.[1] are as follows:

Table

T/K	P/bar
93.251	3.1862
116.325	15.3855
132.595	34.4540

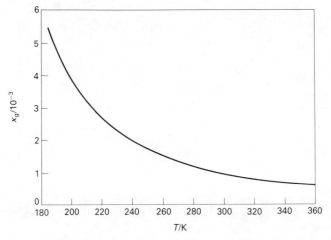

Figure 13.1. The reference line for carbon monoxide at a partial pressure of 1.013 bar.

These data fit the following equation which will be used to estimate the position of reference lines above the critical temperature

$$\ln(P/\text{bar}) = 9.8525 - 907.01/[(T/K) + 11.08]$$

(see Fig. 13.1).

13.2.1 Solubility of carbon monoxide in water

Wilhelm et al.[4] derived an equation for the mole fraction solubility in water at a partial pressure of 1.013 bar from nine data points published by Winkler[2] and Christoff.[3] This equation, plotted in Fig. 13.2, can be written in the form

$$\ln x_g = -171.764 + 8296.9/(T/K) + 23.3376\ln(T/K)$$

standard deviation in $x_g = 0.68\%$
temperature range 273 to 353 K.

According to this equation x_g at 298.15 K is 0.0000172. This is very low compared with the solubility in most organic solvents.

13.2.2 Solubility of carbon monoxide in hydrocarbons

The mole fraction solubility in many hydrocarbons increases with increase in temperature in the temperature range in which measurements have been made (Fig. 13.3).

Makranczy and co-workers[5,6] have measured the solubility in normal

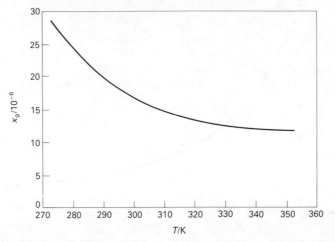

Figure 13.2. The mole fraction solubility in water of carbon monoxide at a partial pressure of 1.013 bar.

alkanes from pentane to hexadecane at 298.2 K and a partial pressure of 1.013 bar. Their measurements correspond to a decrease in mole fraction solubility with increase in chain length. Their value for the solubility in heptane is within about 2.5% of the value published by Gjaldbaek.[7] The general trend towards lower mole fraction solubility with increase in chain length is not supported by data for hexadecane published by Lin and Parcher[8] or by Tremper and Prausnitz[9] (Table 13.1). The solubility in cyclohexane measured by Battino et al.[10] agrees with that measured by Patyi

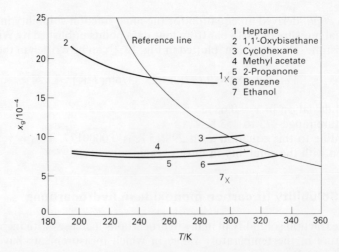

Figure 13.3. Mole fraction solubilities of carbon monoxide, at a partial pressure of 1.013 bar, in hydrocarbons and solvents containing oxygen.

CARBON MONOXIDE, NITROGEN, OXYGEN AND HYDROGEN

Table 13.1 The mole fraction solubility of carbon monoxide at a partial pressure of 1.013 bar

Solvent	T/K	x_g	Ref.
Pentane	298.2	0.001857	5
Hexane	298.2	0.001755	5
	298.2	0.001720	6
Heptane	298.2	0.001687	5
	298.2	0.00173	7
Octane	298.2	0.001553	5
	298.2	0.0001714	51
Nonane	298.2	0.001583	5
Decane	298.2	0.001520	5
	298.2	0.0001662	51
Undecane	298.2	0.001526	5
Dodecane	298.2	0.001541	5
Tridecane	298.2	0.001502	5
Tetradecane	298.2	0.001443	5
Pentadecane	298.2	0.001444	5
Hexadecane	298.2	0.001432	5
	298.2	0.00175	8
	313.2	0.00185	8
	328.2	0.00184	8
	300	0.00186	9
	475	0.00239	9
Hexatriacontane	353.2	0.00289	8
	373.2	0.00279	8
	393.2	0.00291	8
	413.2	0.00304	8
Cyclohexane	283.4	0.000974	10
	298.2	0.000991	10[a]
	298.2	0.001007	6
	308.7	0.001003	10
Methylcyclohexane	284.1	0.001235	11
	298.1	0.001241	11
	313.3	0.001225	11
	298.2	0.001234	11[a]
Bicyclohexyl	300	0.00103	9
	475	0.00138	9
Benzene	298.2	0.000658	6
	298.2	0.000674	7
	298.2	0.000624	13
	298.2	0.000645	12
	333.5	0.000752	12
Methylbenzene	283.7	0.000782	11
	308.3	0.000811	11
	313.3	0.000825	11
	298.2	0.000802	11[a]
	298.2	0.000815	14
	298.2	0.000790	13
	298.2	0.000798	15
Methanol	293.2	0.000377	17
	298.2	0.0003291	16
	298.2	0.000325	13
Ethanol	293.2	0.000485	17
	298.2	0.0004367	16
	298.2	0.000461	13

Table 13.1 (*continued*)

Solvent	T/K	x_g	Ref.
1-Propanol	293.2	0.000550	17
	298.2	0.0005193	16
2-Propanol	293.2	0.000605	17
1-Butanol	293.2	0.000624	17
	298.2	0.0005422	16
2-Butanol	298.2	0.000630	17
2-Methyl-1-propanol	293.2	0.000669	17
2-Methyl-2-propanol	298.2	0.000732	17
1-Pentanol	298.2	0.0005776	16
	298.2	0.000756	13
1-Hexanol	298.2	0.0006313	16
1-Heptanol	298.2	0.0006914	16
1-Octanol	298.2	0.0007463	16
	298.1	0.0008490	51
1-Nonanol	298.2	0.0007901	16
1-Decanol	298.2	0.0008502	16
	298.2	0.0009640	51
1-Undecanol	298.2	0.0008905	16
1-Dodecanol	298.2	0.0009434	16
2-Propanone	298.2	0.000780	12
1,1'-Oxybisethane	298.2	0.001693	12[a]
Acetic acid	298.2	0.000403	13
Methyl acetate	298.2	0.000850	12[a]
Ethyl acetate	298.2	0.001012	13
Propyl acetate	298.2	0.001189	14
Pentyl acetate	298.2	0.001294	13
2-Methylpropyl acetate	298.2	0.001288	13
Benzenamine	298.2	0.000201	13
	298.2	0.000199	15
	298.2	0.000190	14
Pyridine	298.2	0.000386	14
Benzyl cyanide	298.2	0.000358	14
Propanonitrile	298.2	0.000630	14
Nitrobenzene	298.2	0.000371	14
	298.2	0.000393	13
	298.2	0.000390	15
Trichloromethane	298.2	0.000644	13
	298.2	0.000686	15
Tetrachloromethane	298.2	0.000865	12[a]
	298.1	0.000870	18
1,2-Dibromoethane	298.2	0.000300	14
Hexadecafluoroheptane	298.2	0.00388	7
Chlorobenzene	298.2	0.000631	12[a]
Hexafluorobenzene	298.2	0.00215	20
Carbon disulfide	298.2	0.000360	7

[a]Interpolated.

et al.[6] The methyl group in methylcyclohexane enhances the mole fraction solubility.[11] Mole fraction solubility in bicyclohexyl is close to that in cyclohexane at 298.2 K and 1.013 bar.

Solubility in benzene at 298.2 K and 1.013 bar from Gjaldbaek's data[7] is within 2.5% of Patyi's value.[6] It is also within about 4.3% of an interpolated value from Horiuti's data[12] for the temperature range 285 K to 333 K. Measurements by Just[13] at the beginning of the century are probably too low. There is also close agreement between Gjaldbaek's value[14] of the solubility in methylbenzene, at 298.2 K and 1.013 bar, and the value reported by Field *et al.*[11] Older data published by Just and by Skirrow[15] are close to these values.

13.2.3 Solubility in organic solvents containing oxygen, halogens, nitrogen or sulfur

Mole fraction solubilities in normal alcohols at 298.2 K and a partial pressure of 1.013 bar increase with chain length, according to data published by Makranczy *et al.*[16] This trend is supported by data for solubilities in ethanol, 1-propanol and 1-butanol published by Gjaldbaek[17], although the latter values are up to 15% greater than those published by Makranczy. Data for methanol, ethanol and pentanol have also been published by Just.[13]

Horiuti[12] measured solubility in 2-propanone at several temperatures. The mole fraction solubility for a partial pressure of 1.013 bar from these measurements fits the equation:

$$\ln x_g = -26.890 + 723.58/(T/K) + 3.0376\ln(T/K)$$

standard deviation in $x_g = 1.74 \times 10^{-6}$
temperature range 193 to 313 K.

The value of x_g passes through a minimum at about 238 K but maximum change over the temperature range is only about 7%.

Mole fraction solubility in 1,1'-oxybisethane at 298.2 K is higher than that in alcohols and close to that in hydrocarbons. Data for a partial pressure of 1.013 bar from measurements by Horiuti[12] fit the equation:

$$\ln x_g = -26.347 + 853.37/(T/K) + 3.0019\ln(T/K)$$

standard deviation in $x_g = 5.16 \times 10^{-6}$
temperature range 194 to 293 K.

x_g decreases with rise in temperature in this temperature range.

Mole fraction solubilities at 298.2 K and 1.013 bar in alkyl acetates increase with the size of the alkyl group. Horiuti's measurements of the solubility in methyl acetate fit the equation

$$\ln x_g = -21.981 + 514.33/(T/\text{K}) + 2.3142\ln(T/\text{K})$$

standard deviation in $x_g = 7.72 \times 10^{-7}$
temperature range 194 to 313 K.

x_g passes through a minimum at about 222 K.

Solubilities in benzenamine and in nitrobenzene were measured by Gjaldbaek[14] and, at the beginning of the century, by Just[13] and by Skirrow.[15]

Horiuti measured the solubility in tetrachloromethane at 1.013 bar from 253 to 333 K. More recent measurements by Battino et al.[19] from 283 to 308 K are very close to those of Horiuti. The solubility at a partial pressure of 1.013 bar from the two sources of data fits the equation:

$$\ln x_g = -22.762 + 561.46/(T/\text{K}) + 2.4274\ln(T/\text{K})$$

standard deviation in $x_g = 5.5 \times 10^{-6}$
temperature range 253 to 333 K.

Horiuti[12] measured the solubility in chlorobenzene. The mole fraction solubility at a partial pressure of 1.013 bar is fairly close to that in benzene at 298.2 K and fits the equation

$$\ln x_g = -25.570 + 689.58/(T/\text{K}) + 2.7886\ln(T/\text{K})$$

standard deviation in $x_g = 1.56 \times 10^{-6}$
temperature range 233 to 353 K.

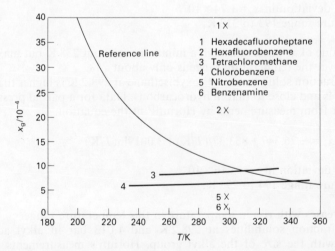

Figure 13.4. Mole fraction solubilities of carbon monoxide. at a partial pressure of 1.013 bar, in solvents containing halogens or nitrogen.

The mole fraction solubilities in hexadecafluoroheptane and hexafluorobenzene are very high compared with those in other halogenated compounds.

The solubility in carbon disulfide has been measured by Gjaldbaek.[7] The measurement is probably more reliable than earlier measurements by Just.[13]

The general pattern of solubilities in organic solvents is shown in Figs 13.3 and 13.4 and Table 13.1.

13.3 SOLUBILITY OF NITROGEN

Nitrogen has the following physical properties:

Melting point	= 63.25 K
Boiling point (1.013 bar)	= 77.34 K
Critical temperature	= 126.2 K
Critical pressure	= 33.9 bar
Critical volume	= 0.0901 dm^3 mol^{-1}
Density at 273.15 K; 1.013 bar	= 1.25051 g dm^{-3}
Relative molecular mass	= 28.0134

Michels et al.[21] have published the following equation for the vapour pressure of liquid nitrogen between 97 K and 125 K:

$$\log_{10}(P/\text{atm}) = -634.337/(T/K) - 15.33647 \log_{10}(T/K) \\ + 0.0332183 T/K + 34.58230$$

Reference lines, to provide a rough guide to solubility, can be based upon an extrapolation of this equation beyond the critical temperature of the gas. The reference line for 1.013 bar is shown in Fig. 13.5.

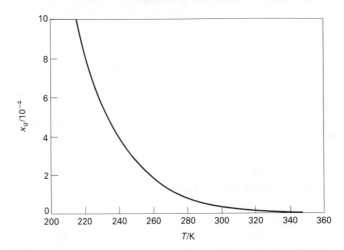

Figure 13.5. The reference line for nitrogen at a partial pressure of 1.013 bar.

Published data on the solubility of nitrogen have been compiled and evaluated in *Solubility Data Series Volume 10*.[22] There is also an excellent critical review of the solubility of nitrogen and air by Battino, Rettich and Tominaga,[23] which has provided the basis for much of the following discussion.

Nitrogen is the chief constituent of air. Wilcock and Battino[24] demonstrated that, within their experimental error of 1%, a mixture of 49.5% of oxygen and 50.5% of nitrogen dissolved to the same extent as calculated from solubilities of the pure gases. For many purposes the solubility of air can therefore be calculated from the partial pressures of the constituent gases and the solubilities of the pure gases.

13.3.1 Solubility of nitrogen in water

Battino *et al.*[23] have derived an equation for the mole fraction solubility in water at a partial pressure of 1.013 bar from 273 to 350 K. This equation is based upon 73 solubility values from nine papers and can be written in the form

$$\ln x_g = -181.5870 + 8632.13/(T/K) + 24.7981\ln(T/K)$$

standard deviation in $x_g = 0.72\%$ at about 300 K

According to this equation the value of x_g passes through a minimum at 340 K.

A similar equation for mole fraction solubilities at a partial pressure of 1.013 bar in the temperature range 350 to 600 K has been derived by Clever and Han.[25] This can be written in the form

$$\ln x_g = -141.2677 + 6921.99/(T/K) + 18.7292\ln(T/K)$$

According to this equation the value of x_g passes through a minimum value at 370 K. Graphical plots of the two equations are shown in Fig. 13.6.

Battino *et al.*[23] have also published a smoothing equation for the solubility in water at temperatures above 350 K and partial pressures to 1000 bar (100 MPa), based on 86 data points from four papers.[26-29] The equation can be written in the form

$$\ln x_g = -107.176 + 4852.4/(T/K) + 13.9321\ln(T/K) + 0.9700\ln(P/\text{bar}) - 0.000483(P/\text{bar})$$

standard deviation in $x_g = 5\%$
temperature range 350 to 503 K; pressure range 25 to 1000 bar.

The equation has been plotted in Fig. 13.7 for pressures to 100 bar.

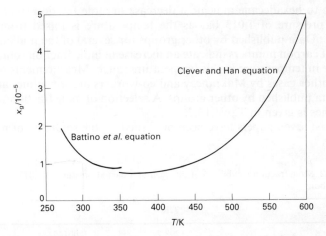

Figure 13.6. The mole fraction solubility in water of nitrogen at a partial pressure of 1.013 bar.

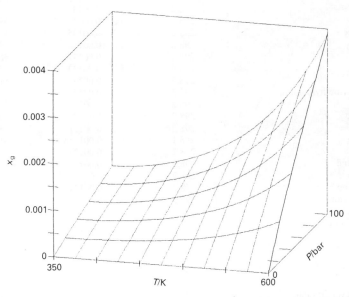

Figure 13.7. The variation with partial pressure and temperature of the mole fraction solubility of nitrogen in water.

13.3.2 Solubility of nitrogen in hydrocarbons

Published data for solubilities of nitrogen in alkanes have been evaluated by Battino[22] and by Battino et al.[23] At a partial pressure of nitrogen of 1.013 bar mole fraction solubilities in normal alkanes tend to decrease with increase in carbon number. Data published by Makranczy et al.[5] for n-alkanes from pentane to hexadecane indicate a decrease in mole fraction solubility for a partial pressure of 1.013 bar as the temperature is raised from 298.2 to 313.2 K. Data published by other groups for several of the n-alkanes in this range of carbon numbers indicate an increase in mole fraction solubility with increase in temperature in this temperature range. Measurements of solubilities of other gases by Makranczy and co-workers are sometimes at variance with data published by other groups. A selection of mole fraction solubilities in alkanes is given in Table 13.2.

At least seven papers have been published on the solubility of nitrogen in

Table 13.2 Mole fraction solubilities of nitrogen at a partial pressure of 1.013 bar in various hydrocarbons

Solvent	T/K	x_g	Ref.
Pentane	298.2	0.00145	5
Hexane	298.1	0.00140	30
	298.2	0.00138	34
	298.2	0.00141	5
	298.2	0.00141	49
Heptane	298.2	0.00135	50
	298.2	0.00138	5
Octane	298.2	0.00131	50
	298.2	0.00133	5
	298.3	0.00131	51
2,2,4-Trimethylpentane	293.2	0.00155	52
	298.2	0.001533	33
Nonane	298.2	0.00128	50
	298.2	0.00131	5
Decane	298.1	0.001214	51
	298.2	0.00125	5
Undecane	298.2	0.00127	5
Dodecane	298.2	0.00123	5
Tridecane	298.2	0.00124	5
Tetradecane	298.2	0.00124	5
Pentadecane	298.2	0.00126	5
Hexadecane	298.2	0.00122	5
	298.2	0.00126	9[a]
Cyclohexane	298.1	0.000753	30
	298.2	0.000768	53
	298.2	0.000761	54
	298.2	0.000755	49
Methylcyclohexane	298.2	0.000946	11
Cyclooctane	298.3	0.000598	51
(Z)-1,2-Dimethylcyclohexane	298.0	0.000884	55
(E)-1,2-Dimethylcyclohexane	298.3	0.001001	55
1,1'-Bicyclohexyl	298.2	0.000740	9
Benzene	298.2	0.000445	23[b]
Methylbenzene	298.2	0.000539	13
	298.2	0.000543	11[c]

[a] Extrapolated.
[b] Smoothed data.
[c] Interpolated.

cyclohexane. There is good agreement in the mole fraction solubility at 298.2 K from data in four of the papers (Table 13.2). Battino et al.[23] have published a smoothing equation based upon three of these sets of data. This may be written in the form

$$\ln x_g = -5.874 - 385.2/(T/K)$$

standard deviation in $\ln x_g = 0.010$
standard deviation in $x_g = 1\%$
temperature range 283 to 333 K.

Data for other cyclic alkanes are shown in Table 13.2.

The solubility in benzene was measured by Horiuti.[12] His measurements were close to those obtained later by Gjalbaek and Hildebrand[30] and Byrne et al.[31] Battino et al.[23] have derived a smoothing equation for the mole fraction solubility at a partial pressure of 1.013 bar from data in these papers. This may be written in the form

$$\ln x_g = -6.0544 - 495.67/(T/K)$$

standard deviation in $\ln x_g = 0.0090$
temperature range 280 to 333 K.

Mole fraction solubilities for other aromatic hydrocarbons are given in Table 13.2. The variation of solubility with temperature in some hydrocarbons is shown in Fig. 13.8.

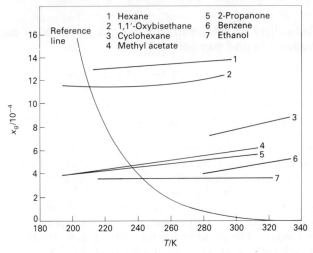

Figure 13.8. Mole fraction solubilities of nitrogen, at a partial pressure of 1.013 bar, in hydrocarbons and organic compounds containing oxygen. The solubility curves for 1,1'-oxybisethane and methyl acetate are based upon smoothing equations derived by the authors. Other curves are based upon equations published by Battino et al. (Ref. 22).

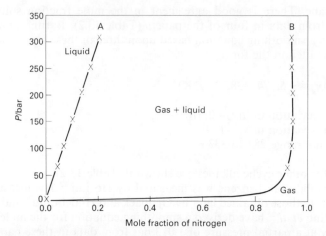

Figure 13.9. Phase equilibria in the nitrogen + benzene system at 398.15 K. Curve A shows the variation of the solubility of nitrogen in benzene with variation in total pressure. Curve B shows the variation of the composition of the gas phase with change in total pressure.

There have been various studies of phase equilibria between aromatic hydrocarbons and nitrogen at pressures above 1.013 bar over a wide temperature range. Phase equilibria between benzene and nitrogen at 398.2 K are shown in Fig. 13.9, which is based on data published by Miller and Dodge.[32] At this temperature the mole fraction concentration of nitrogen in the liquid phase varies linearly with total pressure to at least 307 bar.

13.3.3 Solubility of nitrogen in solvents containing carbon, hydrogen and oxygen

Solubilities in 1-alkanols from C1 to C12 have been measured by various groups. In addition, data for solubility in 2-propanol and 2-methyl-1-propanol have been published. Published data have been evaluated by Battino[22] and by Battino et al.[23] Agreement between different groups is poor for some of the alcohols.

Measurements by Kretschmer et al.[33] and by Katayama and Nitta[34] show that the mole fraction solubilities in methanol, ethanol, 1-propanol and 2-propanol at a partial pressure of 1.013 bar pass through minima at about 273 K. The minimum for 1-butanol[34] is about 253 K.

In general, however, at a partial pressure of 1.013 bar the mole fraction solubilities in alkanols show little change over the temperature ranges which have been studied. The smoothed value[23] of the mole fraction solubility in methanol is 0.000284 at 213.2 K, 0.000270 at 273.2 K and 0.000280 at 323.2 K. Corresponding values for ethanol are 0.000369, 0.000352 and 0.000367.

Horiuti's[12] measurement of the solubility in 2-propanone is in close agreement with that published by Kretschmer et al.[34] Battino et al.[24] have used these data to derive an equation for the mole fraction solubility at a partial pressure of 1.013 bar. This may be written in the form

$$\ln x_g = -22.172 + 377.54/(T/K) + 2.3493\ln(T/K)$$

standard deviation in $x_g = 3.2 \times 10^{-6}$
temperature range 195 to 314 K.

Horiuti[12] also measured solubilities in 1,1'-oxybisethane and in methyl acetate over temperature ranges. Christoff's[3] measurement of the solubility in 1,1'-oxybisethane at 273.2 K is within about 4% of Horiuti's value. Equations for the mole fraction solubilities at a partial pressure of 1.013 bar in these two solvents on the basis of data published by Horiuti are as follows:

1,1'-Oxybisethane

$$\ln x_g = -22.7706 + 553.29/(T/K) + 2.4983\ln(T/K)$$

standard deviation in $x_g = 3.1 \times 10^{-6}$
temperature range 195 to 293 K.

Methyl acetate

$$\ln x_g = -19.1957 + 238.23/(T/K) + 1.9228\ln(T/K)$$

standard deviation in $x_g = 1.6 \times 10^{-6}$
temperature range 194 to 313 K.

Selected values of mole fraction solubilities at a partial pressure of 1.013 bar in other compounds of carbon, hydrogen and oxygen are given in Table 13.3.

13.3.4 Solubility of nitrogen in organic compounds containing halogens

The solubility in trichloromethane at 293.2 K and at 298.2 K was reported by Just.[13] A value at 295.2 K reported by Korosy[36] is about 2% lower than the interpolated value at this temperature from Just's measurements. The solubility in tetrachloromethane was measured by Horiuti[12] for the temperature range 253 K to 333 K. Mole fraction solubilities at a partial pressure of 1.013 bar from these measurements fit an equation published by Battino et al.[23] which may be written in the form

$$\ln x_g = -22.1869 + 407.89/(T/K) + 2.3634\ln(T/K)$$

standard deviation in $x_g = 9.0 \times 10^{-7}$
temperature range 253 to 333 K.

Table 13.3 Mole fraction solubilities of nitrogen at a partial pressure of 1.013 bar in solvents containing carbon, hydrogen and oxygen

Solvent	T/K	x_g	Ref.
Methanol	298.2	0.000273	23[a]
Ethanol	298.2	0.000357	23[a]
1-Propanol	298.2	0.000406	23[a]
2-Propanol	298.2	0.000466	23[a]
1-Butanol	298.2	0.000461	23[a]
2-Methyl-1-propanol	298.4	0.000482	56
1-Pentanol	298.2	0.000544	13
	298.2	0.000495	57
	298.2	0.000516	58
	298.2	0.000546	16
1-Hexanol	298.2	0.000584	58
	298.2	0.000599	16
Cyclohexanol	298.2	0.000266	57
1-Heptanol	298.2	0.000610	58
	298.2	0.000660	16
	298.2	0.000609	59
1-Octanol	298.2	0.000657	59
	298.2	0.000700	16
	298.2	0.000680	58
	298.2	0.000623	51
1-Nonanol	298.2	0.000750	16
1-Decanol	298.2	0.000805	16
	298.1	0.000668	51
1-Undecanol	298.2	0.000849	16
1-Dodecanol	298.2	0.000898	16
2-Propanone	298.2	0.000542	12[b]
	298.2	0.000549	33
Methyl acetate	298.2	0.000587	12[b]
1,1'-Oxybisethane	298.2	0.001255	12[b]
	273.2	0.00116	35
Tetrahydrofuran	298.2	0.000521	60
1,4-Dioxane	298.2	0.000237	60
2,3-Dihydropyran	298.2	0.000493	60
Tetrahydro-2H-pyran	298.2	0.000593	60
1,1'-Oxybispropane	298.2	0.00122	60
1,2-Epoxyethane	298.2	0.000459	61

[a]Smoothed data.
[b]Interpolated or extrapolated.

Mole fraction solubilities increase with rise in temperature.

Horiuti also measured the solubility in chlorobenzene. Mole fraction solubilities at a partial pressure of 1.013 bar fit the equation[23]

$$\ln x_g = -25.0006 + 499.56/(T/K) + 2.7320\ln(T/K)$$

standard deviation in $x_g = 1.8 \times 10^{-6}$
temperature range 253 to 333 K.

Mole fraction solubilities increase with rise in temperature.

Table 13.4 Mole fraction solubilities of nitrogen at a partial pressure of 1.013 bar in solvents containing halogens

Solvent	T/K	x_g	Ref.
Trichloromethane	295.2	0.000425	36
	293.2	0.000427	13
	298.2	0.000445	13
Dichlorodifluoromethane	273.2	0.00159	62
Tetrachloromethane	298.2	0.000641	12[a]
1,2-Dichloro-1,1,2,2-tetrafluoroethane	293.2	0.00809	63
1,1,2-Trichloro-1,2,2-trifluoroethane	298.1	0.001931	64
Chlorobenzene	298.2	0.000427	12[a]
1-Chlorohexane	298.2	0.000819	60
Hexafluorobenzene	298.2	0.00179	20
1-Bromoheptane	298.2	0.000675	59
Undecafluoro(trifluoromethyl)cyclohexane	298.2	0.00336	30
Hexadecafluoroheptane	298.1	0.00393	30
Heptafluorotetrahydro(nonafluorobutyl)furan	298.2	0.00430	65
1,1,1,2,2,3,3,4,4,5,5,6,6-Tridecafluoro-6-[1,2,2,2-tetrafluoro-1-(trifluoromethyl)ethoxy]hexane	298.2	0.00510	65
1,1,2,2,3,3,4,4-Octafluoro-1,4-bis[1,2,2,2-tetrafluoro-1-(trifluoromethyl)ethoxy]butane	298.2	0.00500	65

[a] Interpolated.

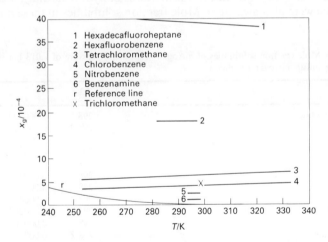

Figure 13.10. Mole fraction solubility of nitrogen, at a partial pressure of 1.013 bar, in solvents containing nitrogen or halogens. The solubility curves are based upon smoothing equations published by Battino et al. (Ref. 22).

Selected values of mole fraction solubilities in other organic compounds containing halogens are given in Table 13.4. Mole fraction solubilities in compounds containing fluorine can be exceptionally high compared with solubilities in other organic solvents (Fig. 13.10).

13.3.5 Solubilities of nitrogen in organic compounds containing nitrogen, sulfur or silicon

Solubilities in several solvents containing nitrogen have been reported. In many cases there are measurements at one temperature only.

Just[13] measured the solubility in nitrobenzene at 293.2 and 298.2 K. A measurement at 298.2 K by Metschl[37] confirms Just's measurement. Just also measured the solubility in benzenamine at 293.2 and 298.2 K. Metschl's measurement at 298.2 K is 13% higher than Just's value in this case.

Kobatake and Hildebrand[38] measured the solubility in 1,1,2,2,3,3,4,4,4-nonafluoro-N,N-bis(nonafluorobutyl)-1-butanamine (perfluorotributylamine, $C_{12}F_{27}N$) over the temperature range from 283 to 303 K. Mole fraction solubilities for a partial pressure of 1.013 bar at 298.2 K in this solvent from measurements by Sargent and Seffl[39] and Powell[40] are 19% and 21% higher than values found by Kobatake and Hildebrand. Whichever measurements are correct it is clear that the mole fraction solubility for a partial pressure of 1.013 bar is high compared with most other organic solvents.

Solubilities in hydrazine, methylhydrazine and 1,1-dimethylhydrazine were measured by Chang et al.[41] Mole fraction solubilities at 298 K and a partial pressure of 1.013 bar, calculated from their measurements, are given in Table 13.5. Solubilities in various mixtures of hydrazine and 1,1-dimethylhydrazine were also measured. Mole fraction solubilities in these mixtures at

Table 13.5 Mole fraction solubilities of nitrogen at a partial pressure of 1.013 bar in solvents containing sulfur, nitrogen or silicon

Solvent	T/K	x_g	Ref.
Carbon disulfide	298.1	0.000223	30
	298.2	0.0002215	38
	298.2	0.000222	40
Sulfinylbismethane	298.2	0.0000833	43
Nitromethane	298.0	0.000201	66
Pyrrolidine	298.2	0.000368	60
Pyridine	298.2	0.000250	60
Piperidine	298.2	0.000422	60
Nitrobenzene	298.2	0.000264	13
	298.2	0.00026	37
Benzenamine	298.2	0.00013	37
	298.2	0.000115	13
1,1,2,2,3,3,4,4,4-Nonafluoro-N,N-bis(nonafluorobutyl)-1-butamine	298.2	0.003490	38
	298.2	0.00415	39
	298.2	0.00424	40
Hydrazine	298.2	0.0000072	41
Methylhydrazine	298.1	0.0000929	41
1,1-Dimethylhydrazine	298.2	0.000376	41[a]
Octamethylcyclotetrasiloxane	298.2	0.00255	48

[a] Extrapolated.

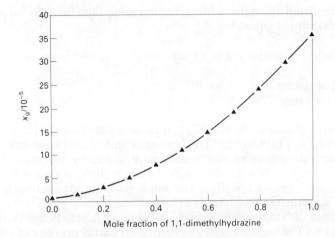

Figure 13.11. Mole fraction solubility of nitrogen in mixtures of hydrazine and 1,1-dimethylhydrazine (partial pressure of nitrogen, 1.013 bar; temperature, 288.15 K).

288.2 K and a partial pressure of nitrogen of 1.013 bar are shown in Fig. 13.11.

Solubilities in 1-methyl-2-pyrrolidinone and in 1,5-dimethyl-2-pyrrolidinone over temperature and pressure ranges were measured by Shakhova et al.[42] The dependence of mole fraction solubility upon partial pressure of nitrogen at 298.2 K for these solvents is shown in Fig. 13.12.

Solubilities in other solvents containing nitrogen are shown in Table 13.5.

Measurements of the solubility in carbon disulfide by Gjaldbaek and Hildebrand,[30] Kobatake and Hildebrand[38] and Powell[40] are consistent with

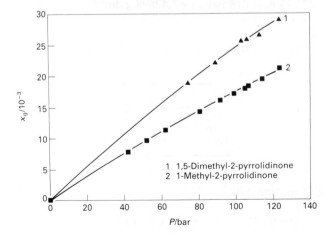

Figure 13.12. Variation with pressure of the mole fraction solubility of nitrogen in 1,5-dimethyl 2-pyrrolidinone and 1-methyl-2-pyrrolidinone at 298 K.

each other within about 0.7% Mole fraction solubilities for a partial pressure of 1.013 bar fit the equation:

$$\ln x_g = -6.4005 - 599.81/(T/K)$$

standard deviation in $x_g = 7.3 \times 10^{-7}$.
temperature range 279 to 303 K.

Dymond[43] measured the solubility in sulfinylbismethane (dimethylsulfoxide) at 298.2 K (Table 13.5). The hydrogen sulfide + nitrogen system was investigated at various pressures and temperatures by Robinson and co-workers.[44,45] A few data for the sulfur dioxide + nitrogen system have been published by Dean and Walls.[46] The sulfur hexafluoride + nitrogen system has been investigated by Miller et al.[47]

Wilcock et al.[48] measured solubility in octamethylcyclotetrasiloxane from 292 to 313 K. The mole fraction solubility for a partial pressure of 1.013 bar is high in this solvent (Table 13.5). Changes in mole fraction solubility at these temperatures are less than the experimental error of the measurements.

13.4 SOLUBILITY OF OXYGEN

Oxygen has the following physical properties:

Melting point	= 54.36 K
Boiling point (1.013 bar)	= 90.18 K
Critical temperature	= 154.8 K
Critical pressure	= 50.3 bar
Critical volume	= 0.078 dm^3 mol^{-1}
Density at 273.15 K, 1.013 bar	= 1.42898 g dm^{-3}
Relative molecular mass	= 31.999

Stull[67] has published the following values of the vapour pressure of liquid oxygen:

T/K	P/bar
90.2	1.013
120.0	10.13
149.1	40.53

These data fit the equation

$$\ln(P/\text{bar}) = 9.6377 - 912.15/[(T/K) + 4.62]$$

This equation can be used to calculate reference lines for oxygen which can

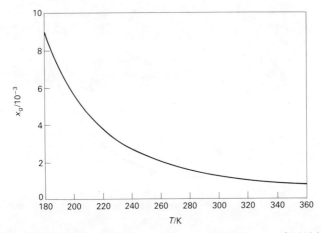

Figure 13.13. The reference line for oxygen at a partial pressure of 1.013 bar.

give an approximate guide to the solubility at temperatures above the critical temperature (Fig. 13.13).

Published data on the solubility of oxygen have been compiled and evaluated in Solubility Data Series, Volume 7.[68] The solubilities of oxygen and ozone have also been critically reviewed by Battino et al.[69]

13.4.1 Solubility of oxygen in water

The solubility of oxygen in water has been measured many times since the pioneering work of Bunsen. The most accurate values of mole fraction solubility below 333K at a partial pressure of 1.013 bar are likely to be those of Benson et al.[70] Battino[68] has published an equation which fits these data. This may be written in the form

$$\ln x_g = -171.2542 + 8391.24/(T/K) + 23.24323 \ln(T/K)$$

standard deviation in $x_g = 0.17\%$
temperature range 273 to 333 K.

Measurements of the solubility in water in the range 373 to 617 K have been made by Stephan et al.[71] and Pray et al.[72] Clever and Han[25] have used these and other data for lower temperatures to produce a smoothing equation for the mole fraction solubility at a partial pressure of oxygen of 1.013 bar. This may be written in the form

$$\ln x_g = -139.485 + 6889.6/(T/K) + 18.554 \ln(T/K)$$

standard deviation in $x_g = 2\%$
temperature range 273 to 617 K.

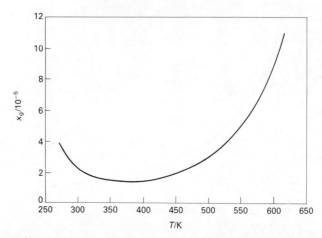

Figure 13.14. The mole fraction solubility in water of oxygen at a partial pressure of 1.013 bar.

The variation in mole fraction solubility, according to this equation, is shown in Fig 13.14.

13.4.2 Solubility of oxygen in hydrocarbons

Battino et al.[68,69] have evaluated the data from the literature for solubility in alkanes. They have prepared smoothed values of mole fraction solubilities.[69]

Mole fraction solubilities at a partial pressure of 1.013 bar show little change with temperature. Mole fraction solubilities in hexane (293 to 298 K) and heptane (293 to 313 K) increase slightly with rise in temperature. Mole fraction solubilities in octane from 283 to 313 K, 2,2,4-trimethylpentane from 243 to 323 K, nonane from 298 to 303 K, and decane from 283 to 313 K decrease with rise in temperature. Mole fraction solubility in 2,2,4-trimethylpentane is higher than in other alkanes.

Measurements of solubilities in straight-chain alkanes from C5 to C16 by Makranczy et al.[5] show a general decrease in mole fraction solubility with increase in chain length at 298.2 and 313.2 K and a partial pressure of 1.013 bar. This is not borne out by Blanc and Batiste[73] who indicate a general increase in mole fraction solubility with increase in chain length from C7 to C18 at 303.2 K. Further work on these systems is required to resolve this discrepancy.

Mole fraction solubilities in cyclic alkanes at temperatures close to 298 K and a partial pressure of 1.013 bar are lower than in straight chain alkanes of the same carbon number (Table 13.6; Fig. 13.15).

Battino et al.[69] have published a smoothing equation for the solubility of oxygen in benzene at a partial pressure of 1.013 bar. This is based upon data published by Byrne et al.,[31] Horiuti,[12] and Schlapfer et al.[74] This equation may be written in the form

Table 13.6 Mole fraction solubilities of oxygen in hydrocarbons at a partial pressure of 1.013 bar

Solvent	T/K	x_g	Ref.
Pentane	298.2	0.00205	5
Hexane	298.2	0.00198	5
	298.2	0.00205	69[a]
Heptane	298.2	0.00217	50
	298.2	0.00194	5
	298.2	0.002055	69[a]
Octane	298.2	0.00206	50
	298.2	0.00190	5
	298.2	0.002177	50
	298.2	0.002120	69[a]
2,2,4-Trimethylpentane	298.2	0.002529	33
	298.2	0.002814	38
	298.2	0.002654	69[a]
Nonane	298.2	0.00209	50
	298.2	0.00189	5
	298.2	0.002110	69[a]
Decane	298.2	0.00185	5
	298.1	0.002200	51
	298.2	0.002178	69[a]
Cyclohexene	298.2	0.00104	60
Cyclohexane	298.2	0.00123	60
	298.2	0.00125	54[b]
Methylcyclohexane	298.2	0.001599	75
(Z)-1,2-Dimethylcyclohexane	298.2	0.001543	55[b]
(E)-1,2-Dimethylcyclohexane	298.1	0.001734	55
Benzene	298.2	0.000810	69[a]
Methylbenzene	298.2	0.000923	69[a] (based on 75)
1,4-Dimethylbenzene	298.2	0.001244	76
1,2-Dimethylbenzene	298.2	0.001118	76
1,3-Dimethylbenzene	298.2	0.001196	76
Ethylbenzene	298.2	0.001220	76
Propylbenzene	298.2	0.001345	76
(1-Methylethyl)benzene	298.2	0.001388	76
1-Methyl-4-propylbenzene	298.2	0.001429	76
Butylbenzene	298.2	0.001440	76
(1-Methylpropyl)benzene	298.2	0.001569	76

[b]Interpolated.
[a]Smoothed data.

$$\ln x_g = -30.1649 + 874.16/(T/K) + 3.53024\ln(T/K)$$

standard deviation in $x_g = 6 \times 10^{-6}$
temperature range 283 to 343 K.

Mole fraction solubility in benzene at 1.013 bar shows a small increase with increase in temperature. Insufficient data are available to see whether methylbenzene behaves in a similar manner. Mole fraction solubilities in the alkylbenzenes at 298.2 K and a partial pressure of oxygen of 1.013 bar increase with increase in the number and size of the side chains (Table 13.6).

13.4.3 Solubility of oxygen in solvents containing carbon, hydrogen and oxygen

Makranczy et al.[16] have reported Ostwald coefficients in straight chain alkanols from methanol to undecanol but values tend to be from about 8% to 27% lower than values from other workers and may not correspond to equilibrium conditions. Battino et al.[69] published a smoothing equation for the mole fraction solubility in ethanol at a partial pressure of 1.013 bar. This is based upon measurements by Kretschmer et al.,[33] Schlapfer et al.,[74] Timofeev,[77] Shchukarev and Tolmacheva,[78] and Naumenko.[79] It can be written in the form

$$\ln x_g = -7.874 + 126.93/(T/K)$$

standard deviation in $x_g = 1.1 \times 10^{-5}$
temperature range 248 to 343 K.

Solubilities in 2-propanone, 1,1'-oxybisethane and methyl acetate have been measured by more than one group and with good agreement. Equations for the mole fraction solubility at a partial pressure of gas of 1.013 bar in each of these solvents have been published by Battino et al.[69] In every case the mole fraction solubility passes through a minimum with rise in temperature.

The equation for 2-propanone, based upon measurements by Horiuti,[12] Kretschmer et al.[33] and Bub and Hillebrand,[80] can be written in the form

$$\ln x_g = -24.3100 + 649.40/(T/K) + 2.6414\ln(T/K)$$

standard deviation in $x_g = 4.7 \times 10^{-6}$
temperature range 195 to 318 K.

The equation for 1,1'-oxybisethane, based upon measurements by Horiuti,[12] Schlapfer[74] and Christoff,[35] can be written in the form

$$\ln x_g = -28.9222 + 970.90/(T/K) + 3.4083\ln(T/K)$$

standard deviation in $x_g = 2.3 \times 10^{-5}$
temperature range 195 to 293 K.

The equation for methyl acetate, based upon measurements by Horiuti[12] and Schlapfer et al.,[74] can be written:

$$\ln x_g = -20.6196 + 466.02/(T/K) + 2.1119\ln(T/K)$$

standard deviation in $x_g = 2.8 \times 10^{-6}$
temperature range 195 to 313 K.

Solubilities in other compounds containing oxygen have been studied less extensively. Available evidence shows that, in the case of ketones, mole fraction solubility increases with the carbon number. Mole fraction solubility in cyclic ethers is about half that in non-cyclic ethers.

Selected mole fraction solubilities at a partial pressure of 1.013 bar in organic compounds containing oxygen are shown in Table 13.7 and Fig. 13.15.

Table 13.7 Mole fraction solubilities of oxygen at a partial pressure of 1.013 bar in various organic compounds containing oxygen

Solvent	T/K	x_g	Ref.
Methanol	298.2	0.0004122	33
Ethanol	298.2	0.000583	69[a]
2-Propanol	298.2	0.0007745	33
1-Butanol	298.2	0.0007894	33
2-Methyl-1-propanol	298.3	0.000854	56
1-Octanol	298.1	0.001132	51
1-Decanol	298.1	0.001219	51
Cyclohexanol	299.2	0.000855	81
2-Propanone	298.2	0.0008399	69[a]
2-Butanone	298.2	0.001011	80
2-Pentanone	298.2	0.001112	80
2-Hexanone	298.2	0.001174	80
Cyclohexanone	298.2	0.000636	60
1,1'-Oxybisethane	298.2	0.001937	69[a]
1,1'-Oxybispropane	298.2	0.00191	60
Tetrahydrofuran	298.2	0.000816	60
1,4-Dioxane	293.2	0.000616	74
	293.2	0.000520	60
	298.2	0.000538	60
Tetrahydro-2H-pyran	298.2	0.000964	60
Methyl acetate	298.2	0.0008905	69[a]
	293.2	0.0008824	69[a]
Ethyl acetate	293.2	0.000871	74

[a]Smoothed data.

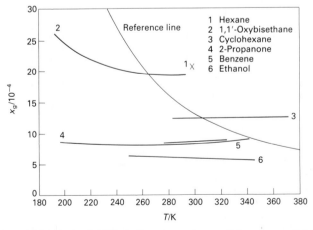

Figure 13.15. Mole fraction solubilities of oxygen, at a partial pressure of 1.013 bar, in hydrocarbons and solvents containing oxygen.

13.4.4 Solubility of oxygen in halogenated solvents

The solubility of oxygen in tetrachloromethane and in chlorobenzene was measured by Horiuti.[12] There is very little change in the mole fraction solubility from 273 to 333 K. The data given by Horiuti correspond to a mole fraction solubility at a partial pressure of 1.013 bar of 0.001203 at 273.2 K, 0.001199 at 303.2 K and 0.001203 at 333.2 K. The mole fraction solubility in chlorobenzene shows a small increase with temperature, namely 0.000778 at 273.2 K and 0.000825 at 353.2 K.

Mole fraction solubilities are exceptionally high in some fluorinated solvents in comparison with solubilities in other organic solvents (Table 13.8 and Fig. 13.16).

Table 13.8 Mole fraction solubilities of oxygen at a partial pressure of 1.013 bar in various organic compounds containing halogens

Solvent	T/K	x_g	Ref.
Tetrachloromethane	298.2	0.001200	12[a]
Chlorobenzene	298.2	0.0007884	12[a]
Hexafluorobenzene	298.2	0.002418	20[b]
1,2-Dibromo-1,1,2,3,3,3-hexafluoropropane; $C_3F_6Br_2$	298.2	0.00303	82
1,1,1,3,3,4,4,5,5,5-decafluoro-2-(trifluoromethyl)-2-pentanol; $C_6HF_{13}O$	298.2	0.00374	83
Fluorobenzene	298.2	0.001508	76
Bromobenzene	298.2	0.000748	76
Iodobenzene	298.2	0.000510	76
1-Chlorohexane	298.2	0.00135	60
1,1,1,2,2,3,3,5,5,5-Decafluoro-4-methoxy-4-(trifluoromethyl)pentane $C_7H_3F_{13}O$	298.2	0.00462	83
1-Bromoheptane	298.2	0.00132	59
1-Bromo-1,1,2,2,3,3,4,4,5,6,6,6-dodecafluoro-5-(trifluoromethyl)hexane $C_7F_{15}Br$	298.2	0.00475	82
1-Chloro-1,1,2,2,3,3,4,4,5,6,6,6-dodecafluoro-5-(trifluoromethyl)hexane $C_7F_{15}Cl$	298.2	0.00490	82
Undecafluoro(trifluoromethyl)cyclohexane; C_7F_{14}	298.2	0.00456	82
Hexadecafluoroheptane	298.2	0.00555	7
Bis(trifluoromethyl)benzene $C_6H_4(CF_3)_2$	298.2	0.00631	82
2-Ethoxy-1,1,1,3,3,4,4,5,5,5-decafluoro-2-(trifluoromethyl)pentane $C_8H_5F_{13}O$	298.2	0.00514	83
Heptafluorotetrahydro(nonafluorobutyl)furan; $C_8F_{16}O$	298.2	0.00560	65
1-Bromo-1,1,2,2,3,3,4,4,5,5,6,6,7,7,8,8,8-heptadecafluorooctane $C_8F_{17}Br$	298.2	0.00566	82
Octadecafluorooctane C_8F_{18}	298.2	0.00534	82
1,1,1,2,2,3,3-Heptafluoro-4,4-bis(trifluoromethyl)heptane $C_9H_7F_{13}$	298.2	0.00494	83
1,1,1,2,2,3,3,5,5,5-Decafluoro-4-propoxy-4-(trifluoromethyl)pentane $C_9H_7F_{13}O$	298.2	0.00509	83
1-Chloro-1,1,2,2,3,3,4,4,5,5,6,6,7,8,8,8-hexadecafluoro-7-(trifluoromethyl)octane; $C_9F_{19}Cl$	298.2	0.00511	82
Eicosafluorononane; C_9F_{20}	298.2	0.00535	82

Table 13.8 (continued)

Solvent	T/K	x_g	Ref.
1,1,1,2,2,3,3,4,4,5,5,6,6-Tridecafluoro-6-[1,2,2,2-tetrafluoro-1-(tri-fluoromethyl)ethoxy]hexane; $C_9F_{20}O$	298.2	0.00660	65
1,1,1,2,2,3,3-Heptafluoro-4,4-bis(trifluoromethyl) octane $C_{10}H_9F_{13}$	298.2	0.00530	83
1,1,1,2,2,3,3-Heptafluoro-6-methyl-4,4-bis (trifluoromethyl)heptane $C_{10}H_9F_{13}$	298.2	0.00502	83
Octadecafluorodecahydronaphthalene; $C_{10}F_{18}$	298.2	0.00390	82
1,1,2,2,3,3,4,4-Octafluoro-1,4-bis[1,2,2,2-tetrafluoro-1-(trifluoromethyl)ethoxy]butane; $C_{10}F_{22}O_2$	298.2	0.00650	65
1,1,1,2,3,3-Hexafluoro-2-(heptafluoropropoxy)-3-[1,2,2-trifluoro-2-(1,2,2,2-tetrafluoromethyl)-1-(trifluoromethyl)ethoxy]propane; $C_{11}HF_{23}O_3$	298.2	0.00683	82
1,1,1,7,7,7-Hexafluoro-2,2,6,6-tetrakis(tri-fluoromethyl)heptane $C_{11}H_6F_{18}$	298.2	0.00478	83
Heptadecafluorodecahydro(trifluoromethyl) naphthalene; $C_{11}F_{20}$	298.2	0.00406	82
Tetradecafluorobis(trifluoromethyl)tricyclo [3,3,1,3,7]decane $C_{12}F_{20}$	298.2	0.00420	82
1,1,1,2,4,4,5,7,7,8,10,10,11,13,13,14,14,15,15,15-Eicosafluoro-5,8,11-tris(trifluoromethyl)-3,6,9,12-tetraoxapentadecane; $C_{14}HF_{29}O_4$	298.2	0.00759	82
1,1,1,2,4,4,5,7,7,8,10,10,11,13,13,14,16,16,17,17,18,18,18-Tricosafluoro-5,8,11,14,-tetrakis (trifluoromethyl)-3,6,9,12,15-pentaoxaoctadecane $C_{17}HF_{35}O_5$	298.2	0.00853	82
1,1,1,2,4,4,5,7,7,8,10,10,11,13,13,14,16,16,17,19,19,20,22,22,23,25,25,26,28,28,29,29,30,30,30-pentacontafluoro-5,8,11,14,17,20,23,26-octakis (trifluoromethyl)-3,6,9,12,15,21,24,27-nonaoxa-triacontane; $C_{29}HF_{59}O_9$	298.2	0.0118	82

[a] Interpolated.
[b] Extrapolated.

13.4.5 Solubility of oxygen in solvents containing nitrogen or sulfur

Relatively few measurements of the solubility of oxygen in such solvents have been reported. The mole fraction solubility in sulfinylbismethane at 298.2 K was measured by Dymond[43] and Baird and Foley[84] with poor agreement (Table 13.9).

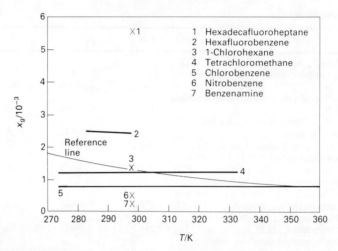

Figure 13.16. Mole fraction solubilities of oxygen, at a partial pressure of 1.013 bar, in solvents containing halogens.

Table 13.9 Mole fraction solubilities of oxygen at a partial pressure of 1.013 bar in organic compounds containing sulfur or nitrogen

Solvent	T/K	x_g	Ref.
Carbon disulfide	298.2	0.000439	7
Sulfinylbismethane	298.2	0.000157	43
	298.2	0.000109	84
Pyrrolidine	298.2	0.000606	60
Pyridine	298.2	0.000458	60
Piperidine	298.2	0.000735	60
Nitrobenzene	298.2	0.000495	76
Benzenamine	298.2	0.000226	76
N-Methylbenzenamine	298.2	0.000291	76
2-Methylbenzenamine	298.2	0.000318	76
3-Methylbenzenamine	298.2	0.000395	76
N,N-Dimethylbenzenamine	298.2	0.000718	76
N-Ethylbenzenamine	298.2	0.000766	76
N,N-Dimethylbenzenamine	298.2	0.000962	76
1,1,2,2,3,3,4,4,4-Nonafluoro-N,N-bis(nonafluorobutyl)-1-butamine $C_{12}F_{27}N$	298.2	0.00520	38

13.5 SOLUBILITY OF HYDROGEN

Hydrogen has the following physical properties:

Melting point	= 16.01 K
Boiling point (1.013 bar)	= 20.37 K

CARBON MONOXIDE, NITROGEN, OXYGEN AND HYDROGEN 301

Critical temperature = 32.99 K
Critical pressure = 12.94 bar
Critical volume = 0.0655 dm^3 mol^{-1}
Density at 273.15 K; 1.013 bar = 0.089883 g dm^{-3}
Relative molecular mass = 2.016

The following vapour pressure data have been published by Stull et al.[67]:

T/K	P/bar
20.7	1.013
31.4	10.132

Extrapolation of these vapour pressure data is unlikely to provide a reliable guide to the solubility of the gas at the temperatures at which most measurements have been carried out.

Published data on the solubility of hydrogen have been compiled and evaluated in Solubility Data Series Volumes 5 and 6.[85]

13.5.1 Solubility of hydrogen in water

Battino and Wilhelm[85] give an equation for the solubility of hydrogen in water at a partial pressure of 1.013 bar which was derived from published data. This can be written in the form

$$\ln x_g = -125.939 + 5528.45/(T/K) + 16.8893\ln(T/K)$$

standard deviation in x_g = 0.52% at about 300 K.
temperature range 273 to 345 K.

x_g has a minimum value at 327.3 K (Fig. 13.17).
Solubility at pressures up to 1013 bar has been measured by Wiebe and Gaddy.[86,87] The variation of mole fraction solubility with pressure at 298.2 K is shown in Fig. 13.18.

13.5.2 Solubility of hydrogen in hydrocarbons

Mole fraction solubilities in alkanes at a partial pressure of 1.013 bar increase with temperature and tend to increase with chain length (Table 13.10).

Solubilities at a partial pressure of 1.013 bar in pentane and other alkanes have been published by Makranczy et al.[5] Their values tend to be lower than those reported by other workers.

Solubility in hexane has been measured by various groups.[5,59,91,92] Clever[85] has published a smoothing equation for mole fraction solubility at a partial

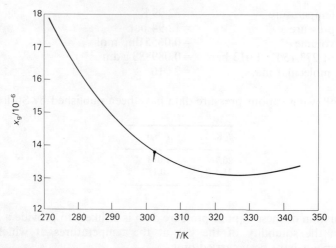

Figure 13.17. The mole fraction solubility in water of hydrogen at a partial pressure of 1.013 bar.

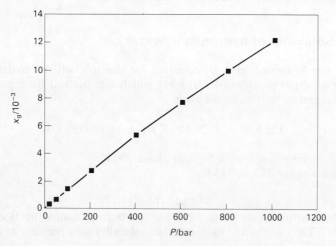

Figure 13.18. The variation with pressure of the solubility of hydrogen in water at 298.15 K.

pressure of 1.013 bar based upon data from different sources. This may be written in the form

$$\ln x_g = -5.8952 - 424.55/(T/K)$$

standard deviation in $x_g = 2.34 \times 10^{-5}$
temperature range 213 to 298 K.

Clever[85] has also published equations for the solubility of hydrogen in

Table 13.10 Solubility of hydrogen in hydrocarbons at a partial pressure of 1.013 bar

Solvent	T/K	x_g	Ref.
Pentane	298.2	0.00067	5
Hexane	298.2	0.000663	85[a]
Heptane	298.2	0.000688	93
	298.2	0.00066	5
	298.2	0.000663	60
Octane	298.2	0.000683	93
	298.2	0.00064	5
2,2,4-trimethylpentane	298.2	0.000781	93
Nonane	298.2	0.000698	50
	298.2	0.00065	5
Decane	298.2	0.00065	5
Undecane	298.2	0.00068	5
Dodecane	298.2	0.00073	5
Tridecane	298.2	0.00074	5
Tetradecane	298.2	0.00068	5
Pentadecane	298.2	0.00070	5
Hexadecane	298.2	0.00072	5
	300.0	0.00090	97
Hexamethyltetracosane	300.0	0.00138	98
Benzene	298.2	0.000260	12[b]
	298.2	0.0002578	93
Methylbenzene	298.2	0.0003167	93
	298.2	0.000318	96
1,3-Dimethylbenzene	298.5	0.000414	110
	305.3	0.000436	110
1,1'-Methylenebisbenzene	300.0	0.000289	97

[a]Smoothed value.
[b]Interpolated.

heptane and octane at a partial pressure of 1.013 bar. These are based upon measurements by Cook et al.[93] and may be written:

Heptane

$$\ln x_g = -5.6689 - 480.99/(T/K)$$

standard deviation in x_g = 4.60 × 10^{-6}
temperature range 238 to 308 K.

Octane

$$\ln x_g = -5.6624 - 484.38/(T/K)$$

standard deviation in x_g = 3.08 × 10^{-6}
temperature range 248 K to 308 K.

Various workers have investigated equilibria of hydrogen with methane, ethane, higher alkanes and alkenes at high pressures. These studies have been compiled and reviewed by Young.[85] Streett and his co-workers studied hydrogen+methane[88] from 92 to 180 K up to a pressure of 1400 bar, hydrogen+ethene[89] from 114 to 247 K up to a pressure of 6000 bar and hydrogen+ethane[90] from 92 to 280 K up to a pressure of 5600 bar.

Horiuti[12] measured the solubility in benzene from 280 to 336 K and Cook et al.[93] from 283 to 308 K. Clever[85] gives an equation for the mole fraction solubility at a partial pressure of 1.013 bar based upon data from these two sources. It may be written:

$$\ln x_g = -5.5284 - 813.90/(T/K)$$

standard deviation in $x_g = 2.89 \times 10^{-6}$
temperature range 280 to 336 K.

Various groups have investigated phase equilibria between benzene and hydrogen at high pressures. Data have been compiled and evaluated by Young.[94] At temperatures close to 298 K the variation in mole fraction solubility with increase in pressure is nearly linear to a pressure of about 500 bar. Part of the phase diagram for 338.7 K and pressures up to 689 bar, based upon measurements by Thompson and Edminster,[95] is shown in Fig. 13.19.

Cook et al.[93] published the solubility in methylbenzene at a partial pressure of 1.013 bar. The value at 298.2 K is close to that given by Saylor and Battino[96] (Table 13.10). Measurements by Cook et al. fit an equation given by Clever[85] which may be written in the form:

$$\ln x_g = -6.0373 - 603.07/(T/K)$$

standard deviation in $x_g = 7.2 \times 10^{-7}$
temperature range 258 to 308 K.

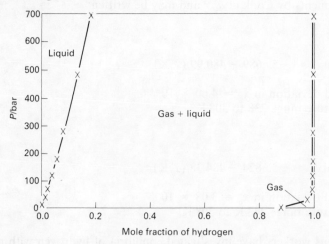

Figure 13.19. Phase equilibria in the benzene+hydrogen system at 338.7 K.

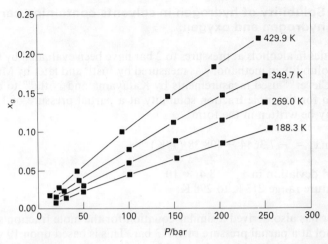

Figure 13.20. The variation with temperature and pressure of the solubility of hydrogen in 9,10-dihydrophenanthrene.

The mole fraction solubilities at 298.2 K and 1.013 bar are in the order: 1,3-dimethylbenzene > methylbenzene > benzene (Table 13.10).

The solubility in 9,10-dihydrophenanthrene in the range 461 to 703 K and 20.3 to 253 bar was measured by Sebastian et al.[99] (Fig. 13.20).

Solubilities in hydrocarbons at a partial pressure of 1.013 bar are shown in Fig. 13.21.

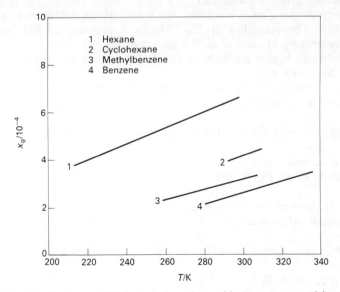

Figure 13.21. Mole fraction solubilities in hydrocarbons of hydrogen at a partial pressure of 1.013 bar.

13.5.3 Solubility of hydrogen in solvents containing carbon, hydrogen and oxygen

Solubilities in alcohols at pressures to 2 bar have been evaluated by Clever.[100]

The solubility in methanol was measured by Just[13] and later by Makranczy et al.[16] Clever[100] used measurements by Katayama and Nitta[101] to derive an equation for the mole fraction solubility at a partial pressure of 1.013 bar. This may be written in the form

$$\ln x_g = -7.3644 - 408.38/(T/K)$$

standard deviation in $x_g = 8.4 \times 10^{-7}$
temperature range 213 K to 298 K.

Clever has also derived a similar equation for the mole fraction solubility in ethanol at a partial pressure of 1.013 bar. This is based upon 19 solubility values measured by Maxted and Moon,[102] Katayama and Nitta[101] and Cargill,[103] and can be written

$$\ln x_g = -7.0155 - 439.18/(T/K)$$

standard deviation in $x_g = 2.0 \times 10^{-6}$
temperature range 213 to 333 K.

Mole fraction solubilities of hydrogen in alcohols at a partial pressure of 1.013 bar are given in Table 13.11.

Various workers have measured the solubility in methanol at elevated pressures.[85] The solubility at 297.55 K to a pressure of 811 bar from measurements by Michels et al.[104] is shown in Fig. 13.22.

Solubility in higher alcohols at elevated pressures has been extensively studied by Brunner.[105]

Horiuti[12] measured the solubility in 1,1'-oxybisethane. Mole fraction solubilities for a partial pressure of 1.013 bar fit the equation

$$\ln x_g = -21.1761 + 126.741/(T/K) + 2.34721\ln(T/K)$$

standard deviation in $x_g = 2.8 \times 10^{-6}$
temperature range 193 to 294 K.

Horiuti[12] also measured solubility in 2-propanone. Mole fraction solubilities for a partial pressure of 1.013 bar fit the equation

$$\ln x_g = -17.8580 - 107.646/(T/K) + 1.77470\ln(T/K)$$

standard deviation in $x_g = 1.7 \times 10^{-6}$
temperature range 191 to 313 K.

Mole fraction solubilities in methyl acetate at a partial pressure of 1.013 bar, based upon Horiuti's measurements, fit the equation

Table 13.11 Mole fraction solubility of hydrogen at a partial pressure of 1.013 bar in solvents containing carbon, hydrogen and oxygen

Solvent	T/K	x_g	Ref.
Methanol	213.2	0.000093	100[a]
	298.2	0.000161	100[a]
Ethanol	213.2	0.000114	100[a]
	298.2	0.000206	100[a]
1-Propanol	298.2	0.000234	101
2-Propanol	298.2	0.000266	107
1-Butanol	298.2	0.000267	101
	298.2	0.000263	16
1-Pentanol	298.2	0.000165	13
	298.2	0.000268	16
1-Hexanol	298.2	0.000301	16
1-Heptanol	298.2	0.000316	16
1-Octanol	298.2	0.000332	16
1-Nonanol	298.2	0.000343	16
1-Decanol	298.2	0.000358	16
1-Undecanol	298.2	0.000362	16
1-Dodecanol	298.2	0.000370	16
Cyclohexanol	299.2	0.000355	80
	298.2	0.000168	108
2-Propanone	298.2	0.0003013	12
1,1'-Oxybisethane	298.2	0.0006251	12
1,1'-Oxybispropane	298.2	0.000643	60
1,4-Dioxane	298.2	0.000193	60
Methyl acetate	298.2	0.0003017	12
Ethyl acetate	293.2	0.000320	13
	298.2	0.000343	13
	294.2	0.000333	102
	303.2	0.000358	102
Pentyl acetate	298.2	0.000473	13

[a]Smoothed data.

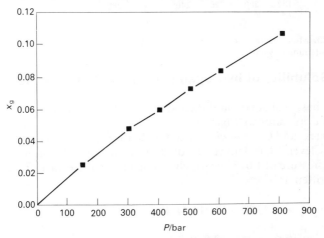

Figure 13.22. The variation with partial pressure of the solubility of hydrogen in methanol at 297.55 K.

$$\ln x_g = -13.3489 - 306.684/(T/K) + 1.10073\ln(T/K)$$

standard deviation in $x_g = 1.7 \times 10^{-6}$
range 195 K to 313 K.

The mole fraction solubility at 298.2 K and 1.013 bar is greater in higher alkyl acetates (Table 13.11).

Schmack and Bittich[106] measured the solubility in 4-methyl-1,3-dioxolan-2-one from 278 to 333 K at a pressure of 1.013 bar and also at pressures of about 10 bar. Shakhova et al.[42] measured the solubility from 283 to 323 K at pressures to a maximum of 141.8 bar.

Solubilities in various solvents containing oxygen are shown in Fig. 13.23.

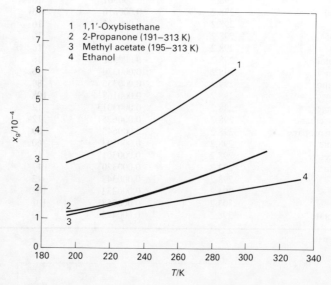

Figure 13.23. Mole fraction solubilities of hydrogen, at a partial pressure of 1.013 bar, in solvents containing oxygen.

13.5.4 Solubility of hydrogen in halogenated solvents

Clever[100] has evaluated published data on the solubility in halogenated solvents at pressures to 2 bar.

Gjaldbaek[7] and Cook et al.[93] have studied the solubility in hexadecafluoroheptane. Clever[100] has derived an equation for the mole fraction solubility at a partial pressure of 1.013 bar which is based upon these measurements. This can be written as

$$\ln x_g = -4.7741 - 532.11/(T/K)$$

standard deviation in $x_g = 2.0 \times 10^{-5}$
temperature range 248 to 323 K.

Phase equilibria between hydrogen and fluorocarbons at high pressures have been studied by Shiau and Ziegler.[109]
Horiuti[12] and Cook et al.[93] measured the solubility in tetrachloromethane at pressures close to barometric. Clever's equation[100] for the mole fraction solubility at a partial pressure of 1.013 bar, based upon both sets of measurements, can be written as

$$\ln x_g = -5.6661 - 707.62/(T/K)$$

standard deviation in x_g = 3.95 × 10^{-6}
temperature range 273 to 331 K.

Tominaga et al.[18] have published a more recent measurement of the solubility in this solvent.

Maxted and Moon[102] measured solubility in trichloromethane. The smoothed value[100] of the mole fraction solubility in this solvent at 298.2 K and a partial pressure of 1.013 bar is 0.000220 compared with a value of 0.000322 in tetrachloromethane.

Solubility in chlorobenzene was measured by Horiuti.[12] The mole fraction solubility at a partial pressure of 1.013 bar fits the equation

$$\ln x_g = -19.1349 - 41.6865/(T/K) + 1.93141\ln(T/K)$$

standard deviation in x_g = 1.0 × 10^{-6}
temperature range 232 to 354 K.

Mole fraction solubilities in halogenated solvents are shown in Table 13.12 and in Fig. 13.24.

Table 13.12 Mole fraction solubility of hydrogen in halogenated solvents at a partial pressure of 1.013 bar

Solvent	T/K	x_g	Ref.
Hexadecafluoroheptane	248.2	0.001004	93
	298.2	0.001401	93
	308.2	0.001492	93
	297.9	0.001415	7
Tetrachloromethane	298.2	0.000326	12
	298.2	0.0003193	93
	298.2	0.000311	18
Trichloromethane	298.2	0.000220	102
1,2-Dichloroethane	298.2	0.000178	91
1,1,2,2-Tetrachloroethane	298.2	0.000406	110
1-Chlorohexane	298.2	0.000446	60
1,1,2-Trichloro-1,2,2-trifluoroethane	298.2	0.000655	111
1-Bromoheptane	298.2	0.000399	59
Chlorobenzene	298.2	0.000256	12

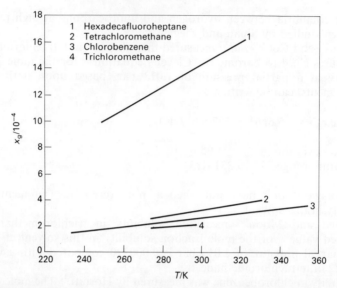

Figure 13.24. Mole fraction solubilities of hydrogen, at a partial pressure of 1.013 bar, in solvents containing halogens.

13.5.5 Solubility of hydrogen in solvents containing nitrogen, sulfur or phosphorus

The solubility in carbon disulfide was measured by Cook et al.[93] from 248 K to 298 K. Measurements by Gjaldbaek[7] at 298 K are within 2.5% of Cook's value. Earlier measurements by Just[13] are probably too low. Clever[85] published a smoothing equation for the mole fraction solubility at a partial pressure of 1.013 bar, based upon Cook's data. This can be written as

$$\ln x_g = -6.2421 - 746.25/(T/K)$$

standard deviation in x_g = 1.19 × 10^{-6}
temperature range 248 to 298 K.

Dymond[43] measured the solubility in sulfinylbismethane at 298.2 K. This was confirmed by Symons[112] who measured the solubility over the range 298 K to 353 K. Clever's equation[100] for the mole fraction solubility at 1.013 bar from these data can be written

$$\ln x_g = -6.0383 - 1027.26/(T/K)$$

standard deviation in x_g = 7.25 × 10^{-6}
temperature range 298 to 353 K.

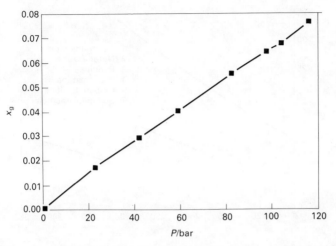

Figure 13.25. The variation with partial pressure of the solubility of hydrogen in tributyl phosphate at 313.15 K.

Solubility in 1-propanamine to elevated pressures has been studied by Moore and Otto[113] and by Brunner.[114] Solubilities at a partial pressure of 1.013 bar from the work by Moore and Otto fit the equation

$$\ln x_g = -18.7230 - 116.677/(T/K) + 1.94693\ln(T/K)$$

standard deviation in $x_g = 2.1 \times 10^{-7}$
temperature range 203 to 303 K.

The solubility in tributyl phosphate at 313.2 K was measured from 22.8 to 116.7 bar by Shakhova et al.[42] (Fig. 13.25).

Solubilities, at a partial pressure of 1.013 bar, in solvents containing nitrogen or sulfur are shown in Table 13.13 and Fig. 13.26.

Table 13.13 Mole fraction solubility of hydrogen at a partial pressure of 1.013 bar in solvents containing sulfur, nitrogen or phosphorus

Solvent	T/K	x_g	Ref.
Carbon disulfide	298.2	0.000160	93
	298.0	0.000164	7
Sulfinylbismethane	298.2	0.000076	43
	298.2	0.000075	112
1-Propanamine	298.2	0.000328	113[a]
Benzenamine	298.2	0.000107	13
Pyridine	298.2	0.000162	60
Nitrobenzene	298.2	0.000156	13

[a]Interpolated.

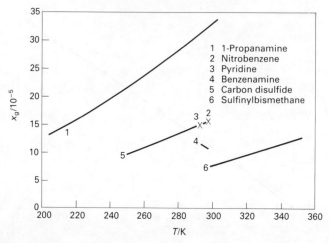

Figure 13.26. Mole fraction solubilities of hydrogen, at a partial pressure of 1.013 bar, in solvents containing nitrogen or sulfur.

REFERENCES

1. Michels, A.; Wassenaar, T.; Zwietering, T. *Physica*, 1952, 18, 160.
2. Winkler, L.W. *Ber.* 1901, 34, 1408.
3. Christoff, A. *Zeit. Phys. Chem.* 1906, 55, 622.
4. Wilhelm, E.; Battino, R.; Wilcock, R. *J. Chem. Rev.* 1977, 77, 219.
5. Makranczy, J.; Megyery-Balog, K.; Rusz, L.; Patyi, L. *Hungarian J. Ind. Chem. Veszprem.* 1976, 4, 269.
6. Patyi, L.; Furmer, I.E.; Makranczy, J.; Sadilenko, A.S.; Stepanova, Z.G.; Berengarten, M.G. *Zh. Prikl. Khim.* 1978, 51, 1296. *Russian J. Appl. Chem.* 1970, 51, 1240.
7. Gjaldbaek, J.Chr. *Acta Chem. Scand.* 1952, 6, 623.
8. Lin, P.J.; Parcher, J.F. *J. Chromatog. Sci.* 1982, 20, 33.
9. Tremper, K.K.; Prausnitz, J.M. *J. Chem. Eng. Data* 1976, 21, 295.
10. Wilhelm, E.; Battino, R. *J. Chem. Thermodyn.* 1973, 5, 117.
11. Field, L.R.; Wilhelm, E.; Battino, R. *J. Chem. Thermodyn.* 1974, 6, 237.
12. Horiuti, J. *Sci. Pap. Inst. Phys. Chem. Res. (Japan)* 1931/32, 17,125.
13. Just, G. *Zeit. Physik. Chem.* 1901, 37, 342.
14. Gjaldbaek, J. Chr.; Andersen, E.K. *Acta Chem. Scand.* 1954, 8, 1398.
15. Skirrow, F. *Zeit. Physik. Chem.* 1901, 37, 342; 1902, 41, 144.
16. Makranczy, J.; Rusz, L.; Balog-Megyery, K. *Hungarian J. Ind. Chem. Veszprem.* 1979, 7, 41.
17. Gjaldbaek, J. Chr. *K. Dan. Vidensk. Selsk. Mat.-Fys. Medd.*, 1948, 24, No. 13.
18. Tominaga, T.; Battino, R.; Gorowara, H.K.; Dixon, R.D.; Wilhelm, E. *J. Chem. Eng. Data* 1986, 31, 175.
19. Gjaldbaek, J. Chr. *Acta Chem. Scand.* 1948, 2, 683.
20. Evans, F.D.; Battino, R. *J. Chem. Thermodyn.* 1971, 3, 753.
21. Michels, A.; Wassenaar, T.; De Graaff, W.; Prins, Chr. *Physica*, 1953, 19, 26.
22. *Solubility Data Series Vol. 10, Nitrogen and Air*, edited by R. Battino, Pergamon, Oxford, 1982.
23. Battino, R.; Rettich, T.R.; Tominaga, T. *J. Phys. Chem. Ref. Data* 1984, 13(2), 563.
24. Wilcock, R.J.; Battino, R. *Nature*, 1974, 252, 614.
25. Clever, H.L.; Han, C.H. *Thermodynamics of Aqueous Systems with Industrial Application*,

ed. S.A. Newman, ACS Symposium Series 133, Amer. Chem. Soc. Washington D.C. 1980, 513.
26. Goodman, J.B.; Krase, N.W. *Ind. Eng. Chem.* 1931, 23, 401.
27. Wiebe, R.; Gaddy, V.L.; Heins, C. *Ind. Eng. Chem.* 1933, 25, 947.
28. Saddington, A.W.; Krase, N.W. *J. Amer. Chem. Soc.* 1934, 56, 353.
29. O'Sullivan, T.D.; Smith, N.O. *J. Phys. Chem.* 1970, 74, 1460.
30. Gjalbaek, J. Chr.; Hildebrand, J.H. *J. Am. Chem. Soc.* 1949, 71, 3147.
31. Byrne, J.E.; Battino, R.; Danforth, W.F. *J. Chem. Thermodyn.* 1974, 6, 245.
32. Miller, P.; Dodge, B.F. *Ind. Eng. Chem.* 1940, 32, 434.
33. Kretschmer, C.B.; Nowakowska, J.; Wiebe, R. *Ind. Eng. Chem.* 1946, 38, 506.
34. Katayama, T.; Nitta, T. *J. Chem. Eng. Data* 1976, 21, 194.
35. Christoff, A. *Z. Phys. Chem.* 1912, 79, 456.
36. Körösy, F. *Trans. Faraday Soc.* 1937, 33, 416.
37. Metschi, J. *J. Phys. Chem.* 1924, 28, 417.
38. Kobatake, Y.; Hildebrand, J.H. *J. Phys. Chem.* 1961, 65, 331.
39. Sargent, J.W.; Seffl, R.J. *Fed. Proc.* 1970, 29, 1699.
40. Powell, R.J. *J. Chem. Eng. Data* 1972, 17, 302.
41. Chang, E.T.; Gokcen, N.A.; Poston, T.M. *J. Phys. Chem.* 1968, 72, 638.
42. Shakhova, S.F.; Zubchenko, Yu.P.; Kaplan, L.K. *Khim. Prom.* 1973, 49, 108.
43. Dymond, J.H. *J. Phys. Chem.* 1967, 71, 1829.
44. Besserer, G.J.; Robinson, D.B. *J. Chem. Eng. Data* 1975, 20, 157.
45. Kalra, H.; Krishnan, T.R.; Robinson, D.B. *J. Chem. Eng. Data* 1976, 21, 222.
46. Dean, M.R.; Walls, W.S. *Ind. Eng. Chem.* 1947, 39, 1049.
47. Miller, H.C.; Verdelli, L.S.; Gall, J.F. *Ind. Eng. Chem.* 1951, 43, 1126.
48. Wilcock, R.J.; McHale, J.L.; Battino, R.; Wilhelm, E. *Fluid Phase Equilib.* 1978, 2, 225.
49. Patyi, L.; Furmer, I.E.; Makranczy, J.; Sadilenko, A.S.; Stepanova, Z.G.; Berengarten, M.G. *Zh. Prikl. Khim.* 1978, 51, 1296.
50. Thomsen, E.S.; Gjaldbaek, J.C. *Acta Chem. Scand.* 1963, 17, 127.
51. Wilcock, R.J.: Battino, R.; Danforth, W.F.; Wilhelm, E. *J. Chem. Thermodyn.* 1978, 10, 817.
52. Baldwin, R.R.: Daniel, S.G. *J. Appl. Chem.* 1952, 2(Apr.), 161; *J.Inst.Petrol. London* 1953, 39, 105.
53. Dymond, J.; Hildebrand, J.H. *Ind. Eng. Chem. Fundam.* 1967, 6, 130.
54. Wilhelm, E.; Battino, R. *J. Chem. Thermodyn.* 1973, 5, 117.
55. Geller, E.B.; Battino, R.; Wilhelm, E. *J. Chem. Thermodyn.* 1976, 8, 197.
56. Battino, R.; Evans, F.D.; Danforth, W.F.; Wilhelm, E. *J. Chem. Thermodyn.* 1971, 3, 743.
57. Gjaldbaek, J.C.; Niemann, H. *Acta Chem. Scand.* 1958, 12, 1015.
58. Boyer, F.L.; Bircher, L.J. *J. Phys. Chem.* 1960, 64, 1330.
59. Ijams, C.C. PhD. *thesis*, 1941, Vanderbilt Univ. Nashville, TN, USA.
60. Guerry, D. PhD. *thesis*, 1944, Vanderbilt Univ. Nashville, TN, USA.
61. Olson, J.D. *J. Chem. Eng. Data* 1977, 22, 326.
62. Steinberg, N.; Manowitz, B.; Pruzansky, J. *US AEC BNL-542 (T-140)* (see *Chem. Abstr.* 1959, 53, 21242g).
63. Williams, V.D. *J. Chem. Eng. Data* 1959, 4, 92.
64. Hiraoka, H.; Hildebrand, J. H. *J. Phys. Chem.* 1964, 68, 213.
65. Tham, M.K.; Walker, R.D.; Modell, J.H. *J. Chem. Eng. Data* 1973, 18, 385.
66. Friedman, H.L. *J. Am. Chem. Soc.* 1954, 76, 3294.
67. Stull, D.R. *Ind. Eng. Chem.* 1947, 39, 517.
68. Battino, R. *Solubility Data Series, Vol. 7, Oxygen and Ozone,* Pergamon Press, Oxford, 1981
69. Battino, R.; Rettich, T.R.; Tominaga, T. *J. Phys. Chem. Ref. Data* 1983, 12(2), 163.
70. Benson, B.B.; Krause, D.; Peterson, M.A. *J. Soln. Chem.* 1979, 8, 655.
71. Stephan, E.L.; Hatfield, N.S.; Peoples, R.S.; Pray, H.A. *U.S. Atomic Energy Commission,* 1956, BMI-1067.
72. Pray, H.A.; Schweichert, L.E.; Minnich, B.H. *Ind. Eng. Chem.* 1952, 44, 1146.
73. Blanc, C.; Batiste, M. *Bull. Cent. Rech. Pau-SNPA* 1970, 4, 235.
74. Schläpfer, P.; Audykowski, T.; Bukowiecki, A. *Schweiz. Arch. Angew. Wiss. Tech.* 1949, 15, 299.

75. Field, L.R.; Wilhelm, E.; Battino, R. *J. Chem. Thermodyn.* 1974, 6, 237.
76. Naumenko, N.K.; Mukhin, N.N.; Aleskovskii, V.B. *Zh. Prikl. Khim.(Leningrad)* 1969, 42, 2522. *J. Appl. Chem. (USSR)* 1969, 42, 2376.
77. Timofeev, W. *Z. Physik. Chem.* 1890, 6, 141.
78. Shchukarev, S.A.; Tolmacheva, T.A. *Zh. Strukt. Khim.* 1968, 9, 21; *J. Struct. Chem.* 1968, 9, 16.
79. Naumenko, N.K. *Candidate's Thesis, Leningrad* 1944 (quoted in refs. 67 and 68).
80. Bub, G.K.; Hillebrand, W.A. *J. Chem. Eng. Data* 1979, 24, 315.
81. Cauquil, G. *J. Chim. Phys.* 1927, 24, 53.
82. Wesseler, E.P.; Iltis, R.; Clark, L.C. *J. Fluorine Chem.* 1977, 9,137.
83. Lawson, D.D.; Moacanin, J.; Scherer, K.V.; Terranova, T.J.; Ingham, J.D. *J. Fluorine Chem.* 1978, 12, 221.
84. Baird, W.R.; Foley, R.T. *J. Chem. Eng. Data* 1972, 17, 355.
85. *Solubility Data Series Volumes 5/6, Hydrogen and Deuterium*, ed. C.L. Young, Pergamon, Oxford, 1981.
86. Wiebe, R.; Gaddy, V.L.; Heins, C. *Ind. Eng. Chem.* 1932, 24, 823.
87. Wiebe, R.; Gaddy, V.L. *J. Am. Chem. Soc.* 1934, 56, 76.
88. Tsang, C.Y.; Clancy, P.; Calado, J.C.G.; Streett, W.B. *Chem. Eng. Commun.* 1980, 6, 365.
89. Heintz, A.; Streett, W.B. *Ber. Bunsenges. Phys. Chem.* 1983, 87, 298.
90. Heintz, A.; Streett, W.B. *J. Chem. Eng. Data* 1982, 27, 465.
91. Waters, J.A.; Mortimer, G.A.; Clements, H.E. *J. Chem. Eng. Data* 1970, 15, 174 and 462.
92. Katayama, T.; Nitta, T. *J. Chem. Eng. Data* 1976, 21, 194.
93. Cook, M.W.; Hanson, D.N.; Alder, B.J. *J. Chem. Phys.* 1957, 26, 748.
94. Young, C.L. ref. 85, page 406.
95. Thompson, R.E.; Edminster, W.C. *Am. Inst. Chem. Engnrs. J.* 1965, 11, 457.
96. Saylor, J.H.; Battino, R. *J. Phys. Chem.* 1958, 62, 1334.
97. Cukor, P.M.; Prausnitz, J.M. *J. Phys. Chem.* 1972, 76, 598.
98. Chappelow, C.C; Prausnitz, J.M. *Am. Inst. Chem. Engnrs. J.* 1974, 20, 1097.
99. Sebastian, H.M.; Simnick, J.J.; Lin, H.-M., Chao, K.-C. *J. Chem. Eng. Data* 1979, 24, 343.
100. Clever, H.L. ref. 85 page 186.
101. Katayama, T.; Nitta, T. *J. Chem. Engng. Data* 1976, 21, 194.
102. Maxted, E.B.; Moon, C.H. *Trans. Faraday Soc.* 1936, 32, 769.
103. Cargill, R.W. *J. Chem. Soc., Faraday Trans. I* 1978, 74, 1444.
104. Michels, A.; de Graaff, W.; van der Somme, J. *Appl. Sci. Res.* 1953, A4, 105.
105. Brunner, E. *J. Chem. Thermodyn.* 1980, 12, 993.; *Ber. Bunsenges. Physik. Chem.* 1979, 83, 715.
106. Schmack, P.; Bittich, H,-J. *Wiss. Zeit. Tech. Hochsch. Chem. Leuna-Merseburg* 1966, 8, (2–3), 182.
107. Puri, P.S.; Reuther, J.A. *Can. J. Chem. Eng.* 1974, 52, 636.
108. Kruyer, S.; Nobel, A.P.P. *Rec. Trav Chim.* 1961, 80, 1145.
109. Shiau, J.F.; Ziegler, W.T. *J. Chem. Eng. Data* 1980, 25, 239.
110. deWet, W.J. *J. S. Afr. Chem. Inst.* 1964, 17, 9.
111. Linford, R.G.; Hildebrand, J.H. *Trans. Faraday Soc.* 1970, 66, 577.
112. Symons, E.A. *Can. J. Chem.* 1971, 49, 3940.
113. Moore, R.G.; Otto, F.D. *Can. J. Chem. Eng.* 1972, 50, 355.
114. Brunner E. *Ber. Bunsenges. Physik. Chem.* 1978, 82, 798.

Chapter 14
SOLUBILITIES OF OTHER GASES

Over 200 substances have boiling points, under barometric pressure, below 298.15 K. On this basis they may be classified as gases. Some of these substances are unstable. A very small proportion of gases have critical temperatures below 298.15 K. There are probably only 18 such gases. Any other gas could exist in the liquid phase at 298.15 K, if it were chemically stable under these conditions. It follows that the vapour pressure of liquefied gas at 298.15 K is a real quantity for the majority of gases. Only in the case of the 18 gases is this vapour pressure an imaginary quantity at this temperature.

There are two important groups of gases which have not been discussed in this book. One group comprises the noble gases of group 0 in the periodic table. The solubilities of these gases have been compiled and evaluated for volumes in the Solubility Data Series.[1-3]

Another group consists of the halocarbons. These have been studied by many groups. A summary of the solubilities of some of these compounds in a wide range of solvents was published by Zellhoefer.[4] Wilhelm, Battino and Wilcock[5] have derived smoothing equations for the solubilities in water of 13 gaseous halocarbons. These equations are based on data published by other workers. Much of the work on the solubilities of halocarbons has been critically evaluated by Gerrard.[6,7]

A reference line can provide a guide to the solubility of a gas for which no solubility data have been published. In the absence of other data, the vapour pressure of a liquefied gas at any temperature may be approximately estimated from the boiling point and heat of vaporization at the boiling point. Such approximate vapour pressures may then be used to prepare reference lines.

Physical data for gases from various sources are tabulated below. There has been no attempt to evaluate the data, and different sources sometimes disagree. The table is presented to give an overall picture of the properties of all gases from which approximate solubilities may be estimated. Readers wishing to obtain further information on the physical properties of gases should consult The Gas Encyclopaedia[8] or other recent compilations.

SOLUBILITY OF GASES IN LIQUIDS

Table 14.1 Physical properties of substances which are gases at 298 K

Molecular formula	M_r	B. pt at 1.013 bar/K	Crit. temp./K	Heat of vap. at b. pt. (1.013 bar) /J mol^{-1}	Vap. press. /bar (T/°C)	Density of gas at 273.15 K and 1.013 bar
He	4.00	4.23	5.25	81.2		0.1785
H_2	2.02	20.37	32.99	904		0.08988
D_2	4.03	23.59	38.35	1226		0.179
Ne	20.18	27.25	44.45	1738.5		0.8999
N_2	28.02	77.34	126.2	5590		1.2505
CO	28.01	81.65	133.15	6042		1.2500
F_2	38.00	85.01	144.15	6544		1.696
Ar	39.94	87.29	150.69	650.2		1.78367
O_2	32.00	90.18	154.75	6828		1.42898
CH_4	16.04	111.54	191.05	8180		0.71683
Kr	83.80	120.25	209.35	9034		3.743
NO	30.01	121.45	180.15	13836		1.3402
F_2O	54.00	128.35	215.15	11100		
NF_3	71.01	144.15	234.15	11594		
CF_4	88.01	145.15	227.65	11996		
O_3	48.00	161.15	261.15	15192		2.141
SiH_4	32.12	161.15	269.15	11014		
N_2F_2	66.02	161.75				
Xe	131.30	165.15	289.75	12644		
C_2H_4	28.05	169.45	282.40	13544		1.26037
PF_3	87.97	171.65		14560		
ClF	54.45	172.35		22343		
BF_3	67.82	172.85	260.90	18901		
SiH_3F	50.10	185.15		17991		
SiF_4	104.09	178.15	287.25			4.67
$SiHF_3$	86.09	178.75				
B_2H_6	27.67	180.65	289.85	14435		
GeH_4	76.63	183.15	308	14062	26 (0)	3.43
N_2O	44.02	183.7	309.65	16560	50 (20)	1.9775
C_2H_6	30.07	184.55	305.45	14703	38.4 (21)	1.356
PH_3	34.00	185.45	324.45	14598	38.6 (21)	1.5307
COF_2	66.01	188.15	288.15	16084	30 (10)	
HCl	36.47	188.20	324.68	16159	43.2 (21)	1.6392
PF_5	125.97	188.55	288.15	16740		
C_2H_2	26.04	189.12s	308.4			1.1747
F_3CNO	99.10	189.15				
$CH_2=CF_2$	64.04	190.15	303.25	15870	36.7 (21)	
CHF_3	70.02	191.05	298.75	18085	44.8 (21)	
$CClF_3$	104.47	191.75	301.95	15506	32.6 (21)	
NF=NF	66.00	192.15				
CO_2	44.01	194.4s	304.3		58.3 (21)	1.977
C_2F_6	138.02	194.95	292.85	16150	18.5 (0)	
CH_3F	34.03	194.80	317.75	17569	39.6 (25)	
$CF_2=CF_2$	100.00	197.15	306.15			
SiH_2F_2	68.10	197.15				
F_3CNF_2	121.00	198.15				
$B(CH_3)_3$	55.92	199.92				
NO_2F	65.00	200.75				
$CH_2=CHF$	46.14	201.15	327.85	17157	25.5 (21)	
NF_2NF_2	104.00	203.15	309.15			

Table 14.1 (continued)

Molecular formula	M_r	B. pt at 1.013 bar/K	Crit. temp./K	Heat of vap. at b. pt. (1.013 bar) /J mol^{-1}	Vap. press. /bar (T/°C)	Density of gas at 273.15 K and 1.013 bar
$SiClF_3$	120.56	203.15				
F_3CCN	95.00	205.15				
HBr	80.92	206.43	363.15	17606	23.1 (21)	3.6163
SF_6	146.07	209.35 (s)	318.7	2343s	21 (19)	6.6132
BH_3CO	41.84	210.15				
AsF_3	131.92	210.15				
PSF_3	119.94	210.55				
Rn	222.00	211.15	377.5	16736	8.9 (0)	9.96
AsH_3	77.95	210.67	373.05	16686	18.38 (21)	
H_2S	34.08	212.88	373.55	18673	18 (20)	1.5392
$CBrF_3$	148.93	215.35	340.15	17686		
CH_3SiH_3	43.11	215.65				
F_2O_2	70.00	216.15				
NOF	49.00	217.15				
SO_2F_2	102.07	217.95		19230		
AsF_5	169.92	220.15		20753		
SnH_4	122.72	221.15		18410		
CH_2F_2	52.00	221.55	351.55	18754	8 (0)	
COS	60.08	222.95	378.15	18506	12.1 (21)	
$SiHClF_2$	102.56	223.15				
$F_3CC=CH_2$	94.04	224.85				
C_3H_6	42.08	225.45	364.95	18418	10.4 (21)	1.9149
$C_2H_3F_3$	84.04	225.55	346.25	19352	11 (20)	
PF_2Cl	104.44	225.85				
ClO_2F	86.47	226.35				
SeF_6	192.96	226.55				
NO_3F	81.00	227.25				
CH_3SiH_2F	45.13	229.15				
SOF_2	86.00	229.35				
C_3H_8	44.10	231.08	369.95	18778	8.6 (21)	2.020
NO_3F	81.10	231.15				
F_3CF_2NO	137.01	231.15				
H_2Se	80.98	231.15	411.15	19874	12 (30)	
$CHClF_2$	86.48	232.35	369.15	20225	9.4 (21)	
SF_4	108.07	232.75	364.15	21687	10.6 (21)	
POF_3	103.97	233.35				
TeF_6	241.60	234.25 (s)	356.4	28221(s)		
C_2ClF_5	154.48	234.45	353.15	19456	8.0 (21)	
S_2F_2	102.00	234.75				
F_2CCCF_2	112.03	235.15				
CH_3CH_2F	48.00	236.05	375.35	20444	7.5 (20)	
$CF_3CF_2CF_3$	188.03	236.45	345.05	19623	7.91 (21)	
GeF_4	148.59	236.6 (s)				
HI	127.92	237.75	425.95	19761	62 (10)	5.7245
$ClCF_2NO$	115.48	238.15				
F_3CCF_2NF	152.02	238.15				
CH_2CCH_2	40.07	238.65	393.85	18636	8.0 (25)	
Cl_2	70.94	239.04	417.15	20355	6.9 (21)	3.212
F_3CNO_2	115.01	239.55				
NH_3	17.03	239.80	405.4	23351	8.9 (21)	0.77141

Table 14.1 (*continued*)

Molecular formula	M_r	B. pt at 1.013 bar/K	Crit. temp./K	Heat of vap. at b. pt. (1.013 bar) /J mol^{-1}	Vap. press. /bar (T/°C)	Density of gas at 273.15 K and 1.013 bar
cyclo C$_3$H$_6$	42.08	240.29	398.30	20054	6.2 (21)	1.88
SiCl$_2$F$_2$	137.00	241.45				
D$_3$N	20.05	242.25				
SiH$_3$Cl	66.58	242.65				
CCl$_2$F$_2$	120.93	243.36	385.15	19962	5.9 (21)	
CF$_3$CFCF$_2$	150.03	244.15			7.0 (21)	
CF$_3$COCF$_3$	166.03	245.15	357.25		6.2 (20)	
C$_2$ClF$_3$	116.48	245.25	378.95	20773	5.3 (21)	
F$_3$CPH$_2$	102.00	246.65				
(CH$_3$)$_2$O	46.07	248.33	400.05	21518	5.2 (21)	
CH$_3$CHF$_2$	66.05	248.45	386.65	21556	5.4 (21)	
CH$_3$Cl	50.49	248.93	416.25	21401	5.1 (21)	
NHF$_2$	53.00	249.55				
Si$_2$OF$_6$	186.18	249.85				
CH$_3$C=CH$_2$	40.07	249.95	402.39	21861	5.2 (21)	
CH$_2$N$_2$	42.00	250.15				
CH$_2$CHSiH$_3$	48.16	250.35				
CF$_3$I	196.00	250.65				
COSe	106.97	251.45				
C$_2$N$_2$	52.04	251.98	399.70	23355	5.2 (21)	
GeF$_3$Cl	165.06	252.55				
B(CH$_3$)$_3$	55.92	252.95				
BrF	98.91	253.15				
F$_3$C(SF$_3$)	158.01	253.15				
HCHO	30.03	253.15				
(CH$_3$)$_2$SiH$_3$	60.17	253.55				
SF$_5$Cl	162.47	254.05				
Si$_2$F$_6$	170.17	254.65				
SiHCl$_2$F	119.01	254.75				
C$_2$HClF$_2$	98.48	254.55	400.55	22471	29 (10)	
SbH$_3$	124.77	255.95				
PBrF$_2$	148.88	257.05				
ClO$_4$F	118.47	257.25				
(H$_3$Si)$_2$O	78.22	257.25				
NO$_2$Cl	81.47	258.15				
SCl$_4$	133.88	258.15				
ClN$_3$	77.47	258.15				
Si$_2$H$_6$	62.22	258.65	424	21387		
CH$_3$PH$_2$	48.02	259.15				
CH$_2$=CHCl	62.50	259.25	431.55	20803	3.3 (21)	
PFCl$_2$	120.91	259.30				
C$_2$H$_5$SiH$_3$	50.17	259.45				
ClF$_5$	130.45	260.05	415.75	22245	3.4 (20)	
F$_3$CCF$_2$CF$_2$NO	199.03	261.15				
BrCF$_2$NO	159.92	261.15				
CH(CH$_3$)$_3$	58.12	261.45	408.05	21292	3.1 (21)	
ClCH$_2$F	68.50	263.05				
SO$_2$	64.07	263.15	430.65	24943	3.3 (21)	2.927
CH$_3$CClF$_2$	100.50	263.95	410.25	22426	3.0 (21)	

Table 14.1 (continued)

Molecular formula	M_r	B. pt at 1.013 bar/K	Crit. temp./K	Heat of vap. at b. pt. (1.013 bar) /J mol^{-1}	Vap. press. /bar (T/°C)	Density of gas at 273.15 K and 1.013 bar
$(CH_3)_2SiHF$	78.16	264.15				
SeO_2F_2	148.96	264.75				
$CH_2=C(CH_3)_2$	56.11	266.25	417.85	22121	2.7 (21)	
CH_3NH_2	31.06	266.70	430.05	25983	3.0 (20)	
$C_2H_5CH=CH_2$	56.11	266.92	419.55	21895	2.6 (21)	
cyclo C_4F_8	200.04	267.11	388.45	23146	2.7 (21)	
ClO_2F	86.47	267.15				
NOCl	65.47	267.35	440.65	25389		
$CH_2=CHCH=CH_2$	54.09	268.74	425.15	22585	2.5 (21)	
$FCH_2CH=CH_2$	60.07	270.15				
$BrFC=CF_2$	160.93	270.15				
GeF_2Cl_2	181.53	270.35				
$B_2H_4(CHS)_2$	55.72	270.55				
C_3H_7F	62.09	270.65				
H_2Te	129.62	271.15		19210		
C_4F_{10}	238.03	271.45	386.45	23370	22 (20)	
C_4H_{10}	58.12	272.65	425.15	22393	2.0 (21)	2.7032
(E)-$CH_3CH=CHCH_3$	56.11	274.04	428.15	22736	2.0 (21)	
$PH(F_3C)_2$	170.00	274.15				
$H_2C_3F_6$	152.05	274.35				
$C_2F_4=C_2F_4$	200.04	274.35			2.0 (21)	
SiH_3Br	111.01	275.05				
AsH_3CH_3	91.97	275.15				
Cl_2O	86.94	275.15				
$(CH_3)_3N$	59.11	276.02	433.25	22949	1.9 (21)	
CH_3Br	94.95	276.61	467.15	23928	1.9 (21)	4.3379
$(CClF_2)_2$	170.93	276.70	418.85	23430	1.9 (21)	7.9216
(Z)-$CH_3CH=CHCH_3$	56.11	276.87	433.15	23322	1.9 (21)	
IF_7	259.90	277.65 (s)				
$B_2H_4(CH_3)_2$	55.72	278.05				
NO_2Cl	81.47	278.15				
BrCl	115.38	278.15				
$CH_2=CHC\equiv CH$	52.08	278.3				
C_3O_2	68.03	279.15				
$CH_3OCH=CH_2$	58.08	279.15				
CH_3SH	48.10	279.40	469.95	24573	2.0 (21)	
$(CH_3)_2NH$	45.08	280.02	437.65	26485	1.8 (21)	
$COCl_2$	98.93	280.71	455.15	24656	1.7 (21)	
$CH_3OC_2H_5$	60.10	280.75				
$C_2H_5C\equiv CH$	54.09	281.25	463.65	24514	1.6 (21)	
SiH_2Cl_2	101.04	281.65	449.45	25205	0.75 (0)	
$CHCl_2F$	102.93	282.05	451.65	24950	1.6 (21)	
$C(CH_3)_4$	72.15	282.65	433.75	22753	1.5 (21)	
ClO_2	67.47	283.05		25941		
PCl_2F_3	158.91	283.15				
B_2H_5Br	106.57	283.15				
C_2H_4O	44.05	283.88	468.15	25527	1.5 (21)	

Table 14.1 (*continued*)

Molecular formula	M_r	B. pt at 1.013 bar/K	Crit. temp./K	Heat of vap. at b. pt. (1.013 bar) /J mol^{-1}	Vap. press. /bar (T/°C)	Density of gas at 273.15 K and 1.013 bar
ClF_3	92.46	284.40	447.15	27543		1.5 (21)
$SOClF$	102.47	285.35				
C_2H_5Cl	64.52	285.55	460.35	24702		1.4 (21)
BCl_3	117.19	285.65	451.95	23962		1.3 (20)
$ClCN$	61.48	285.85		27366		
$CH_2=CHBr$	106.96	288.85		25971		1.4 (25)
SiH_3PH_2	64.10	285.85				
cyclo C_4H_8	56.11	285.66	459.95	24196		0.9 (10)
B_4H_{10}	53.32	289.15	443	26187		1.55 (20)
$C_2H_5NH_2$	45.09	289.73	456.35	27171		1.17 (20)
$(CF_3)_3P$	238.01	290.45				
WF_6	297.85	290.65	443	26187		1.55 (20)
$NO_2(OCl)$	97.47	291.15				
HF	20.01	292.66	461.15	7497		1.05 (20)
$C_3H_7CH=CH_2$	70.14	293.21	444.65	24125		1.16 (21)
CH_3CHO	34.05	293.95				
$(F_3C)_2PCl$	204.47	294.15				
N_2O_4/NO_2	92.0/46.0	294.15	431.15			1.5 (30)
$CH_2=CClCH$	76.54	295.65				
$(CF_3)_3PO$	254.01	296.75				
CCl_3F	137.38	296.92	471.15	24995		0.92 (21)
CBr_2F_2	209.84	297.65	471.35			0.83 (21)
HCN	27.02	298.85	456.75	25217		
Thoron $^{220}Rn_{86}$	\multicolumn{6}{l}{Consider as radon for solubility purposes.[2]}
Actinon $^{219}Rn_{86}$						

(s) Sublimes.

REFERENCES

1. Clever, H.L. *Solubility Data Series, Vol. 1, Helium and Neon,* Pergamon Press, Oxford, 1979.
2. Clever, H.L. *Solubility Data Series, Vol. 2, Kryton, Xenon and Radon,* Pergamon Press, Oxford, 1979.
3. Clever, H.L. *Solubility Data Series, Vol. 4, Argon,* Pergamon Press, Oxford, 1980.
4. Zellhoefer, G.F. *Ind. Eng. Chem.* 1937, 29, 548.
5. Wilhelm, E.; Battino, R.; Wilcock, R.J. *Chem. Rev.* 1977, 77, 219.
6. Gerrard, W. *Solubility of Gases and Liquids,* Plenum, New York, 1976.
7. Gerrard, W. *Gas Solubilities, Widespread Applications,* Pergamon Press, Oxford, 1980.
8. *Gas Encyclopaedia,* L'Air Liquide, Elsevier, Amsterdam, 1976.

INDEX OF SOLVENTS

This index includes all the components of liquid phases, other than the solute gas, which are mentioned in the text. Components which are listed in tables but do not appear in the text have not been included.

Acetaldyhde 150, 254
Acetic acid 63, 202, 226, 256
Acetic anhydride 226, 256
Acetonitrile 65, 232
2-Aminoethanol 71, 222, 245, 261
Ammonia 129, 152, 167, 222, 245
Ammonium
 chloride 222
 salts 43
 sulfate 222

Barium chloride 176
Benzaldehyde 255
Benzenamine 69, 97, 101, 229, 258, 280, 290
Benzene 39, 48, 91, 99, 103, 105, 118, 135, 143, 148, 178, 193, 224, 249, 285, 294, 304
Benzeneacetonitrile 65, 260
1,2-Benzenediol 64
Benzenemethanol 60, 107, 251
Benzenesulphonic acid, butyl ester 212
Benzonitrile 40, 65, 232
1,1'-Bicyclohexyl 91, 249, 279
Boric acid, tripentyl ester 215
Bromobenzene 73, 105, 208, 229, 257
1-Bromobutane 207, 228
Bromoethane 207, 228
1-Bromohexane 207
1-Bromooctane 72, 207, 228
1-Butanamine 71
Butane 98, 110, 140, 223
Butanoic acid 256
1-Butanol 94, 103, 196, 225, 279, 286
2-Butoxymethanol 60

Cadmium salts 43
Calcium chloride 221
Carbon dioxide 129, 222
Carbon disulfide 75, 233, 262, 281, 291, 310
Carbonyl sulfide 129
Chlorobenzene 72, 105, 126, 136, 151, 207, 211, 229, 257, 280, 288, 298, 309
1-Chlorobutane 207
Chlorodifluoromethane 125
2-Chloroethanol 95
1-Chlorohexane 207
(Chloromethyl)benzene 258
1-Chlorooctane 72, 96
2-Chlorophenol 63, 73
Cyclohexane 47, 48, 118, 178, 223, 249, 271, 276, 285
Cyclohexanol 252
Cyclohexanone 62
Cyclohexene 91
Cyclooctane 249

Decahydronaphthalene 48
Decane 47, 98, 103, 107, 118, 134, 140, 160, 178, 223, 249, 294
1-Decanol 251
1,2-Dibromoethane 181, 228, 257
N,N-Dibutylbutanamine 71
1,2-Dichlorobenzene 73, 208
Dichlorodifluoromethane 125
1,1-Dichloroethane 205
1,2-Dichloroethane 72, 96, 204
2,2-Dichloroethanol 95
Dichloromethane 180, 204, 210

1,2-Dichloropropane 72
Diethanolamine 222, 261
N,N-Diethylbenzenamine 69
N,N-Diethylethenamine 98
Diglycolamine 261
9,10-Dihydrophenanthrene 305
1,3-Dimethoxybenzene 61
1,4-Dimethoxybenzene 61
1,2-Dimethoxyethane 60
Dimethoxymethane 60
N,N-Dimethylbenzenamine 69, 101, 230
1,2-Dimethylbenzene 178, 193, 224, 250
1,3-Dimethylbenzene 48, 103, 121, 178, 193, 224, 250, 305
1,4-Dimethylbenzene 150, 178, 193
1,2-Dimethylcyclohexane 249
N,N-Dimethylformamide 40, 65, 152, 179, 232, 260
1,1-Dimethylhydrazine 290
N,N-Dimethylmethanamine 102
2,2-Dimethylpropane 118
Dimethyl sulfate 75
Dimethylbenzene 166
1,1'-Dimethylhydrazine 97
1,4-Dioxane 39, 61
Dodecane 46, 140, 223
Dodecanol 251
Dotriacontane 118

Eicosane 118, 140, 160, 246
Ethane 118, 223, 303
1,2-Ethanediol 40, 59, 95, 107, 196, 226, 252
2,2'[1,2-Ethanediylbis(oxy)]bisethanol 226
Ethanol 57, 93, 103, 122, 150, 196, 225, 246, 251, 279, 286, 296, 306
Ethanolamine 71, 222, 245, 261
Ethene 118, 304
Ethoxybenzene 60
Ethyl acetate 63, 150, 255
N-Ethylbenzenamine 69, 230
Ethylbenzene 178, 224

Fluorobenzene 208
Fluorosulfuric acid 215
Formamide 68
Formic acid 63
2-Furanmethanol 60

Heptadecane 134
Heptane 134, 140, 160, 177, 276, 303
1-Heptanol 57
1,1,2,3,4,4-Hexachloro-1,3-butadiene 181
Hexadecafluoroheptane 125, 181, 257, 281, 308
Hexadecane 47, 118, 134, 192, 223, 246, 276, 284
1-Hexadecene 191
Hexafluorobenzene 258, 281
Hexamethylphosphoric triamide 262
2,6,10,15,19,23-Hexamethyltetracosane 246
Hexane 46, 91, 98, 102, 105, 110, 118, 135, 140, 223, 246, 294, 301
Hexanedinitrile 98
Hexanoic acid 63, 226
1-Hexanol 56
Hexatriacontane 246
Hydrazine 97, 290
Hydrochloric acid 44, 116, 175, 221
Hydrogen iodide 221
Hydrogen sulfide 129, 245, 292

Iodobenzene 73, 105, 208, 211, 229, 257
1-Iodohexane 207

Methane 223, 303
Methanol 39, 57, 91, 103, 122, 124, 150, 161, 225, 251, 286, 296, 306
Methoxybenzene 60
1-Methoxybutane 199
2-Methoxythanol 60, 150, 255
1,1'-(Methoxymethylene)bisbenzene 60
2-Methoxyphenol 64
Methyl acetate 63, 124, 135, 145, 255, 279, 287, 296, 306
N-Methylbenzenamine 40, 69, 101, 230
Methylbenzene 48, 60, 103, 121, 135, 150, 166, 178, 193, 224, 249, 295, 304
2-Methylbutane 118
Methylcyclohexane 223, 249, 279
4-Methyl-1,3-dioxolan-2-one 63, 226, 256, 308
1,1'-Methylenebisbenzene 122, 250
(1-Methylethyl)benzene 224, 250

Methylhydrazine 97, 290
1-Methylnaphthalene 122, 250
1-Methyl-2-nitrobenzene 65, 211
3-Methylpentane 118
3-Methylphenol 125, 256
2-Methylpropane 118, 140, 223
2-Methyl-1-propanol 251, 286
Methylpyridine 69
1-Methyl-2-pyrrolidinone 69, 95, 127, 230, 260, 291
1-Methyl-3-nitrobenzene 211
1-Methyl-4-(1-methylethenyl)cyclohexene 249

Naphthalene 250
Nitrobenzene 64, 179, 211, 230, 260, 280, 290
Nitromethane 65
Nitrous oxide 129
1,1,2,2,3,3,4,4,4-Nonafluoro-N,N-bis(nonafluorobutyl)-1-butanamine 259, 290
Nonane 98, 294

Octacosane 246
Octamethylcyclotetrasiloxane 128, 262, 292
Octane 46, 140, 178, 192, 223, 249, 294, 303
1-Octanol 59, 103, 161, 198, 225, 251
2-Octanone 62
1,1'-Oxybisbenzene 201
1,1'-Oxybisbutane 60
1,1'-Oxybisethane 60, 124, 150, 200, 255, 279, 287, 296, 306
2,2'-Oxybisethanol 59, 95
1,1'-Oxybisoctane 200, 226
1,1'-Oxybis(3-methylbutane) 201
2,2'-Oxybispropane 60
Oxybispropanol 255

Pentachloroethane 206
Pentane 45, 134, 140, 223, 246, 276, 284, 301
1-Pentanol 251
Pentyl acetate 255
Perchloric acid 221
Phenol 63, 256
2-Phenoxyethanol 60
Phenylmethyl acetate 63
Phenylmethyl benzoate 63

Phosphoric acid
  tributyl ester 63, 75, 127, 232, 262, 311
  triethyl ester 75, 96, 128, 262
  trimethyl ester 262
  tripropyl ester 75, 128, 262
  tris(methylphenyl) ester 75
  tris(2-methylpropyl) ester 75, 128, 262
Phosphorous acid
  dibutyl ester 96
  triethyl ester 96
  triphenyl ester 96
Poly(ethylene glycols) 226
Potassium
  carbonate 222, 245
  chloride 43, 176, 244
  nitrate 43, 244
  salts 43
1-Propanamine 311
Propane 140, 223
1,2-Propanediol 59
Propanenitrile 259
1,2,3-Propanetriol 59, 95, 107, 246, 254
Propanoic acid 63, 256
1-Propanol 93, 103, 123, 251, 279, 286
2-Propanol 94, 286
2-Propanone 39, 62, 95, 124, 135, 145, 150, 226, 246, 254, 279, 286, 296, 306
Propyl acetate 255
Propylbenzene 224
Pyridine 40, 69, 101, 230, 258

Quinoline 69, 98, 101, 127, 230, 259

Silicic acid
  tetraethyl ester 214
  tetramethyl ester 214
  tetrapropyl ester 214
  tetra(4-methyl-2-pentyl) ester 214
Sodium
  bicarbonate 222
  carbonate 222
  chloride 43, 116, 176, 221, 244
  iodide 254, 260
  nitrate 243
  perchlorate 221
  salts 43
  sulfate 43, 221

Sulfinylbismethane 73, 262, 292, 299, 310
Sulfur 233
Sulfur dioxide 129, 245, 292
Sulfur hexafluoride 292
Sulfuric acid 43, 212
Sulfuryl chloride 75

1,1,2,2-Tetrabromoethane 228
1,1,2,2-Tetrachloroethane 181, 206, 228
Tetrachloroethene 181
Tetrachloromethane 71, 96, 102, 103, 107, 125, 136, 151, 180, 204, 210, 257, 287, 298, 309
Tetradecane 178, 223
Tetrahydrofuran 61, 151, 255
Tetrahydrothiophene, 1,1-dioxide 74, 262
Tetrahydro-1,1-thiophene 233
Tetrahydro-2-furanmethanol 61
1,2,3,4-Tetrahydronaphthalene 48, 250
Tribromomethane 228

1,1,2-Trichloroethane 206
2,2,2-Trichloroethanol 72, 95
Trichloroethene 72
(Trichloromethyl)benzene 258
2,2,3-Trimethylpentane 118
2,2,4-Trimethylpentane 46, 178, 247, 294
Trichloromethane 71, 96, 102, 103, 107, 180, 204, 210, 228, 257, 287, 309
1,3,5-Trimethylbenzine 48, 103, 121, 224

Undecanol 296
Urea 246

Water 40, 41, 88, 115, 134, 138, 147, 156, 166, 174, 188, 219, 242, 243, 266, 270, 273, 275, 282, 293, 301

Zinc chloride 221
Zinc sulfate 43

SUBJECT INDEX

(T) after a page number refers to a Table. (F) refers to a Figure.

Absorption coefficient 6
Acetylene, see Ethyne
Adsorption of gas 16
Air, solubility 282
Ammonia
  ionization 88
  phase equilibria with methane 129
  physical properties 86
  reference line 87
  solubility in
    alchohols 91, 92(T), 94(F), 95(F)
    halogen compounds 92(T), 95, 96
    hexanedinitrile 29(F)
    hydrocarbons 91, 92(T), 93(F), 94(F)
    nitrogen compounds 92(T), 96, 96(F), 97(F), 98(F)
    oxygen compounds 91, 92(T), 94(F), 95(F), 96(F)
    phosphorus compounds 92(T), 96
    water 88, 88(F), 89(F)
  vapour pressure 87
Apparatus 10

Bubble point 18, 143
Bunsen coefficient 6
Butane
  physical properties 114, 155
  reference line 113(F), 156(F)
  solubility in
    alcohols 161, 162(T), 165(F)
    alkanes 160, 161(F), 162(T), 164(F)
    aromatic hydrocarbons 162, 165(F)
    halogen compounds 163(T), 165(F)

    hydrocarbons 161(F), 162, 162(T), 165(F)
    nitrous compounds 163(T), 165(F)
    oxygen compounds 161, 162(T), 165(F)
    water 157, 157(F), 159(F)
  vapour pressure 155

Carbon dioxide
  ionization 242
  phase equilibria
    benzene 249
    methane 129, 130(F)
  physical properties 241
  reference line 241
  removal from industrial gases 245
  solubility in
    alcohols 246, 251, 252(F), 253(T), 257(F)
    aldehydes 253(T), 254
    alkanes 246, 247(F), 248(T), 251(F)
    alkanolamines 245
    amines 258, 260(T), 261(F)
    ammonia 245
    aqueous organic solutions 246
    aromatic hydrocarbons 248(T), 249, 250(F), 251(F)
    carboxylic acids 253(T), 256, 257(F)
    N,N-dimethylformamide 260, 260(T), 261(F)
    esters 253(T), 255, 257(F)
    ethanol 246, 251, 252(F), 253(T)
    ethers 253(T), 255, 256(F), 257(F)
    haloalkanes 257, 258(T), 259(F)

SUBJECT INDEX

Carbon dioxide (*cont.*)
  solubility in, (*cont.*)
    halogen compounds 257, 258(T), 259(F)
    hydrocarbons 246, 247(F), 248(T), 250(F), 251(F)
    hydrogen sulfide 245
    ketones 246, 253(T), 254, 256, 257(T)
    1-methyl-2-pyrrolodinone 260, 260(T), 261(F)
    nitrobenzene 260, 260(T), 261(F)
    nitrogen compounds 246, 258, 260(T), 261(F)
    non-aromatic hydrocarbons 246, 247(F), 248(T), 251(F)
    octamethylcyclotetrasiloxane 262, 262(F), 263(T)
    oxygen compounds 246, 251, 252(F), 253(T), 254(F), 255(F). 256(F), 257(F), 258(T), 260, 261(F), 262, 263(T)
    phenols 253(T), 256
    phosphorus compounds 262, 262(F), 263(T)
    2-propanone 246, 253(T), 254, 254(F), 255(F)
    salt solutions 243, 244(F), 254, 255(F)
    sulfur compounds 262, 262(F), 263(T)
    water 242, 243(F)
  vapour pressure 241
Carbon monoxide
  physical properties 274
  reference line 274, 275(F)
  solubility in
    alkanes 275, 276(F), 277(T)
    alkanols 277(T), 279
    aromatic hydrocarbons 276(F), 277(T), 279
    carbon disulfide 278(T), 281
    esters 276(F), 278(T), 279
    halogen compounds 278(T), 280, 280(F)
    hydrocarbons 275, 276(F), 277(T)
    nitrogen compounds 278(T), 280, 280(F)
    oxygen compounds 276(F), 277(T), 279, 280(F)
    water 275, 276(F)

  vapour pressure 274
Chlorine
  hydrate 174
  hydrolysis 174
  physical properties 173
  reference line 22(F), 23(F), 274(F)
  reference surface 23(F)
  solubility in
    alkanes 177, 178(F)
    aromatic hydrocarbons 178, 179(F), 179(T)
    halogen compounds 180, 180(T), 181(F), 182(F)
    hydrocarbons 177, 178(F), 179(T)
    hydrochloric acid 175, 176(F)
    liquids, alphabetical order 183
    liquids, increasing mole fraction 184
    liquids, increasing weight ratio 185
    nitrogen compounds 179
    salt solutions 176 177(F)
    water 174, 175(T), 175(F), 176(F)
  vapour pressure 173
Chromatography 15, 128, 134, 135, 226, 230, 251, 260
Compressibility factor 15
Critical temperature (*see physical properties of individual gases*) 22, 115, 133, 274, 275, 281, 293, 315

Degassing of solvent 9
Dew point 18, 143
N,N-Dimethylmethanamine
  ionization 88
  phase equilibria
    hexane 105, 106(F)
    methanamine 101(F), 102
    tetrachloromethane 107, 109(F)
  physical properties 86
  reference line 87(F)
  solubility in
    alcohols 107, 107(F), 108(T), 109(T)
    halogen compounds 107, 108(T), 109(F), 110
    hydrocarbons 105, 106(F), 108(T)
    nitrogen compounds 107, 108(T)
    oxygen compounds 107, 107(F), 108(T), 109(T)
    water 89, 90(F)

SUBJECT INDEX

vapour pressure 87
Dry method 13
Duhem–Margules equation 26

Enthalpy of solution 31
Entropy of solution 31
Ethanamine
  ionization 88
  phase equilibria
    butane 110
    hexane 110
  physical properties 86
  reference line 87(F)
  solubility in hexane 110, 110(F)
Ethane
  phase equilibria
    benzene 143
    hydrogen 303
  physical properties 114, 138
  reference line 113(F), 139(F)
  solubility in
    2-propanone 141(F), 145, 145(T), 146(F)
    alcohols 141(F), 143, 144(T), 144(F), 145(T)
    halogen compounds 141(F), 142(T), 145(T)
    hydrocarbons 140, 140(F), 142(T), 142(F), 143(F)
    methyl acetate 145, 145(T), 145(F)
    nitrogen compounds 141(F), 145(T)
    organic solvents 140, 140(F), 141(F), 142(T), 142(F), 143(F), 144(T), 144(F), 145(T), 145(F), 146(F)
    oxygen compounds 141(F), 143, 144(T), 144(F), 145(T), 145(F), 146(F)
    sulfur compounds 142(T), 145(T)
    water 139, 139(F)
  vapour pressure 138
Ethene
  phase equilibria with hydrogen 304
  physical properties 114, 133
  reference line 113(F), 133
  solubility in
    alcohols 135, 136(F), 137(T), 138(F)
    alkyl phosphates 137(T)
    halogen compounds 136, 137(T), 138(F)
    hydrocarbons 134, 136(F), 137(T)
    nitrobenzene 137(T)
    organic solvents 134, 136(F), 137(T), 138(F)
    oxygen compounds 135, 136(F), 137(T), 138(F)
    2-propanone 135, 137(T), 138(F)
    water 134, 134(F)
  vapour pressure 133
Ethylene, see Ethene
Ethyne
  physical properties 114, 146
  reference line 113(F), 147(F)
  solubility in
    alcohols 149(F), 150, 154(T)
    ammonia 152, 154(F), 155(T)
    aqueous solutions 148
    aromatic hydrocarbons 148, 148(F), 150(F), 154(T)
    N,N-dimethylformamide 152, 153(F), 155(T)
    esters 149(F), 150, 154(T)
    halogen compounds 149(F), 151, 155(T)
    hydrocarbons 148, 148(F), 150(F), 154(F)
    organic solvents 148, 149(F), 150(F), 151(F), 152(F), 153(F), 154(T)
    oxygen compounds 149(F), 150, 151(F), 152(F), 153(F), 154(T)
    2-propanone 149(F), 150, 151(F), 152(F), 154(T)
    water 147, 148(F)
  vapour pressure 146
Experimental methods 7, 40

Fugacity 15, 25, 26

Gas burette 10, 11(F)
Gibbs free energy of solution 31
Girdler–Sulfide process 218
Group contribution methods 37

Halocarbons 315
Henry's law 5, 24, 175
Henry's law constant 5, 15, 24, 27, 30, 32, 74, 128, 134, 135, 226, 230, 260, 262

SUBJECT INDEX

Higher amines, solubility 111
High pressure 17, 57, 65, 98, 116, 118, 121, 124, 125, 127, 129, 135, 140, 143, 150, 151, 152, 158, 161, 220, 223, 226, 230, 249, 282, 301, 303, 304, 306, 308, 311
Hydrocarbons (*also see* Methane, Ethane, etc.)
  physical properties 114
  solubility in
    1,1-dichloroethane 169(F)
    methanol 168(F)
    methylbenzene 167(F)
    2-propanone 167(F)
    tetrachloromethane 168(F)
    water 165, 166(T), 166(F)
Hydrogen
  phase equilibria
    benzene 304, 304(F)
    ethane 303
    ethene 304
    fluorocarbons 309
  physical properties 300
  solubility in
    alcohols 306, 307(T), 307(F), 308(F)
    alkanes 301, 303(T), 305(F)
    amines 311, 311(T), 312(F)
    aromatic hydrocarbons 303(T), 304, 304(F), 305(F)
    esters 306, 307(T), 308(F)
    ethers 306, 307(T), 308(F)
    haloalkanes 308, 309(T), 310(F)
    halogen compounds 308, 309(T), 310(F)
    hydrocarbons 301, 303(T), 304(F), 305(F)
    nitrobenzene 311(T), 312(F)
    nitrogen compounds 310, 311(T), 312(F)
    oxygen compounds 306, 307(T), 307(F), 308(F), 310, 311(T)
    2-propanone 306, 307(T), 308(F)
    sulfur compounds 310, 311(T), 312(F)
    tributyl phosphate 311, 311(F)
    water 301, 302(F)
  vapour pressure 301
Hydrogen bromide
  physical properties 187
  reference line 188
  solubility in
    alcohols 196, 197(T), 199(F)
    alkanes 192, 192(F), 193(T)
    aromatic hydrocarbons 194, 194(F), 195(F)
    carboxylic acids 203(T)
    ethers 200, 201(T), 202(F)
    halogen compounds 209(T), 210, 210(F)
    nitrobenzene 211, 211(T)
    nitrogen compounds 211, 212(T)
    non-aromatic hydrocarbons 192, 192(F), 193(T)
    sulfur compounds 213(T), 215
    tripentyl borate 213(T), 215
    water 189(F), 191
  vapour pressure 187
Hydrogen chloride
  hydrates 188
  physical properties 187
  reference line 188
  solubility in
    alcohols 196, 196(F), 197(T)
    alkanes 191, 191(F), 192(T)
    aromatic hydrocarbons 193, 193(F), 194(T)
    boron compounds 213(T), 214
    carboxylic acids 202, 203(T)
    esters 202, 203(T)
    ethers 199, 200(F), 201(T), 202(F)
    1-ethoxybutane 8, 9(F)
    halogen compounds 202, 204, 204(F), 205(F), 207(F), 209(T)
    nitrobenzene 211, 212(T)
    nitrogen compounds 211, 212(T)
    non-aromatic hydrocarbons 191, 191(F), 192(T)
    phosphorus compounds 213(T), 214
    silicon compounds 214, 214(F), 215(F)
    sulfur compounds 212, 213(T)
    water 188, 189(F)
  vapour pressure 187
Hydrogen fluoride
  physical properties 187
  reference line 188
  solubility in
    alcohols 198(T)
    benzene 195, 195(F)

SUBJECT INDEX

ethers 201, 201(T)
fluorosulfuric acid 214(T), 215
octane 192
water 190(F), 191
vapour pressure 187
Hydrogen iodide
physical properties 187
reference line 188
solubility in
alcohols 196, 198(T), 199(F)
carboxylic acids 203(T)
ethers 200, 201(T), 202(F)
halogen compounds 211, 211(T), 212, 213(T)
sulfur compounds 214(T), 215
water 190(F), 191
vapour pressure 187
Hydrogen sulfide
hydrate formation 220
ionization 219
phase equilibria
ethane 223
methane 129, 131(F), 223
propane 223
physical properties 218
reference line 23(F), 219
removal from effluent gases 222
solubility in
alcohols 225, 226(F), 227(T)
alkanes 223, 224(F), 225(T)
alkanolamines 222
alkyl phosphates 232, 233(T)
amines 229, 231(T), 232(F)
ammonia 222
aromatic hydrocarbons 224, 225(T)
benzene 224, 225(T)
carbon dioxide 222
carboxylic acids 226, 227(F), 227(T)
N,N-dimethylformamide 231(T), 232
haloalkanes 227, 228(F), 229(T)
halogen compounds 227, 228(F), 229(T), 230(F)
hydrochloric acid 221
liquids, alphabetical order 234
liquids, increasing mole fraction 235
liquids, increasing weight ratio 237

4-methyl-1,3-dioxolan-2-one 226, 227(T)
1-methyl-2-pyrrolidinone 230, 231(F), 231(T)
nitrogen compounds 229, 231(F), 231(T), 232(F)
non-aromatic hydrocarbons 223, 224(F), 225(T)
oxygen compounds 225, 226(F), 227(F), 227(T), 230, 231(F)
salt solutions 220
sulfur compounds 233, 233(F)
water 219, 220(F)
vapour pressure 219

Ideal solution 21

Krichevsky-Il'inskaya equation 27
Kuenen coefficient 6

Methanamine
ionization 88
phase equilibria
N,N-dimethylmethanamine 101(F), 102
tetrachloromethane 102
physical properties 86
reference line 87(F)
solubility in
alcohols 99, 100(T), 101(F)
halogen compounds 100(T), 102
hydrocarbons 98, 99(F), 100(T), 101(F)
nitrogen compounds 100(T), 101
oxygen compounds 99, 100(T), 101(F)
water 89, 89(F)
vapour pressure 87
Methane
phase equilibria
ammonia 129
carbon dioxide 129, 130(F)
carbonyl sulfide 129, 129(F)
carbonyl sulfide 129, 129(F)
hydrogen 303
hydrogen sulfide 129, 131(F)
nitrous oxide 129, 132(F)
quinoline 127, 127(F)
sulfur dioxide 129
physical properties 114
reference line 23(F), 113(F), 115

Methane (*cont.*)
  solubility in
    alcohols   122, 123(T)
    alkanes   118, 119(T), 119(T), 120(F)
    alkyl phosphates   127
    amines   126(T)
    aromatic hydrocarbons   118, 119(T), 121(F), 122(F)
    halogen compounds   125, 126(T), 126(F), 127(F)
    hydrochloric acid   116, 118(F)
    methyl acetate   123(T), 124
    3-methylphenol   125, 125(F)
    1-methyl-2-pyrrolidinone   126(T), 127, 128(F)
    nitrobenzene   126(T)
    nitrogen compounds   125, 126(T), 128(F), 129, 132(F), 133(F)
    non-aromatic hydrocarbons   118, 119(T), 119(F), 120(F)
    oxygen compounds   122, 123(T), 124(F), 125(F), 126(T), 127, 128(F)
    2-propanone   123(T), 124
    salt solutions   116, 117(F)
    sulfur compounds   126(T), 129, 129(F), 130(F), 131(F), 132(F)
    water   115, 116(F), 117(F)
  vapour pressure   115
N-Methylmethanamine
  ionization   88
  phase equilibria
    hexane   102, 102(F)
    tetrachloromethane   103, 105(F)
  physical properties   86
  reference line   87(F)
  solubility in
    alcohols   103, 104(T)
    halogen compounds   103, 104(T), 105(F)
    hydrocarbons   103, 103(F), 104(T)
    nitrogen compounds   103, 104(T)
    oxygen compounds   103, 104(T)
    water   89, 90(F)
  vapour pressure   87
2-Methylpropane
  physical properties   114, 155
  reference line   113(F), 156(F)
  solubility in
    alcohols   162, 162(T)
    alkanes   160, 162(T)
    halogen compounds   163(T), 165(F)
    hydrocarbons   160, 162(T)
    nitrogen compounds   163(T), 165(F)
    oxygen compounds   162, 162(T), 165(F)
    water   157, 157(F)
  vapour pressure   155
Mole fraction solubility   2, 24, 31, 33
Mole ratio solubility   2, 25

Nitric oxide
  physical properties   265
  reference line   270, 272(F)
  solubility in
    organic solvents   271, 272(T), 273(F)
    salt solutions   271
    water   270, 271(F)
  vapour pressure   270
Nitrogen
  phase equilibria with benzene   286
  physical properties   281
  reference line · 281, 281(F)
  solubility in
    alkanes   283, 284(T), 285(F)
    alkanols   285(F), 286, 288(T)
    amines   289(F), 290, 290(T)
    aromatic hydrocarbons   284(T), 285, 285(F), 286(F)
    ethers   285(F), 287, 288(T)
    haloalkanes   287, 288(T), 289(F)
    halogen compounds   287, 288(T), 289(F), 290(T), 292
    hydrocarbons   283, 284(T), 285(F), 286(F)
    methyl acetate   285(F), 287, 288(T)
    1-methyl-2-pyrrolidinone   290(T), 291, 291(F)
    nitrobenzene   289(F), 290, 290(T)
    nitrogen compounds   289(F), 290, 290(T), 291(F)
    octamethylcyclotetrasiloxane   290(T), 292
    oxygen compounds   285(F), 286, 288(T), 289(T), 290, 290(T), 292
    2-propanone   285(F), 286, 288(T)

sulfur compounds 290(T), 291
  water 282, 283(F)
  vapour pressure 281
Nitrogen dioxide
  dimerization 265, 273
  organic solvents 273
  phase equilibria 273
  physical properties 265
  reaction with water 273
Nitrous oxide
  phase equilibria with methane 129, 132(F)
  physical properties 265
  reference line 266(F)
  solubility in
    organic solvents 268, 268(T), 269(F), 270(F)
    salt solutions 267(F), 268
    water 266, 267(F)
  vapour pressure 265
Noble gases 23(F), 315

Ostwald coefficient 3, 32, 34
Oxygen
  physical properties 292
  reference line 292, 293(F)
  solubility in
    alkanes 294, 295(T), 297(F)
    alkanols 296, 297(T), 297(F)
    amines 300(F), 300(T)
    aromatic hydrocarbons 294, 295(T), 297(T)
    esters 296, 297(T)
    ethers 296, 297(T), 297(F)
    fluorocarbons 298, 298(T), 300(F)
    haloalkanes 298, 298(T), 300(F)
    halogen compounds 298, 298(T), 300(F)
    hydrocarbons 294, 295(T), 297(F)
    ketones 296, 297(T), 297(F)
    nitrobenzene 300(F), 300(T)
    nitrogen compounds 299, 300(F), 300(T)
    oxygen compounds 296, 297(T), 297(F), 298(T), 299, 300(F), 300(T)
    sulfur compounds 299, 300(T)
    water 293, 294(F)
  vapour pressure 292

Partial molar enthalpy of solution 32
Partial pressure 5, 7, 17, 20, 25, 26, 33, 35
Phase rule 1
Propanamines, solubility in hexane 111
Propane
  physical properties 114, 155
  reference line 113(F), 156(F)
  solubility in
    alcohols 161, 162(T), 164(F)
    alkanes 159, 160(F), 162(T), 164(F)
    aromatic hydrocarbons 162(T), 164(F)
    halogen compounds 163(T), 164(F)
    hydrocarbons 159, 160(F), 162(T), 164(F)
    nitrogen compounds 163(T), 164(F)
    oxygen compounds 161, 162(T), 164(F)
    water 157, 157(F), 158(F)
  vapour pressure 155

Raoult's law 8, 20, 35
Rate of absorption 9, 33
Reference line (*see individual gases*) 21, 34, 35, 274, 315
Reference surface 22, 23(F), 35
Regular solution theory 37

Salt effects 42, 116 117(F), 176, 177(F), 220, 243, 244(F), 254, 260, 268
Scaled particle theory 37
Sechenov equation 176, 220, 244
Short range ordering 35
Solubility data
  evaluation 32
  sources 30
Solubility
  determining factors 35
  empirical equations 31
  mixtures of gases 282
  parameters 24
  prediction 35, 223
Sources of error 32
Statistical–mechanical methods 37
Structure breaking effect 35

Structure making effect 36
Sulfur dioxide
  addition compounds 39, 69
  ionization 42
  phase equilibria with methane 129
  physical properties 39
  reference line 23, 39, 40(F)
  solubility in
    alkanes 44, 45(T), 46(F), 47(F)
    alkanols 55, 56(F), 57(T), 58(F), 59(F), 60(F)
    amines 62(T), 69, 70(F)
    aromatic hydrocarbons 48, 49(F), 50(T), 51(F), 52(T), 53(F), 54(T), 55(F), 56(F)
    benzene 48, 49(F), 50(T), 51(F)
    carboxylic acids 63, 65(F)
    N,N-dimethylformamide 62(T), 65, 68(F)
    esters 63, 66(F)
    ethers 57(T), 60, 61(F), 63(F)
    haloalkanes 67(T), 71, 71(F), 72(F), 73(F)
    halobenzenes 67(T), 72, 74(F)
    halogen compounds 67(T), 71, 71(F), 72(F), 73(F), 74(F)
    hydrochloric acid 44
    ketones 28(F), 57(T), 62, 64(F)
    liquids, alphabetical order 76
    liquids, increasing mole fraction 79
    liquids, increasing weight ratio 82
    nitrobenzene 62(T), 64, 67(F), 68(F)
    nitrogen compounds 62(T), 64, 67(F), 68(F), 69(F), 70(F)
    non-aromatic hydrocarbons 44, 45(T), 46(F), 47(F)
    oxygen compounds 28(F). 53, 56(F), 57(F), 58(F), 59(F), 60(F), 61(F), 63(F), 64(F), 65(F)
    phenols 63, 66(F)
    phosphorus compounds 75
    2-propanone 28(F)
    pyridine 62(T), 69, 69(F)
    salt solutions 42, 43(F)
    sulfur compounds 73
    sulfuric acid 43, 44(F)
    water 41, 42(F)
  vapour pressure 39
Supersaturation 9

Thin film methods 14

Wet method 12

Xenon
  reference line 23(F)

LIBRARY USE ONLY

MAY 8 1991